Behavioral Technologies for Fish Guidance

Funding for the publication of these proceedings was provided by

Electric Power Research Institute

Funding for planning the symposium was provided by

Flash Technology Corporation of America

Oak Ridge National Laboratory

Behavioral Technologies for Fish Guidance

Edited by
Charles C. Coutant
Senior Research Ecologist, Oak Ridge National Laboratory
Environmental Sciences Division, Building 1505, Oak Ridge, Tennessee 37831-6036

American Fisheries Society Symposium 26

Proceedings of the Symposium
Behavioral Technologies for Fish Guidance
held at Charlotte, North Carolina, USA
30–31 August 1999

American Fisheries Society
Bethesda, Maryland
2001

The American Fisheries Society Symposium series is a registered serial. Suggested citation formats follow.

Entire book

Coutant, C. C., editor. 2001. Behavioral technologies for fish guidance. American Fisheries Society, Symposium 26, Bethesda, Maryland.

Chapter within the book

Goetz, F. A., J. J. Dawson, T. Shaw, and J. Dillon. 2001. Evaluation of low-frequency sound transducers for guiding salmon smolts away from a navigation lock. Pages 91–114 *in* C. C. Coutant, editor. Behavioral technologies for fish guidance. American Fisheries Society, Symposium 26, Bethesda, Maryland.

Printed in the United States of America.

Cover design by Gail Sweeden, Creative Media Solutions, Oak Ridge National Laboratory

Library of Congress Catalog Card Number 2001091420
ISBN 1-888569-24-7
ISSN 0892-2284

American Fisheries Society
5410 Grosvenor Lane, Suite 110
Bethesda, Maryland 20814-2199
USA

TABLE OF CONTENTS

Preface

The use of fish behavior to guide animals away from hazards of water withdrawals is a topic of considerable current interest. Regulatory concerns for fish protection at water developments have stimulated this interest. For example, there are new guidelines pending from the U.S. Environmental Protection Agency for cooling water intakes for steam-electric generating stations, enhanced enforcement of fish protection rules at water diversions (often for irrigation) by the National Marine Fisheries Service, and relicensing actions for nonfederal hydropower facilities by the U.S. Federal Energy Regulatory Commission. Public relations difficulties have arisen for many water users because of widely publicized fish losses. The interest in behavioral guidance technologies is broadly spread among regulators, those being regulated, suppliers of engineering and biological services to the affected industries, and fish researchers. Experiences in the United States are paralleled in many other countries.

The symposium *Behavioral Technologies for Fish Guidance* was initially conceived by Ron Brown, organized by Ron and Chuck Coutant, and held on August 30–31, 1999 at the Annual Meeting of the American Fisheries Society in Charlotte, North Carolina to address this interest. The meeting was well attended by a diverse group of biologists and engineers involved with various technologies for fish guidance and diversion. This book contains the majority of the papers presented at the symposium. The papers have benefited from comments and questions received at the meeting and those stimulated by preprints distributed among presenters. All papers received at least two peer reviews, usually one from a symposium participant and one from someone knowledgeable about the topic who was not at the meeting.

This collection of papers complements several other recent works related to joint efforts of biologists and engineers to improve the technical design and efficacy of fish passage facilities, particularly at dams. Behavioral technologies comprise a subset of fish passage approaches, in which cues to the normal behavior of fish are manipulated to repel fish away from hazardous water withdrawals or attract them to safer alternative routes (or both). The book *Fish Migration and Fish Bypasses*, edited by Mathias Jungwirth, Stefan Schmutz, and Steven Weiss (1998), provided a foundation for the needs of riverine fish populations, regional overviews of European and North American state-of-the-art information on fish passage designs, and case studies of conventional and "nature-like" designs for passage provision. Two American Fisheries Society books edited by Mufeed Odeh (1999, 2000), *Innovations in Fish Passage Technology* and *Advances in Fish Passage Technology*, chronicle the rapid proliferation of innovative ideas and their site-specific testing, development, and deployment. Taken together, these three volumes and the current one offer the novice a rapid education on the subject and the professional an indispensable compendium of the latest information.

Behavioral systems are generally attractive financially for a user of a water resource. Guidance systems based on fish behavior are generally less costly and easier to implement than are structural systems. However, the notion of merely hanging a simple device off an intake to frighten fish away has proven elusive and often misguided. Nevertheless, behavioral systems can be tested relatively inexpensively, and if they fail to significantly reduce fish losses, then more costly structural alternatives can be considered.

In this volume, you will find the principal behavioral technologies described and discussed, with case studies (not necessarily in the following order). Strobe lights and sound as repellants of American eels are presented by Paul Patrick and coauthors R. S. Poulton and Ron Brown, along with the novel idea that sound might be used as a conditioned attraction response (as well as a repellant) to aid passage. Gene Ploskey and Peter Johnson also evaluated strobe lights and sound (infrasound), but for effectiveness with juvenile salmon, as did Bob Mueller and coauthors Duane Neitzel and Brett Amidan. The success of strobe lights for displacing salmon smolts at ship locks in Seattle, Washington is reported by Peter Johnson and coauthors Fred Goetz and Gene Ploskey. The efficacy of strobe lights was further demonstrated by Melo Maiolie and coauthors B. Harryman and B. Ament, when applied to free-ranging kokanee (landlocked sockeye salmon) in lakes. Strobe lights, drop lights, and infrasound were compared for guidance abilities at an irrigation diversion by Steve Amaral and coauthors Fred Winchell and Todd Pearsons. Low-frequency sound transducers were also evaluated at navigation locks by Fred Goetz and coauthors James Dawson, Travis Shaw, and Jeffrey Dillon. Because sound, particularly infrasound, is often misunderstood in its application to fish guidance, Olav Sand and coauthors Per Enger, Hans Karlsen, and Frank Knudsen provide a mini-review of their recent work with infrasound on juvenile salmonids and European silver eels.

The efficacy of a structural barrier as a behavioral guidance device for encouraging migration of salmon smolts to a bypass at a Snake River dam was evaluated by Noah Adams and four coauthors. Because the fine-scale behavior of smolts is the essence of behavioral guidance, Bob Johnson and coauthors Mary Ann Simmons, Carver Simmons and Susan Blanton report on a new sonar technique for obtaining this information and its testing at the same behavioral guidance structure on the Snake River.

Another guidance concept, that of mimicking natural riverine turbulence as a cue to attraction of downstream-migrating juvenile salmonids to bypasses, is presented by Chuck Coutant. Jaci Savino and coauthors Dave Jude and Melissa Kostich report on tests of electrical barriers to deter downstream movement of the exotic round gobi.

Although behavioral technologies have most often focused on single sensory cues, Chuck Coutant discusses the accelerating trend toward considering multisensory or hybrid systems that include more than one sensory system to accommodate the varied responses of multiple species at a site or to combine repulsion from a hazard with attraction to a preferred route. The history of field testing of many behavioral devices by a principal funding organization, the Electric Power Research Institute (EPRI), is chronicled by Ned Taft and coauthors Doug Dixon and Chuck Sullivan.

We recognize that the quality of the written paper often depends on the constructive criticism and thoughtful suggestions of peer reviewers. Therefore, we thank the following individuals for their reviews: Stephen V. Amaral, James J. Anderson, D. Hugh Barwick, David H. Bennett, Mark S. Bevelhimer, G. Russell Danner, Dennis J. Dunning, Albert E. Giorgi, Frederick A. Goetz, Michael J. Hansen, Charles Hanson, Gary E. Johnson, Peter N. Johnson, Melo A. Maiolie, Robert P. Mueller, Richard S. Nemeth, Paul H. Patrick, Gene R. Ploskey, Arthur N. Popper, Kate Puckett, William S. Rainey, Dennis W. Rondorf, David R. Sager, Jacqueline F. Savino, Dana Schmidt, William D. Swink, and Edward P. Taft.

Also, we thank the organizations that supported publication of this book, either financially or with in-kind contributions. EPRI, through Project Manager Douglas A. Dixon, was the underwriter of publication. Oak Ridge National Laboratory and Flash Technology Corporation of America provided time and materials for Chuck Coutant and Ron Brown, respectively, for organizing the symposium and preparing the manuscripts for publication.

We close this preface with a hope for the future, that the "technological fixes" for fish protection, which so often are maligned by fisheries professionals as inadequate for accommodating the biology of the species to be protected, can evolve to make productive use of normal attractive and repulsive behavioral patterns of the fish rather than attempting to subvert or replace them with purely structural passage systems. Designing *with* nature rather than to replace it should have a bright future for resource protection and enhancement.

Chuck Coutant and Ron Brown,
Symposium organizers

References

Jungwirth, M., S. Schmutz, and S. Weiss, editors. 1998. Fish migration and fish bypasses. Fishing News Books, London.

Odeh, M., editor. 1999. Innovations in fish passage technology. American Fisheries Society, Bethesda, Maryland.

Odeh, M., editor. 2000. Advances in fish passage technology. American Fisheries Society, Bethesda, Maryland.

Symbols and Abbreviations

The following symbols and abbreviations may be found in this book without definition. Also undefined are standard mathematical and statistical symbols given in most dictionaries.

A	ampere
AC	alternating current
Bq	becquerel
C	coulomb
°C	degrees Celsius
cal	calorie
cd	candela
cm	centimeter
Co.	Company
Corp.	Corporation
cov	covariance
DC	direct current; District of Columbia
D	dextro (as a prefix)
d	day
d	dextrorotatory
df	degrees of freedom
dL	deciliter
E	east
E	expected value
e	base of natural logarithm (2.71828…)
e.g.	(exempli gratia) for example
eq	equivalent
et al.	(et alii) and others
etc.	et cetera
eV	electron volt
F	filial generation; Farad
°F	degrees Fahrenheit
fc	footcandle (0.0929 lx)
ft	foot (30.5 cm)
ft^3/s	cubic feet per second (0.0283 m^3/s)
g	gram
G	giga (10^9, as a prefix)
gal	gallon (3.79 L)
Gy	gray
h	hour
ha	hectare (2.47 acres)
hp	horsepower (746 W)
Hz	hertz
in	inch (2.54 cm)
Inc.	Incorporated
i.e.	(id est) that is
IU	international unit
J	joule
K	Kelvin (degrees above absolute zero)
k	kilo (10^3, as a prefix)
kg	kilogram
km	kilometer
l	levorotatory
L	levo (as a prefix)
L	liter (0.264 gal, 1.06 qt)
lb	pound (0.454 kg, 454g)
lm	lumen
log	logarithm
Ltd.	Limited
M	mega (10^6, as a prefix); molar (as a suffix or by itself)
m	meter (as a suffix or by itself); milli (10^{-3}, as a prefix)
mi	mile (1.61 km)
min	minute
mol	mole
N	normal (for chemistry); north (for geography); newton
N	sample size
NS	not significant
n	ploidy; nanno (10^{-9}, as a prefix)
o	ortho (as a chemical prefix)
oz	ounce (28.4 g)
P	probability
p	para (as a chemical prefix)
p	pico (10^{-12}, as a prefix)

Pa	pascal
pH	negative log of hydrogen ion activity
ppm	parts per million
qt	quart (0.946 L)
R	multiple correlation or regression coefficient
r	simple correlation or regression coefficient
rad	radian
S	siemens (for electrical conductance); south (for geography)
SD	standard deviation
SE	standard error
s	second
T	tesla
tris	tris(hydroxymethyl)-aminomethane (a buffer)
UK	United Kingdom
U.S.	United States (adjective)
USA	United States of America (noun)
V	volt
V, Var	variance (population)
var	variance (sample)
W	watt (for power); west (for geography)
Wb	weber
yd	yard (0.914 m, 91.4 cm)
α	probability of type I error (false rejection of null hypothesis)
β	probability of type II error (false acceptance of null hypothesis)
Ω	ohm
μ	micro (10^{-6}, as a prefix)
′	minute (angular)
″	second (angular)
°	degree (temperature as a prefix, angular as a suffix)
%	per cent (per hundred)
‰	per mille (per thousand)

American Fisheries Society Symposium 26:1–11, 2001

Responses of American Eels to Strobe Light and Sound (Preliminary Data) and Introduction to Sound Conditioning as a Potential Fish Passage Technology

P. H. Patrick and J. S. Poulton
Environmental Technologies Department, Kinectrics (formerly Ontario Power Technologies)
800 Kipling Ave, Toronto, Ontario M8Z 5S4, Canada

R. Brown
Flash Technology Corporation of America, PO Box 681509
Franklin, Tennessee 37068, USA

Abstract.—Laboratory experiments were conducted on the responses of both juvenile and adult American eels *Anguilla rostrata* to white strobe light and sound. All eels showed a strong avoidance of strobe light, although the response time varied for eel size-class. Juvenile eel avoidance was immediate whereas adult animals displayed marked avoidance only after several minutes exposure to the light source. In all tests, there were significant reductions in numbers occupying lit areas relative to control conditions. After avoidance occurred, there was no evidence of either juvenile or adult animal acclimating to the light source. An avoidance response was evident for different flashing rates. Future research activities are proposed. Preliminary experiments were conducted on the responses of both juvenile and adult eels to sound. We were able to demonstrate the concept that American eels may be attracted to sound. This is a new application since most studies involve using sound as a repellent. Another acoustical technique that involves "sound conditioning" is also introduced as a method that may also have application in upstream/downstream bypass applications.

Recently there is concern about the decline of American eels *Anguilla rostrata* in the northeast. Reasons for this decline are not well known and research activities are currently being focused to better address this issue (Casselman 1999). It is also difficult to determine how much decline has occurred since quantitative data are not readily available. One measure to assess population trends in eels is to compare the numbers of upstream migrants that have used the eel ladder at Ontario Power Generation's Saunders hydroelectric facility located on the St. Lawrence River. Millions of eels have passed over the original ladder since operation began in 1974, which is described in detail by Whitfield and Kolenosky (1978) and Eckersley (1980). Recently, however, eel migration has been very low and just a fraction of numbers that occurred in the late 70's and early 80's. Concern over general eel decline has also been expressed by fishermen, fisheries managers and other scientists, which has resulted in the establishment of a working group by the Atlantic States Marine Fisheries Commission (ASMFC) to develop a Fisheries Management Plan (FMP) for the American eel fishery in order to protect and restore this species (ASMFC 1999).

One reason for the decline may be an inability to pass by large hydropower projects (Euston et al. 1998). Diversion bypass facilities may be needed. At present there is no industry-accepted technology available for diverting eels. Research is ongoing on both behavioral as well as physical means of diverting eels. Behavioral methods have typically included lights (Patrick et al. 1982; Hadderingh and Symthe 1997), sound (Jerko et al. 1989; Sand 1999) and electric fields (Gleeson 1997). Results have been mixed with some systems being more effective than others.

The objective of this report is to provide further quantitative data on the behavioral responses of American eels to strobe light, and, to a lesser extent, sound. We also investigated the feasibility of using sound as an attractant for American eels. Finally, sound conditioning is introduced as a technique, which may have application in some fish bypass situations. This work was initially presented at a symposium on "Behavioral Technologies for Fish Guidance" as part of the 1999 American Fisheries Society Annual Meeting. This paper includes both previously reported and new research results.

Methods

Experiments focused on both laboratory and field trials on the responses of both yellow (juvenile) and adult eels in response to strobe light. Some of the work on juvenile eels has already been published (Patrick et al. 1982), but is included for direct comparison with recent tests involving adult animals. Work involving sound includes data on both American eels as well as other species.

Test Animals

In the earlier strobe light work, yellow eels, 30–55 cm in length, were netted in late August at the R. H. Saunders dam on the St. Lawrence River. Juvenile eels were also collected by Hydro Quebec in the St. Lawrence River (courtesy of Richard Verdon) for our tests involving sound. Adult eels that exceeded 50 cm in length were also netted on the St. Lawrence, and were graciously provided by Dr. Kevin McGrath (New York Power Authority and Kleinschmidt Associates) for both the strobe and sound work. Fish were acclimated under natural photoperiod conditions in large 5.5-m tanks for two weeks prior to testing, and were fed earthworms, crayfish, and small fish.

Strobe Light Tests

Chambered Experiments

Short-term experiments were conducted at Kinectrics (formerly Ontario Power Technologies) to determine the initial responses of both juvenile and adult eels to strobe light, and to assess the role of flash frequency in eliciting a response. Experiments involving juvenile eels were conducted in a 2.4 m diameter circular tank which was divided equally into three areas, each approximately 1.5 m^2. In the design, one chamber was illuminated with white (xenon) strobe light, which was the sole source of illumination during the testing period. The spectral distribution curve for the xenon strobe (Tandy Electronics) had the highest energy output in the violet-blue-green regions (400–570 nm) of the spectrum (Patrick et al. 1982), which is typical of other strobe light sources. For each experiment, the strobe light was projected downwards in the tank producing concentric zones of varying light intensities. Light intensities, measured with a Quantum/Photometer (Lambda model LI-185), ranged from approximately 7.4 $\mu E/m^2/s$ (333 lux) near the light source to approximately 0.3 $\mu E/m^2/s$ (18 lux) at the edge of the light zone. Light recorded near the outer wall was less than 0.1 $\mu E/m^2/s$ (2.0 lux). The illumination chamber had no influence on the other two chambers, which were in darkness during the testing period. During control tests, the light was not measurable in the experimental chamber (<0.01 $\mu Em^2/s$; <0.1 lux).

For juvenile eels, nine replicates (15 eels/replicate) were performed for each 5- and 15-min illumination period at the following flash frequencies: 66, 200, 484, 748, and 1090 flashes/min. After each selected time period, chutes, which divided the chambers, were closed, and the number of eels in each chamber recorded. Observations were also made on the approximate position of each eel within each chamber, relative to the light source. Twenty-eight control tests were conducted randomly at various times throughout the study.

To determine whether juvenile eels showed a behavioral adaptation to strobe light, eel movement was monitored over five 12-h periods. The number of eels recorded in the lighted chamber over time, behavior, and extent of time that eels spent in the lighted area were determined.

For adult eels, the experimental design was similar to tests involving juvenile eels but with these exceptions. The test facility was larger at 5.5-m in diameter. The strobe light intensities were about an order of magnitude higher with a single strobe unit provided by Flash Technology (Franklin, Tennessee). Ten replicates were conducted for each experimental treatment (5 and 15 min). Twenty controls were also randomly conducted to determine the distribution and behavior in the tank with the light source off. Thirty eels were used for each test. The 15-min duration tests also included two different flashing rates (300 and 450 flashes/ min). No long-term acclimation experiments were conducted with adult eels.

Parametric statistics such as analysis of variance (ANOVA) were used to compare mean numbers in the experimental treatment versus the control. Before conducting this analysis, the assumption of equal or homogeneous variances was tested using a Bartlett's test. In cases, where variances were not equal (even after data transformed), nonparametric statistics were also considered for comparative purposes. However, nonparametric statistics are not as robust as parametric ones.

Field Studies

Field studies were conducted at the entrance of the eel ladder constructed by the Ontario Ministry of

Natural Resources located at the Saunders GS facility on the St. Lawrence (Figure 1). Details of the experimental protocol are given in Patrick et al. (1982) and are summarized below. In the experimental design, strobe lights were placed near the entrance of the ladder (Figure 1), and the number of eels passing over the ladder with the strobe lights on was evaluated relative to control conditions (light source off). A total of 11 tests were conducted (6 experimental and 5 control) at night and typically lasted 12 h (1800–0600) in duration. Strobe lights were placed either on the ladder or underwater with the light being projected towards the mouth of the ladder. Strobe light intensities ranged from 2.3 $\mu Em^2/s$ (156 lux) directly in front of the light source to less than 0.01 $\mu Em^2/s$ (5 lux) at the entrance of the eel ladder. The strobe flash frequency exceeded 800 flashes/min. During the testing period, the ladder terminated into a common collecting trough (4.0 m × 0.3 m × 0.3 m), which diverted the migrating eels into a collecting tank (4.0 m × 1.0 m × 1.0 m). The collecting tank was divided into four sizing chambers using 2.5, 1.9, 1.6, and 0.6 cm mesh size screens. Eels collected from the collecting tank during both experimental and control tests were counted and released.

Sound Attraction Tests

Preliminary tests were conducted on the responses of both adult and juvenile eels to complex signals of sound. There is already published material on the auditory sensitivities of European eels *Anguilla anguilla* (Jerko et al. 1989), but our experience has suggested that simple tones will not be effective in

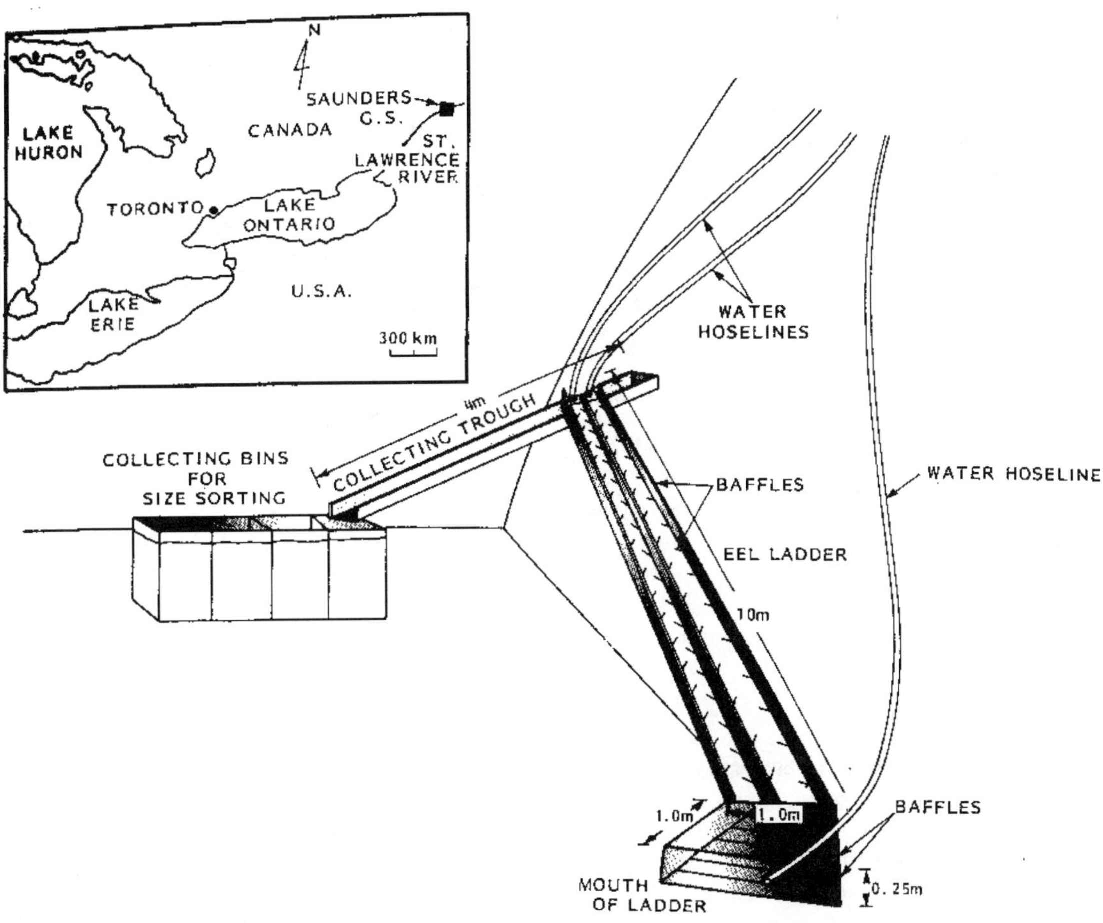

Figure 1. Location and description of eel ladder for collecting eels at Saunders GS.

eliciting a marked behavioral response. The actual signals were complex audible sounds and were based on earlier experiments that involved observations of eels and other migrants responding to various environmental cues. The device we used to project sounds is a modification of an in-house assembly that was capable of producing resonant frequencies of complex signals (below 1000 Hz) up to 190 μPa, 1 m from the source. The intent of these tests was to determine if a positive response occurred under laboratory conditions. If attraction occurs, more rigorous experimental protocols and expanded tests facilities would be developed for future testing.

Tests were conducted in a circular tank similar to the adult eel strobe light tests (5.5-m diameter). The tank was modified to minimize sound reverberation by using sound-absorbing material on the chamber walls, as well as in the tank infrastructure. Acoustical measurements were mapped throughout the tank at surface, middepth and bottom with the sound projector on as well as off (ambient conditions). The tank approached semianechoic conditions although there was still considerable reverberation from surface and bottom reflections. Still, there were well-defined gradients of acoustic energy, especially when comparing the experimental chamber relative to other sections of the tank. It was felt that this facility could be used for preliminary trials investigating sound as an attractant. The projector was placed in one chamber, and the distribution and behavior of eels in each chamber was assessed for both control (no sound) and experimental conditions (sound on). A total of five, independent replicated tests were conducted for a given set of acoustical signals. Five controls were also conducted to determine eel behavior and distribution in the absence of a sound source. Each test typically lasted approximately 15 min. Ten eels (mixed adult/juvenile) were used in these tests. Emphasis was placed on using different animals for each test.

Sound Conditioning Tests

Acoustical conditioning was evaluated as a technique to condition fish to feed using sound (Figure 2). The concept here is to use sound as a stimulus to attract fish towards a location whereby the fish are rewarded with food (i.e., a positive reinforcement). Although this technique was developed as a method to improve growth and survival in fingerling fish in aquaculture applications (Patrick et al. 1992; Patrick 1999), it is believed that this approach may have application in some fish bypass situations involving eels and other species.

Details of the experimental protocol are described elsewhere but have involved both laboratory and field trials (Patrick et al. 1992; Patrick 1999). The initial laboratory work involved four 1000-L circular tanks. One tank served as a control whereas the other three were used as experimental tanks. All four tanks were isolated using a sound proof curtain, which was effective in reducing ambient noise and acoustical cross talk. The initial conceptual tests involved walleye *Stizostedion vitreum*, and lake sturgeon *Acipenser fulvescens*. Laboratory tests focused on fingerlings (100–500/tank), which were fed Bio Kyowa feed (1,700 and 2,700 micron) at 30-min intervals throughout each 24-h day at a ration of 3–5% body weight. Conditioning was defined when the majority of fish in the tank approached the feeder area when the sound was on, and before the feed was dispensed. An assessment of conditioning, fish mortality, length, and size was monitored weekly over a 80-d period. The feed-sound system was fully automated with a microprocessor-based system delivering an appropriate acoustical signal at predetermined time periods prior to feeding. Usually thirty seconds after the sound was introduced, fish were fed using an automated feeder. A similar procedure was used in field tests, although the fish were juvenile and adult (>0.5 kg).

Results and Discussion

Strobe Light

Laboratory Studies

Juvenile eels. For all strobe flash frequencies, the mean number of eels in the illuminated chamber over a 5-min period was at least six times lower than in the same chamber (C1) under control conditions, indicating an avoidance response (Table 1; $P < 0.01$). The mean number of eels in the lighted chamber did not vary statistically among strobe frequencies used ($P > 0.05$), suggesting that strobe frequency did not influence eel avoidance. Following 15-min exposure to each strobe frequency, almost all fish were observed outside the lighted chamber, suggesting a strong avoidance response. Eels that entered the lighted chamber were generally positioned at the edge of the light zone where intensities were lowest (0.3 $\mu E/m^2/2$; <10 lux).

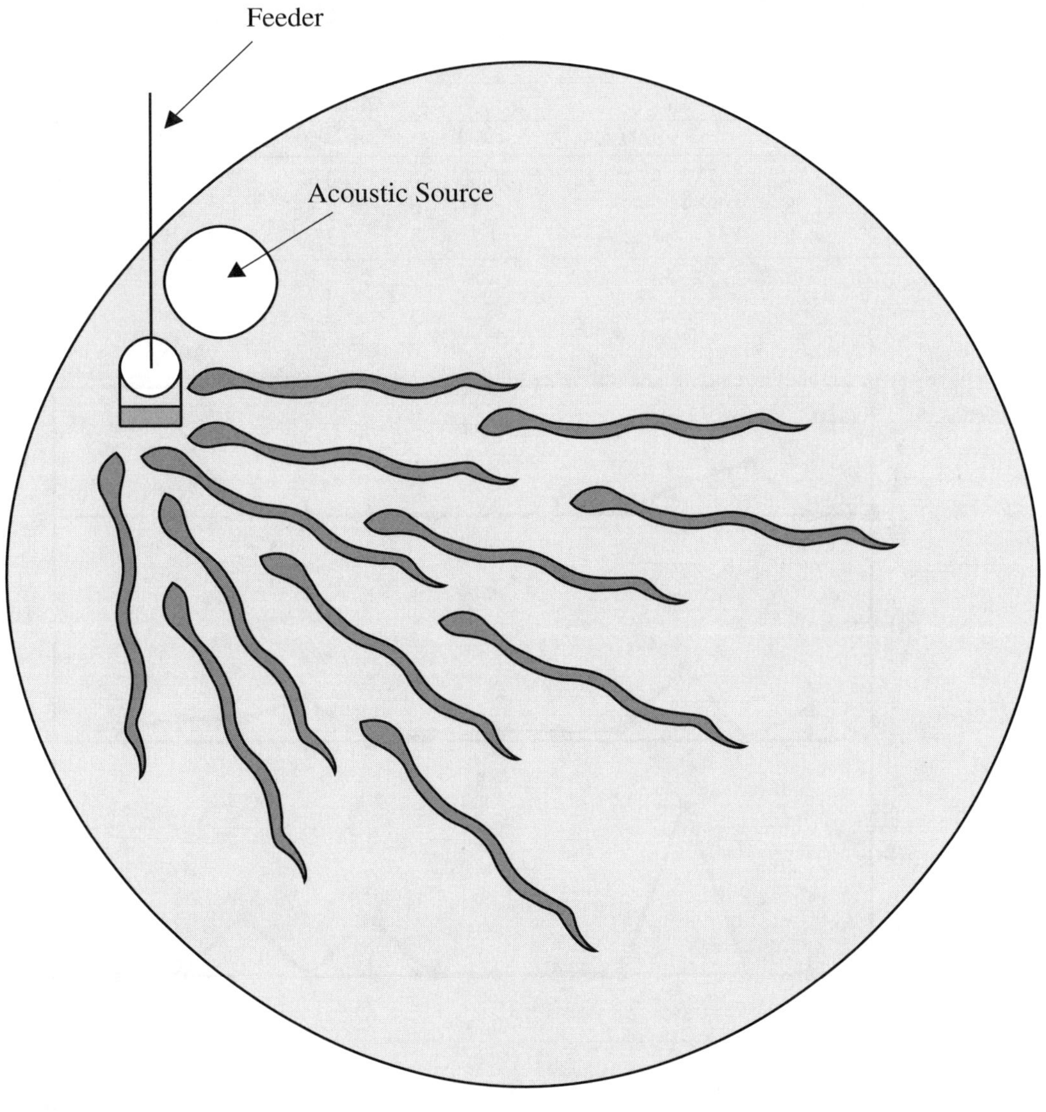

Figure 2. Concept of fish conditioning using acoustics.

Prior to the strobe being turned on, eels were generally inactive in the experimental tank. Once the strobe was turned on, an increase in activity was observed, as eels showed a strong avoidance of the lit area. A marked avoidance occurred over a 12-h period (Figure 3). There was a 94% reduction in total number of eels found in the experimental zone (mean = 17) when compared with occupancy under control conditions (mean = 289). For each replicate there was no apparent increase in fish numbers recorded in the lighted zone over time. Throughout each 12-h monitoring period, the mean time that eels spent in the lighted chamber ranged from 7.4 to 22.2 s, with an overall mean of 13.8 s. There was no evidence of increased time spent in the lighted area for each hour of monitoring, suggesting no behavioral acclimation to the light source.

Adult Eels. Similar to juvenile eels, adult eels displayed an avoidance response to strobe light (Figures 4 and 5). After 5-min exposure, there were statistically fewer fish in the lighted zone compared with the control ($P < 0.01$). However, the response time was considerably slower than observed for juvenile eels even though the light intensities in the tank were at least an order of magnitude high-

Table 1. Mean number of American eels found in control and experimental chamber.

Chamber number	Controls		66		220		484		770		1090	
	Experimental strobe frequency (flashes/min-1)											
	$\bar{x}$	SD(x)	$\bar{x}$	SD(x)	$\bar{x}$	SD(x)	$\bar{x}$	SD(x)	$\bar{x}$	SD(x)	$\bar{x}$	SD(x)
C1 lit in exp	4.55	1.81	0.67*	0.71	0.33*	0.50	0.33*	0.50	0.22*	0.44	0.55*	0.73
C2 not lit	5.69	2.98	5.00	2.69	7.00	1.41	6.55	3.13	9.11	1.82	12.11	1.90
C3 not lit	4.75	2.78	9.33	2.56	7.67	1.22	8.11	3.10	5.67	1.87	2.33	1.50

*Significant at $P < 0.01$.

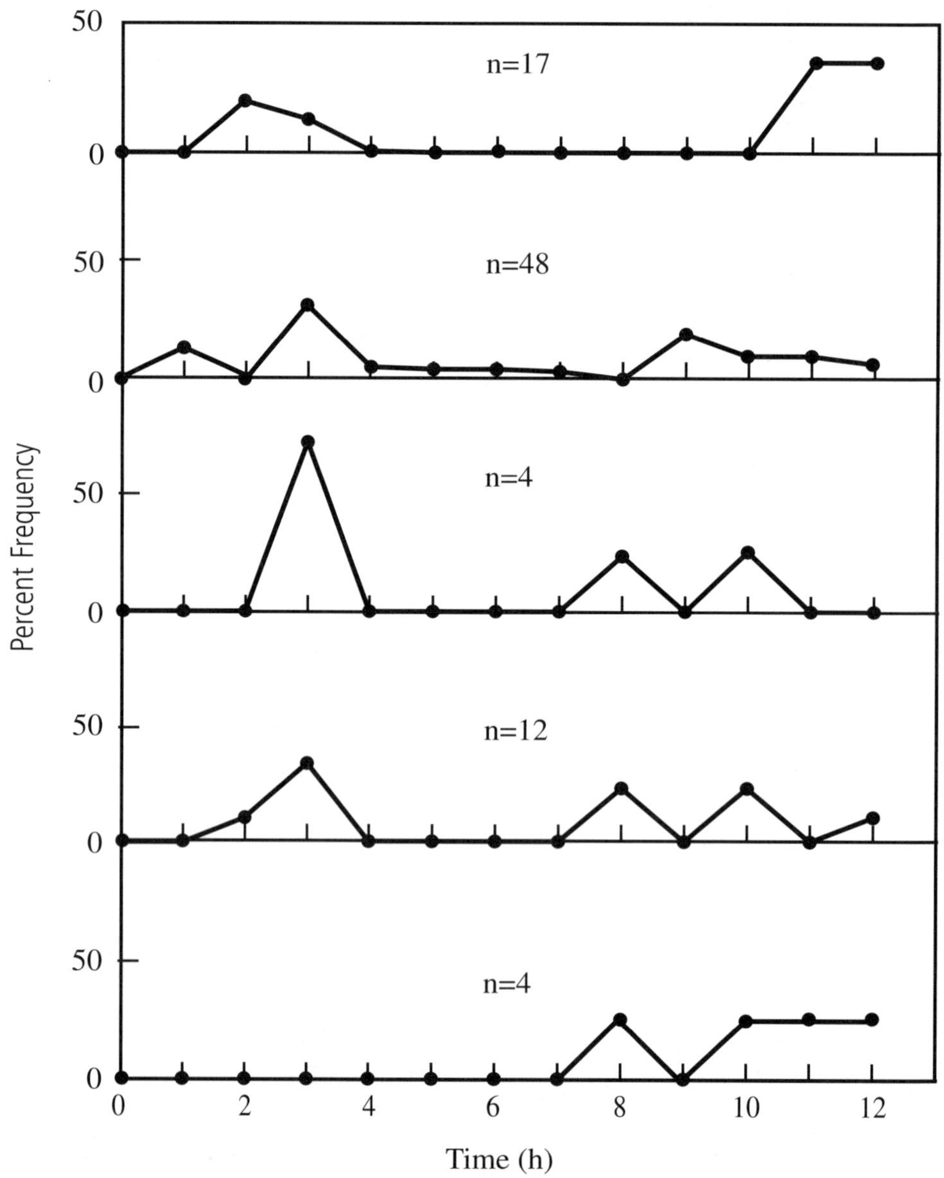

Figure 3. Percent frequency of eels recorded in lighted chamber over time.

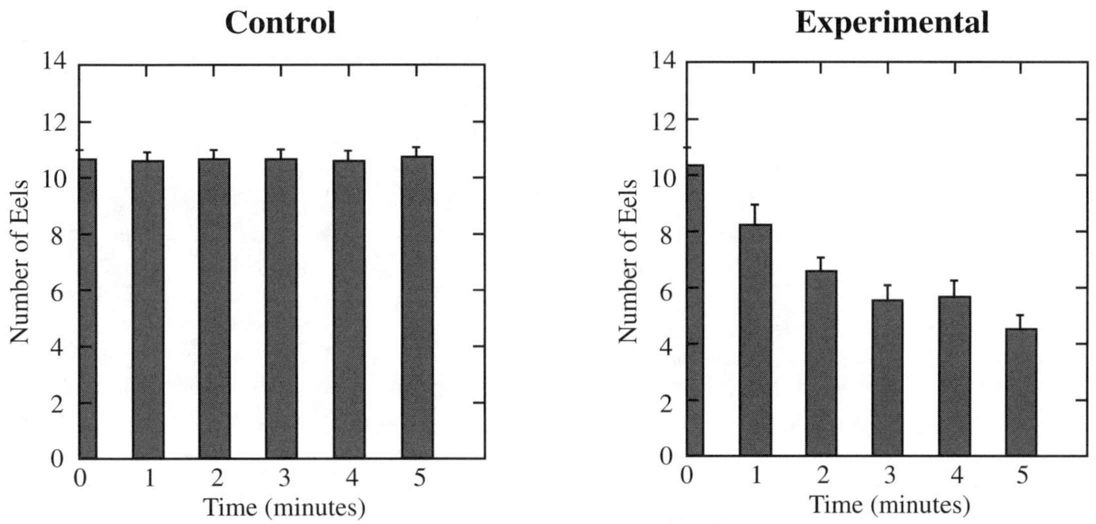

Figure 4. Number of adult eels in lighted chamber vs. control chamber—initial response tests.

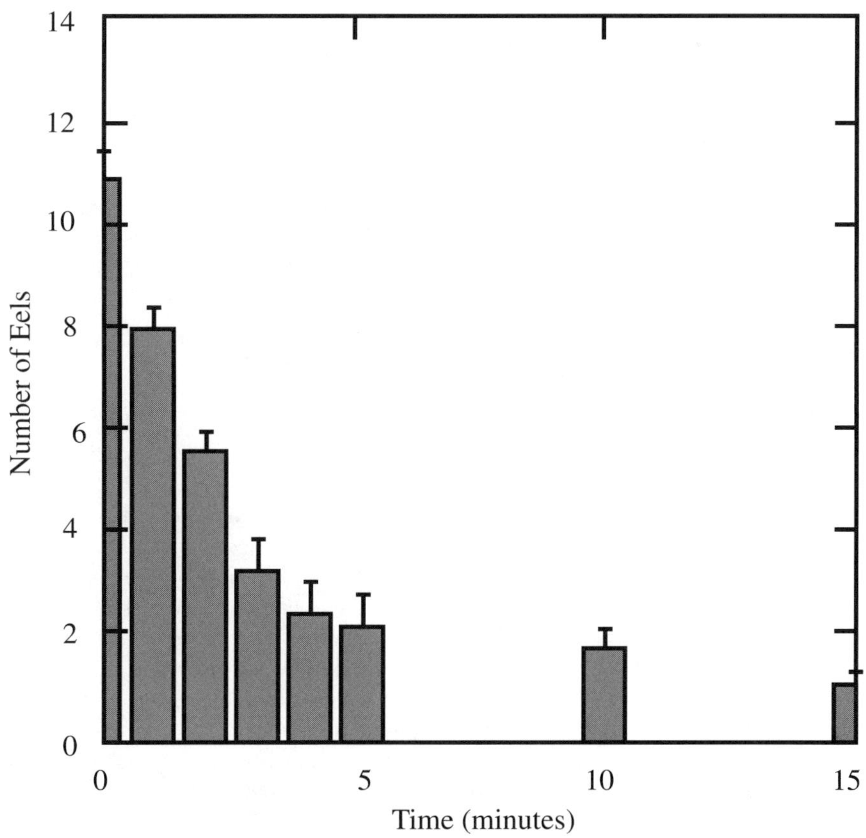

Figure 5. Number of adult eels in lighted chamber over time.

er than in the juvenile laboratory tests. Adult eels appear to be less sensitive to strobe light than juvenile eels. It took several minutes for eels to move out of the lighted area compared with the above-described results for juvenile eels (taking into account the larger test chamber). For instance, after one minute of strobe exposure, there was only about a 20% reduction of adult eels in the lighted chamber compared with more than 55% after 5-min of exposure. This compared with more than an 85% reduction for juvenile eels after a 5-min period. For adult eels, maximum reduction of approximately 90% occurred 15-min after exposure (Figure 5). This was statistically significant compared with control conditions ($P < 0.01$). There was no further reduction several hours after exposure to the light source and no initial evidence of acclimation (although long-term tests were not conducted). Animals that stayed in the lighted area were positioned so that their heads were turned away from the light source. There was also no significant difference in avoidance response as a function of flash rate (300 versus 450 flashes/min; $P < 0.01$). This was consistent with those found in tests involving juvenile eels.

Field Studies

Field experiments focused on the effectiveness of strobe light as a method to reduce movement of eel passage over the fish ladder at the Saunders GS facility (Figure 1). Tests involved the use of lights sequentially turned "on" and "off" during subsequent 24-h periods. During the study, eleven tests were conducted involving alternate strobe/control combinations. This procedure resulted in considerable temporal variability in eel counts up the ladder during the testing period, which was not the historical norm (Eckersley 1980). In successive tests, there were reductions ranging from 65.3% to 91.6% in the numbers passing up the ladder when strobes were on, compared with control conditions with no light. This reduction was statistically significant over the entire testing period ($P < 0.01$). Observations by divers indicated that juvenile eels were positioned in the shadowed areas where light intensities were less than 0.1 $\mu E/m^2/s$ (<5 lux). This was in agreement with the laboratory results in which strong avoidance was noted to occur above this light intensity.

There was a noticeable reduction in eels passing up the ladder for both control and experimental periods as the testing progressed, especially compared with eel movement from the previous year. This temporal variability was not related to variations in water temperature. It may have been attributed to an avoidance of the area after several unpleasant encounters with strobe light. Controls may therefore have included eels that had been repelled during the strobe "on" cycles. Eels may have acquired a learned behavior to avoid the eel ladder after previous exposure to strobe light.

Strobe light effectiveness appeared to vary with size of eels. During strobe tests, eels in the 38–52 cm range were observed in the collecting bins with a lower frequency (relative to controls) than were eels in the 22–28 cm range. These results suggest that smaller eels are less sensitive to strobe light. Still, these results may be somewhat biased since smaller eels were observed to use the baffles of the ladder for cover more frequently than larger eels as they move up the ladder. Still, these results are interesting since differences in behavioral sensitivity to strobe light may occur in at least three size classes of eels (juveniles 22–28 cm, 38–52 cm, and adults larger than 52 cm; although avoidance still occurs).

These results are important since the implementation of strobe light alone or as part of a fish bypass system must take into account the behavioral differences of both juvenile (upstream migrants) and adult fish (downstream migrants) to strobe light. Although both size classes of eel avoid strobe light, their response times differed. These sensitivity differences will likely become increasingly important in the field, especially under variable flow conditions.

Sound Attraction Results

Preliminary testing was conducted on the use of low frequency signals as an attractant (Figure 6). Although data are limited, we believe we have found a complex signal of low frequency sound (<1000 Hz) that attracts eels. Over a 15-min period, there was evidence of both juvenile and adult eel attraction to the sound projector. Eels in the experimental chamber (mean = 3.7) displayed an obvious orientation towards the projector with some individuals actually "bumping" against it. Three other eels were typically observed orienting within 2 m from the projector (i.e. up to 70% of the eels in the tank were responding to some extent to the sound source). Under control conditions, there was never more than one individual observed with-

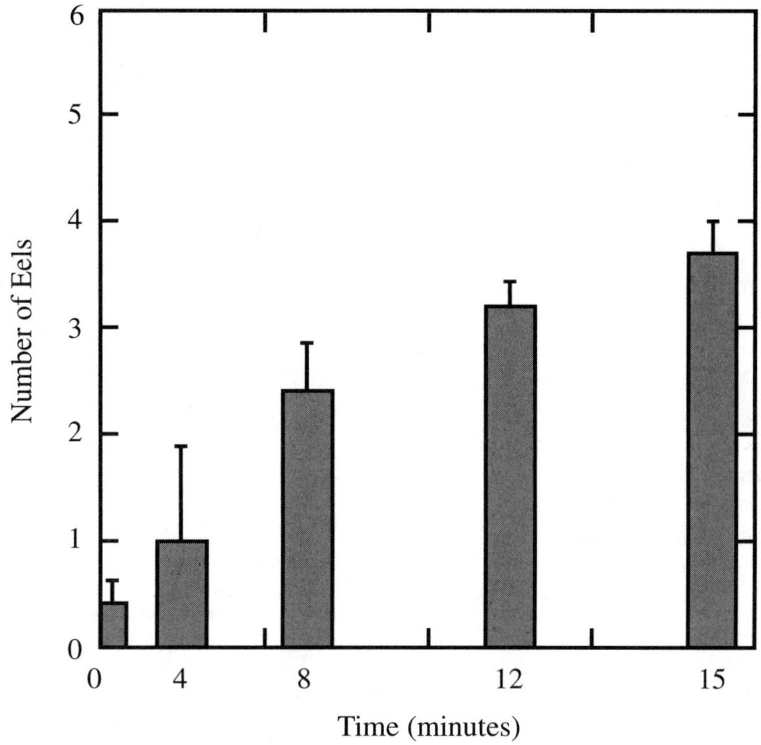

Figure 6. Mean number of adult eels within 1m of the experimental sound projector during attraction tests.

in 1 m of the projector over a 15-min period. Individuals in the test chamber under control conditions were usually positioned away from the projector. These preliminary data have provided evidence that eels may be attracted to sound, which is unique since most sound based studies involve using sound as a repellent (e.g., Haymes and Patrick 1986; Popper and Carlson 1998). More work is required to better define eel response (quantitatively) and the acoustic characteristics, especially under ambient noise conditions in the field.

Sound Conditioning Results

Considerable work on the feasibility of using sound to condition fish to feed has been conducted on several species of fish such as walleye and lake sturgeon. Fish conditioning to sound typically occurred within 10 d. This technique has not yet been tested on American eels.

We tested conditioning in fish by monitoring numbers and distributions in the tank using different low frequency acoustical tones. During typical control conditions fish were generally distributed evenly in the tank, and there was no preference for the area where the feeder was located. During experimental tests, fish immediately swam to the feeder and were subsequently fed (data replicated). These "conditioned" fish tended to have better survival and growth rates than corresponding fish grown under more traditional methods (Patrick et al. 1992). Growth improvements typically ranged from 21% to 29%, whereas survival improvement was more variable, and ranged from 3% to 50% depending on species tested. Additional tests were conducted on memory retention to determine at what point conditioning would be lost. Results, initially from both the laboratory and later in the field, indicated that fish would not lose this conditioning for periods up to 3 weeks. Occasional reinforcement is necessary to extend this time period. How this technique may have potential application in a fish bypass is as follows:

- At an operating hydroelectric facility, sound could potentially be used to complement other existing measures such as light or attraction flows in diverting previously conditioned fish towards a bypass. The intent here is to enhance the performance of an existing system, which may only be operating at 50% effectiveness or less.

- Conditioned fish (fingerlings) would be released (as part of standard practice). Present data suggests that fish may retain conditioning for periods lasting several weeks. Within this time period, one should be able to calculate or estimate the distance migrating fish will travel. In some situations, we believe there may be enough time for fingerlings to migrate from one hydroelectric facility to another. Gradually, fish will lose their conditioning behavior without positive reinforcement (food).
- The potential application of this technology is preferred where there are concerns about cumulative effects of fish mortality, and fish migrants have to pass by more than one hydroelectric or other industrial water user intake. Since hatcheries exist at or near many power plant installations (as part of fisheries enhancement programs), sound conditioning can easily be added as part of the rearing process. This technique likely will also result in an improved growth and survival of these fish during their early development.

Conclusions

1. This behavioral symposium has focused on the use of behavioral stimuli such as light, sound, and turbulent flows in fish diversion applications. Our work has been consistent with many papers in the symposium, which show the effects of various systems in eliciting fish behavioral responses. Both juvenile and adult American eels avoid strobe light, although the avoidance response times for adult eels are slower than those observed for younger fish. This behavioral difference must be taken into account when designing laboratory and/or field studies using strobe lights alone or combined with other systems.
2. Preliminary experiments using sound suggest that some complex signals may be effective as an attractant. At this point, all we have demonstrated is the concept that eels may be attracted by sound, which is unique since most sound-based studies involve using sound as a repellent. More work is required to better define eel response (quantitatively) and the acoustic characteristics, especially under field conditions.
3. Sound conditioning is introduced as a technique that may have application in some fish bypass applications.

Acknowledgment

The authors gratefully acknowledge Dr. Kevin McGrath (NYPA), Kleinschmidt Associates, and Richard Verdon (Hydro Quebec) for providing eels for laboratory studies. Also, thanks to Dr. Charles Coutant for discussions regarding the potential use of sound conditioning for fish passage applications. Funding for this project was provided by Ontario Power Generation. We also thank the two reviewers for their comments of the manuscript.

References

ASMFC. 1999. Fishery Management Plan for American Eel. Fishery Management Report of the Atlantic States Marine Fisheries Commission, Draft Report, Washington, DC.

Casselman, J. M. 1999. American Eels—Another Bad News Story? AFS Newsletter Canadian Aquatic Resources Section. Volume 8, Number 2.

Eckersley, M. 1980. Operation of the Eel Ladder of the Moses-Saunders Dam, Cornwall: 1971–1979. Ontario Ministry of Natural Resources Report, Kingston, Ontario.

Euston, E. T., D. D. Royer, and C. L. Simons. 1998. American Eels and Hydro Plants: Clues to Eel Passage. Hydro Review 17:94–103 (August), Kansas City, Missouri.

Gleeson, L. 1997. Use of Electric Fields to Direct Eel Movement; Eel Passage. Fish Passage Workshop, Milwaukee, Wisconsin, May 6–8, 1997, Alden Research Laboratories, Holden, Massachusetts.

Hadderingh, R. H., and A. G. Smythe. 1997. Deflecting Eels from Power Plant Stations with Light. Fish Passage Workshop, Milwaukee, Wisconsin, May 6-8, 1997.

Haymes, G. T., and P. H. Patrick. 1986. Exclusion of Adult Alewife (*Alosa pseudoharengus*), using a Low Frequency Sound for Application at Water Intakes. Canadian Journal Fisheries and Aquatic Sciences 45:855–862.

Jerko, H., I. Tutunem-Rise, Z. P. S. Enger, and O. Sand. 1987. Hearing in the Eel (*Anguilla anguilla*). Journal of Comparative Physiology 165:455–459.

Patrick, P. H., R. S. Sheehan, and B. Sim. 1982. Effectiveness of a Strobe Light Eel Exclusion Scheme. Hydrobiologia 94:269–277.

Patrick, P. H., E. A. McLeod, and J. J. Kowalewski. 1992. Advanced Aquaculture Techniques—Acoustical Conditioning of Fish. Ontario Hydro Research Division Report, 92-64-K, Toronto.

Patrick, P. H. 1999. The Blue Revolution and Sustainability; at a Crossroads. Chapter 18 *in* J. Leach and R. Claudi, editors. Non-indigenous Freshwater Organisms, Lewis Publishers, New York.

Popper, A. N., and T. J. Carlson. 1998. Application of the Use of Sound to Control Fish Behavior. Transactions of the American Fisheries Society 127:673–707.

Whitfield, R. E., and D. P. Kolenosky. 1978. Prototype Eel Ladder in the St. Lawrence River. Progressive Fish-Culturist 40:152–154.

American Fisheries Society Symposium 26:13–25, 2001

Evaluation of Strobe Lights for Vertically Displacing Juvenile Salmon near a Filling Culvert Intake at the Hiram M. Chittenden Locks, Seattle, Washington

Peter N. Johnson
AScI Corporation, Fisheries Engineering Team
Post Office Box 40, North Bonneville, Washington 98639, USA

Frederick A. Goetz
U.S. Army Corps of Engineers, Seattle District Environmental Resources Section
Post Office Box 3755, 4733 East Marginal Way S., Seattle, Washington 98124, USA

Gene R. Ploskey
U.S. Army Engineer Research and Development Center
Waterways Experiment Station, Fisheries Engineering Team, Post Office Box 40
North Bonneville, Washington 98639, USA

Abstract.—In May 1998, we evaluated the use of strobe lights for vertically redistributing salmon smolts from in front of a culvert used to fill a 15.2 m deep navigation lock chamber in Seattle, Washington. Flows into the culvert during low tides were up to 62.3 cm/s, with maximum instantaneous water velocities of 1.7 m/s. We installed ten strobe light flash heads around the perimeter and in front of the north-filling culvert. Fish densities near the culvert entrance were monitored hydroacoustically. Strobe lights were evaluated using paired on and off treatments during daytime (N = 32). In control treatments before fills, most juvenile salmon were within 5 m of the intake floor. With strobe lights turned on, the fish distribution shifted upward, resulting in a 7-m displacement of the peak densities towards the surface. Mean fish densities at the depth of the culvert (8–13 m) decreased by more than 96% during strobe light-on treatments as compared with control treatments. Based on paired t-tests, mean fish densities in 1-m strata below a depth of 9 m were all significantly lower (α = 0.05) during on treatments than during off treatments. In control treatments during fills, the distribution of juvenile salmon was relatively uniform, with slightly greater mean densities 8–10 m from the surface. During strobe light-on treatments, juvenile salmon were distributed higher in the water column, resulting in a 3-m displacement of the peak densities towards the surface. Mean fish densities at the depth of the culvert decreased by 87% during strobe light-on treatments as compared with control treatments. Based on paired t-tests, mean fish densities in all 1-m strata below a depth of 8 m were significantly lower (α = 0.05) during on treatments than during off treatments. Nighttime tests were also conducted, but nighttime densities were too low for meaningful strobe light evaluation.

Since the 1970s, Lake Washington salmon and steelhead *Oncorhynchus mykiss* populations have undergone a steep decline. The Lake Washington steelhead stock plummeted from a run size of 3,000 adult fish in the early 1980s to the lowest recorded level, 70 fish, in 1994 (S. Foley, Washington Department of Fish and Wildlife, personal communication). Sockeye salmon *O. nerka* populations have experienced a significant decline since building to initial highs in the late 1960s and 1970s from introduction in 1936. The mean spawner return ratio during the last 11 brood years, for which full return data are available, is 0.79 (Washington Department of Fish and Wildlife, Olympia, unpublished data, 1997). Of the four sockeye salmon stocks in Lake Washington, all are considered diminished (Washington Department of Fish and Wildlife 1993). From 1989 to 1996, naturally spawning chinook salmon *O. tshawytscha* have averaged about 700 returning adults for the Cedar River and North Lake Washington tributaries combined. The Lake Washington chinook salmon population is one of 13 populations within the Puget Sound Evolutionary Significant Unit recently listed as threatened. Coho salmon *O. kisutch* escapement peaked at more than 30,000 fish in 1970, but declined to less than 2,000 fish in 1992 (Fresh 1994).

Although declines in salmon and steelhead populations have been prevalent throughout the Northwest, Lake Washington declines appear to be more dramatic for all species than are nearby populations in Puget Sound basins. For example, comparison of Lake Washington coho salmon hatchery returns to two parallel watersheds with coho salmon hatcheries, Skykomish and Green Rivers, indicates that Lake Washington coho salmon have been returning at unusually low numbers. From 1978 to 1994, Lake Washington coho salmon have averaged less than 40% of the average rate of return (smolt-adult survival) of the two other river system hatchery runs. Over a more recent period, 1990–1994, Lake Washington coho salmon returns have averaged less than 10% of the other watersheds' average return rate (U.S. Army Corps of Engineers 1998). This exceptionally low rate of return, in combination with the severe decline in all Lake Washington anadromous fish runs, precipitated resource agency and tribal biologists to investigate the potential causes for the declines.

Since 1995, the Lake Washington Ship Canal at the Hiram M. Chittenden Locks (the Locks) has been the focus of pilot studies and research efforts investigating behavioral aspects of out-migrating juvenile salmon. These studies have been a cooperative effort between many resource agencies including the Washington Department of Fish and Wildlife, Muckleshoot Indian Tribe, King County Land and Water Resources, U.S. Army Engineers Seattle and Portland Districts, and the U.S. Army Engineers Waterways Experiment Station (WES). Since the Locks is the major bottleneck for smolt passage out of the Lake Washington system, agency biologists have assessed the need for smolt passage facilities at the Locks and smolt behavior investigations have been undertaken in recent years (Washington Department of Fish and Wildlife 1997; Ploskey and Johnson 2001, this volume; Dillon and Goetz 1999; Goetz et al. 2001, this volume). Presently, migrating juvenile salmon pass the Locks by one of three major routes: the spillway, lock gates, or filling culverts. Passage through the filling culverts is considered deleterious to migrants. As the large lock chamber is filled prior to lockage, the filling culverts draw water, entraining smolts in the vicinity of the culvert entrance. More than 80% of the filling culvert surface area is covered with sharp-edged barnacles that lacerate and descale smolts as they pass through the culvert into the lock chamber. In addition, there are a number of bends where entrained smolts can impact either concrete or the barnacle-laden surface. Lacerations, wounds, and overall descaling incurred during entrainment through the lock filling system reduce the survival of out-migrant salmon (Washington Department of Fish and Wildlife 1997; Goetz et al. 2001, this volume). There are no data indicating that the spillway or the lock gates are harmful passage routes for out-migrant juvenile salmon. Guidance of smolts away from the culvert entrance and toward alternative passage routes should improve overall survival.

Strobe lights have been demonstrated to elicit avoidance behavior for juvenile salmon under controlled conditions. Avoidance by hatchery-reared subyearling chinook salmon and steelhead was observed in a flume under static conditions (Puckett and Anderson 1987; Anderson et al. 1988). Nemeth and Anderson (1992) reported distinct differences in the behavior of coho and chinook salmon smolts when exposed to strobe lights; coho salmon typically hid while chinook salmon actively swam. In a recent series of studies conducted in a cylindrical net pen in Salmon Bay near the Locks, Ploskey and Johnson (2001, this volume) found that yearling coho and subyearling coho and chinook salmon exhibit avoidance to strobe light stimuli and characterized the nature of the response as an immediate avoidance reaction to the strobe light. Repeated testing showed that smolts and juveniles swam to the portion of the net pen least illuminated by the flash heads. Amaral et al. (1998a) conducted tests in an experimental cage and reported that chinook salmon exhibited strong avoidance to strobe lights during nighttime hours, but little or no reaction was observed under daylight and early evening conditions. Young-of-the-year wild and hatchery-reared fall chinook salmon (43–50 mm TL) exhibited strong and moderate avoidance responses, respectively, when exposed to pulsed light in an experimental test tank (Mueller et al. 2001, this volume).

In field tests, McKinley and Patrick (1988) reported that strobe lights were 56% effective in diverting downstream migrating sockeye salmon away from turbine intakes at Seton Hydroelectric Station in British Columbia. Strobe lights effectively reduced entrainment of Atlantic salmon smolts into two of four turbine intakes at Maine's Weldon Dam (Bernier 1995). Kokanee exhibited strong avoidance of strobe lights in tests conducted in 1997 and 1998 at Spirit Lake, Idaho and at Lake Pend Orielle, Idaho (Maiolie et al. 2001, this volume).

Preliminary results suggest a strobe light barrier system installed at a highway overpass has effectively reduced entrainment of juvenile chinook salmon into the Burbank 3 intake channel on the Columbia River (Amaral et al. 1998b). Johnson and Ploskey (1998) compared smolt behaviors in terms of swimming direction and orientation with and without strobe light stimuli in a juvenile bypass channel, and found that 95% of all observed smolts modified their swimming behavior during strobe light-on treatments. By contrast, although strobe lights appeared to alter the vertical distribution of fish in front of Rocky Reach Dam, the lights were reported to be ineffective at improving fish guidance efficiency (Anderson et al. 1988).

Our objective was to evaluate the effectiveness of strobe lights for vertically displacing migrant juvenile salmon near a filling culvert intake at the Locks with intention of developing a permanent light deployment by the year 2000. Vertical displacement would reduce risk of entrainment and encourage out-migration via lock chambers or spillway. Strobe light evaluation tests were conducted during daytime and nighttime periods, before and during fill events.

Methods

Site Description

We conducted the strobe light evaluation at the Locks located at the outlet of the Lake Washington Ship Canal in Seattle, Washington. The Locks primarily function as a navigation lock for vessels passing between the freshwater Lake Washington system and the salt water of Shilshole Bay and Puget Sound. From north to south, the Locks consists of a large lock, small lock, a spillway, and an adult fish ladder (Figure 1). The large lock chamber is 24.4 m wide by 251.5 m long by 15.2 m deep and accommodates vessels with drafts as deep as 9.1 m. The chamber is divided lengthwise into two half-chambers (upper and lower) by an intermediate (middle) gate. The head differential from upstream to downstream of the lock chambers

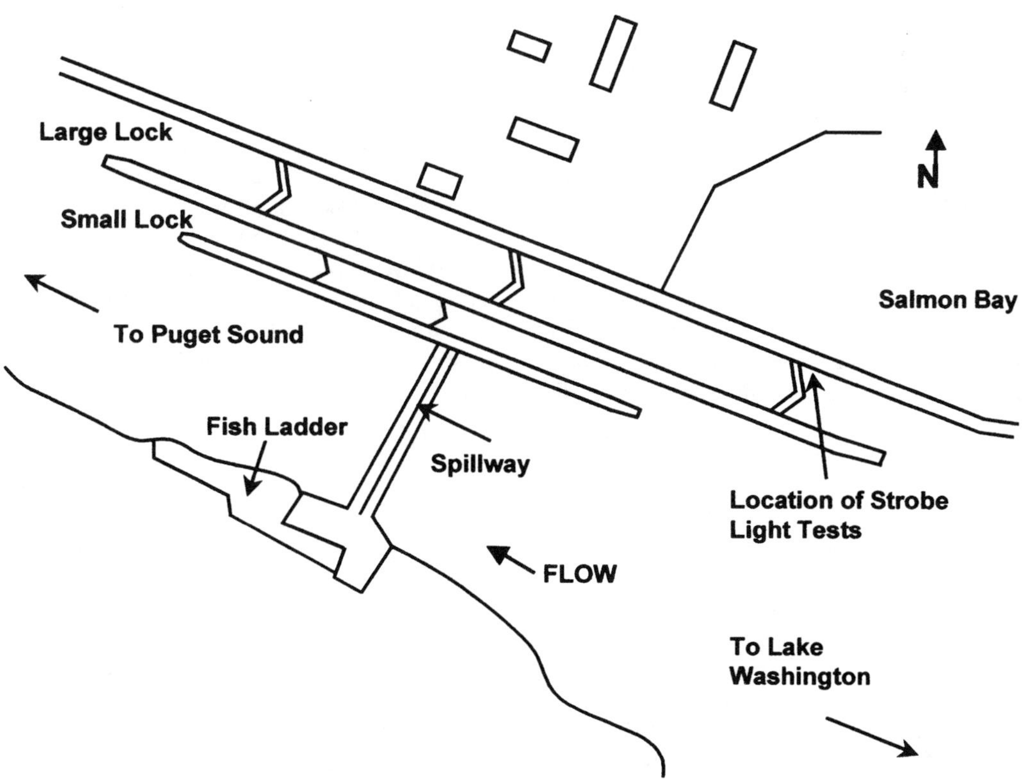

Figure 1. Plan view of the Hiram M. Chittenden Locks Project showing all major structures and location of strobe light evaluations.

varies from 1.8 to 7.9 m depending on the tidal elevation and the level of Lake Washington. Tide levels measured just downstream from the Locks in Shilshole Bay fluctuate about 3.7 m over the course of each tidal cycle. Except for annual pumpouts for maintenance purposes, the locks are operable 24 h/d throughout the year.

The large lock chamber is filled via two culverts that originate as 4.9 m wide by 4.3 m tall openings located at the floor of the channel in both walls just upstream of the upper miter gates. The culverts route water north and south, respectively, for about 4.5 m before turning westward (90° angle) and constricting to 2.6-m wide by 4.3-m tall. The culverts continue laterally along the chamber walls, emptying into the chamber through a series of 22 (1.2-m wide by 0.6-m high) filling ports. The chamber is filled by opening three fixed-wheel, vertical lift Stoney gate valves just west of the upper miter gates and just east of the middle miter gates. For all strobe light tests, the "slow-continuous" valve-opening procedure was used, resulting in the Stoney gate valves opening completely in 4 min and 15 s. Maximum inflow at low tide (= greatest hydraulic head) into each culvert is approximately 62.3 cm/s with inflows greater than 51.0 cm/s lasting up to 3 min.

Strobe Lights

During the annual maintenance pumpout in November 1997, we installed ten 35.6-cm-diameter strobe heads manufactured by Flash Technology, Inc., in the vicinity of the north wall filling culvert. We placed seven lights along the perimeter of the culvert and three in a diagonal pattern south and east of the culvert entrance (Figure 2). Floor-mounted lights were recessed into the concrete sill to prevent damage from cables and chains hanging from vessels entering the lock chamber. All floor-mounted strobe heads were oriented up approximately 45°. The three lights at the floor of the culvert entrance were aimed in a southward direction and the three in the diagonal line were aimed in a southeastward direction. Wall-mounted lights were placed along the side-wall edges of the culvert entrance and aimed in horizontally at approximately 45°.

The three lights at the floor of the culvert entrance were AGL 901 series strobe lights. This is a new model that allows for flexibility in flash rate, light intensity and sequencing via a laptop computer. The other seven lights were older and were limited to a flash rate of 300 flashes/min and operation at medium (600 W) power. We set the flash rate and light intensity values of the AGL 901 series lights to match the older lights in order to equalize stimuli from all of the strobe heads. During strobe light treatments, all lights were flashing at the same time (i.e. we did not sequence the lights).

Turbidity Measurement

On 11 June we deployed a model 8-CBA Little Giant water pump along the vertical midline of the culvert entrance to pump water to a basin on the walkway deck to test water clarity. The pump was lowered in successive distances of 1.25 m to obtain a vertical profile of turbidity in the strobe light testing environment. We measured turbidity with a LaMotte model 2008 nephelometer calibrated to a 0.5 nephelometric turbidity unit (NTU) standard. We acquired at least three replicate samples per station, and up to five samples when differences of up to 0.15 NTU's were observed among the first three samples obtained.

Velocity Measurement

On 22 and 23 September, we deployed a Marsh McBirney model 511 electromagnetic current meter from a 23-m aluminum pole assembly suspended from a crane. We lowered the probe along the vertical midline of the culvert entrance to obtain instantaneous maximum velocity estimates at nine depths. Once the probe was positioned at each depth, the valves were opened to fill the lock chamber and we recorded maximum water velocity estimates for two dimensions: (1) flow toward or away from the culvert entrance, and (2) lateral flow west (downstream from culvert) and east (upstream from culvert) along the wall.

Fish Monitoring

We used a pair of 6° single-beam Biosonics, Inc., transducers to monitor fish densities near the culvert entrance during strobe light and control tests. The transducers were deployed from an aluminum pole that could be easily raised for vessel traffic when not in use. The transducers were aimed downstream 25° from vertical which resulted in the acoustic beams crossing the culvert opening at the center of the culvert entrance floor (Figure 2). We operated the transducers using a Biosonics, Inc., 420 kHz ES 2000 echosounder, multiplexor, echo signal processing card, and 120 MHz personal computer. Transducers were fast-multiplexed at 10 pings/s each.

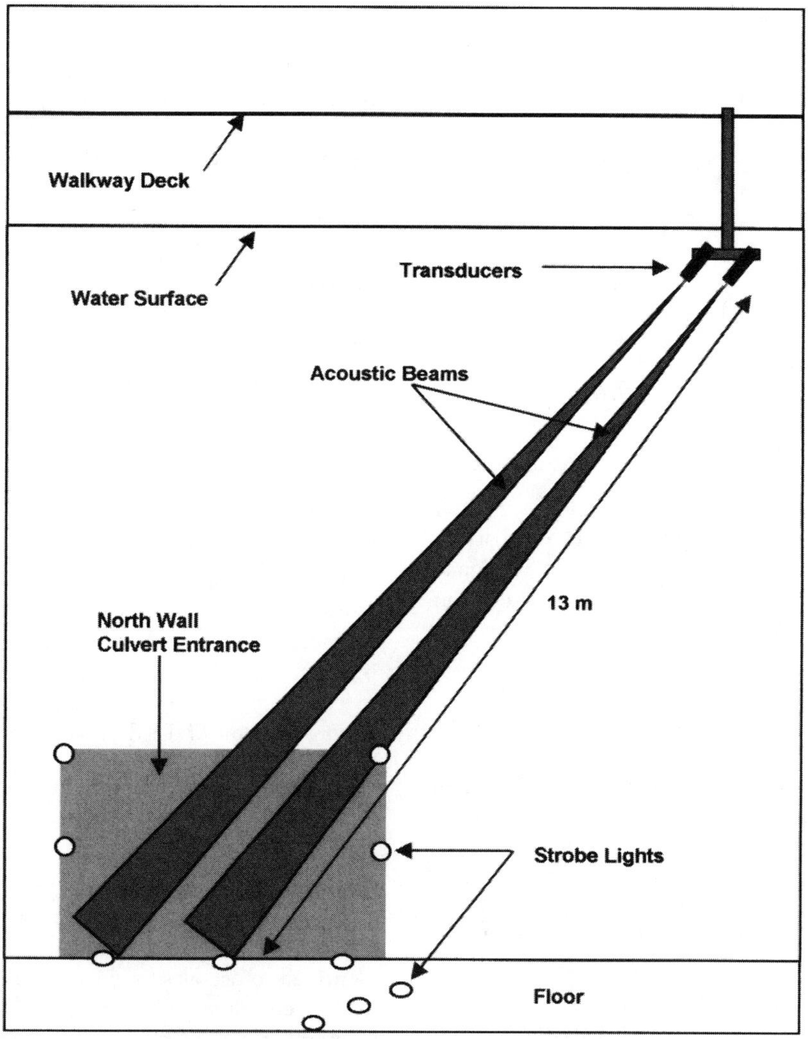

Figure 2. Light placement and hydroacoustic monitoring beams for strobe light evaluations along the north wall of the lock chamber entranceway.

Acoustic Signal Processing

Due to the large numbers of fish in the acoustic beams, we were unable to use single target tracking to count fish during strobe light testing. Instead we used echo integration, which measures the total reflected energy from the ensonified volume. We programmed the software to amplify acoustic signal returns based on a 20 Log R (range) Time Varied Gain. We integrated fish densities every 15 s for each of 11 1-m range strata. Fish density estimates reported herein are aggregated as range strata from the transducers. Table 1 lists depth increments that are equivalent to the range strata from the transducers.

The echo integration technique results in average relative intensity. To convert the relative intensity to values of fish density, it is necessary to calculate an echo integrator scaling constant based upon calibration parameters, data collection parameters, and an estimate of the reflectivity of an average-sized individual fish (MacLennan 1990). This constant (A) is defined as

$$A = 1/(c \times \tau \times \pi \times p^2 \times g^2 \times b_{av}^{\ 2} \times \sigma_{bs}),$$

where c is the speed of sound in water (1500 m/s); τ is the pulse length in s, π is pi (3.1416); p^2 is a pressure term calculated from system source level; g^2 is a gain term calculated from system receive

Table 1. Conversion of range from transducers to depth intervals.

Range (m)	Depth (m)
2.0	1.8
3.0	2.7
4.0	3.6
5.0	4.5
6.0	5.4
7.0	6.3
8.0	7.3
9.0	8.2
10.0	9.1
11.0	10.0
12.0	10.9
13.0	11.8

sensitivity, b_{av}^2 is the average squared beam pattern (based on transducer directivity), and σ_{bs} is sigma, or back scattering cross section (acoustic size) of an average fish.

Strobe Light Tests

We evaluated strobe light effectiveness by examining differences in fish densities among paired strobe light-on and strobe light-off treatments before and during large lock fill events. We ran a total of 32 pairs of strobe-on and strobe-off tests during daylight hours from 9 May through 20 May and 30 pairs of nighttime tests from 20 May through 2 June. Before the beginning of each paired test, we randomly chose either a test treatment (strobe light on) or a control treatment (strobe light off). At least 10 min before a filling event, we turned the strobe lights on or kept them off and started collecting acoustic information. We then recorded the time when the fill event started and continued to collect hydroacoustic data through the end of the fill event while maintaining the same treatment. The opposite treatment then followed to complete the pair. Average fill times were approximately 9 min, but varied slightly with tidal elevation (lock volume).

During testing, all lockages were performed using the full lock chamber. Under normal boat traffic conditions only 3–4 full lockages occur per day. To meet the needed number of testing pairs, additional lockages were performed when boats were not present using "false lockages." The lockmasters performed these additional, false lockages at prescribed times where they opened the Stoney gate valves and filled the lock chamber just as if a vessel were locking through.

Acoustic Data Processing

We summed the fish density estimates for both acoustic beams and all 15-s integration intervals by range strata for each test period before the fill event, and during each fill event. We expanded the density estimates for prefill test periods to 10 min in order to standardize in time. For example, if we collected 9 min and 20 s of acoustic data for a given before-fill test period, then the summed total density estimates by depth strata for that period would be increased by a factor of 1.07 (10 min/sampling time). Likewise, we adjusted density estimates collected during fill-event test periods to account for time differences among paired treatments.

Purse Seining the Large Lock Chamber

The Washington Department of Fish and Wildlife and the Muckleshoot Indian Tribe conducted a series of purse-seine sets in the upper large lock chamber as part of an effort to obtain estimates of injury rates associated with fill events as well as species composition data through the migration season. A total of 13 d of sampling consisting of approximately four purse-seine sets/d occurred from 6 May through 11 June. A 206-m-long by 11-m-deep purse seine with 1–2 cm (stretch) webbing was deployed along the perimeter of the upper lock chamber 10–20 min following a fill event immediately after cessation of seiche in the lock chamber. Fish were identified by species with juvenile salmon identified to age-class and evaluated for injury following entrainment through the lock conduits. Captured smolts were removed from the lock chamber before the next fill. Strobe light testing did not occur during periods of purse seining.

Results

Physical Environment

Mean turbidity values by depth strata ranged from 1.0 to 1.41 NTUs, with a slight increase in turbidity with depth. Two-dimensional velocity measurements acquired along the vertical midline of the culvert entrance are shown in Table 2. The maximum instantaneous velocity estimates obtained for flow moving into the culvert was 1.7 m/s and laterally along the wall was 1.2 m/s. Based on observed discharge values, Waller et al. (1999) computed velocities in the culvert during slow-continuous valve operation to range from 4.3 to 7.6 m/s.

Species Composition

Juvenile sockeye salmon dominated the catch during the period when strobe light evaluations were conducted. Table 3, provided by D. Seiler, Washington Department of Fish and Wildlife, lists summarized catch data of juvenile salmon captured in purse-seine efforts by date.

Biological Assumptions and Average Fish Size

We relied on the purse-seine catch data collected from inside the lock chamber to provide an estimate of species composition since species cannot be identified with hydroacoustic methods. From 6 May to 2 June (the time period of purse seining bracketing the time period of strobe light investigations) juvenile sockeye salmon represented 74% of the catch. Based on this, we assumed that outmigrating salmon, especially juvenile sockeye, dominated the fish composition we sampled during strobe light tests.

We used a sigma value of 6E–5 to scale the fish density estimates collected during strobe light tests. This value equates with a length of about 14 cm (approximate average size of juvenile sockeye salmon collected in purse-seine samples; D. Seiler, Washington Department of Fish and Wildlife, personal communication) and an echo strength of –42.2 dB.

Strobe Light Tests Prior to Fill Events

Distributions of fish near the filling culvert differed significantly among daytime test and control treatments (Figure 3). During control treatments, the vertical distribution of fish was skewed toward the floor at the depth of the culvert, i.e., mean fish densities were greatest within the 11–13 m range

Table 2. Instantaneous maximum velocity estimates for two dimensions along vertical midline of culvert entrance: (1) flow toward or away from the culvert entrance, and (2) lateral flow west (downstream from culvert) and east (upstream from culvert) along the wall. Velocities are listed by depth increment, and tide levels above mean sea level are included. The symbol ' – ' denotes no net flow in that direction.

Depth (m)	Away from culvert (mps)	Toward culvert (mps)	Downstream from culvert (mps)	Upstream from culvert (mps)	Tide (m)
1.5	0.12	0.30	0.43	0.09	1.2
3.3	0.06	0.12	0.64	–	1.2
4.4	0.30	0.30	0.61	–	1.2
5.6	0.55	0.18	0.55	–	3.2
6.8	0.43	0.46	0.64	–	1.1
8.0	–	0.58	0.67	–	1.2
9.2	–	1.10	1.22	0.12	1.8
10.7	–	0.85	0.12	0.55	3.5
12.2	–	1.68	0.24	0.73	1.4

Table 3. Daily total number of individuals and species composition of juvenile salmonids captured in purse seine sets in the locks during the period of strobe light testing, Hiram M. Chittenden Locks, 1998.

		Composition (%)			
Date	Total number	Sockeye	Coho	Chinook	Steelhead
6-May	990	86.4	12.2	1.4	0.0
7-May	2934	92.2	7.0	0.3	0.5
12-May	303	62.7	30.4	6.6	0.3
13-May	695	86.5	10.5	3.0	0.0
14-May	823	84.8	6.7	8.5	0.0
20-May	424	43.9	10.8	45.3	0.0
21-May	138	83.3	8.0	8.7	0.0
27-May	77	45.5	22.1	32.5	0.0
28-May	343	9.9	1.2	88.9	0.0
4-Jun	631	4.1	8.7	87.2	0.0

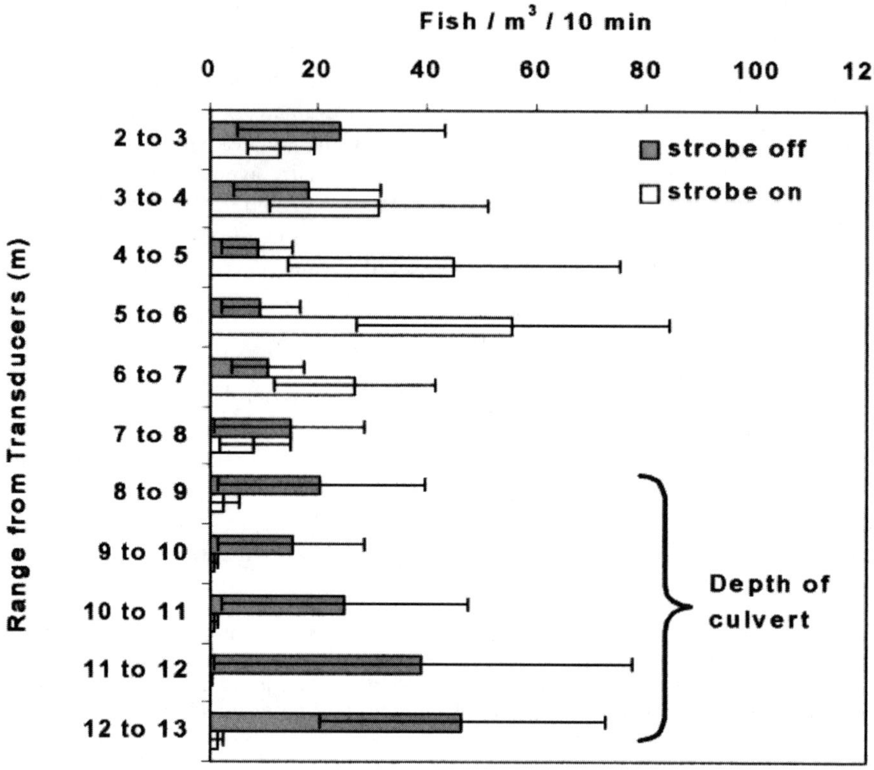

Figure 3. Mean fish densities among control and test treatments 10 min before daytime fill events. Error bars indicate 95% confidence limits about the mean.

strata. During strobe light treatments, the peak distribution of fish shifted towards the surface, with most fish concentrated between 5 and 6 m from the transducers. Mean fish densities were greatly reduced at the depth of the culvert entrance. Two-sample paired *t*-tests indicated mean fish densities were significantly lower (α = 0.05) during strobe light treatments than during control treatments in several range strata at the elevation of the culvert entrance, but overall mean densities for the entire sample volume did not differ (Table 4). Mean fish densities at ranges of 4–6 m from transducers were significantly greater (α = 0.05) during strobe light treatments as compared with control treatments. On average, at-risk fish (proportion of fish within 5 m of the floor) decreased from 63% during control treatments to 2.9% during test treatments.

Vertical distribution patterns during nighttime control treatments were similar to those observed for daytime control treatments, with most of the fish concentrated near the surface or close to the floor (Table 5). During nighttime strobe light-on treatments, greater densities of fish were observed near the surface and 8–9 m from the transducer. Fish densities were significantly lower (α = 0.05) during strobe light treatments as compared with control treatments from 11 to 13 m in range from the transducers. Nighttime fish densities prior to fill events for both treatments combined were reduced by 92% as compared with daytime fish densities (compare Tables 4 and 5).

Strobe Light Tests During Fill Events

Distributions also differed significantly among daytime test and control treatments sampled during fill events (Figure 4). Fish were generally distributed towards the floor, with peak concentrations at 8–10 m from the transducers during control treatments. The fish distribution shifted toward the surface during strobe light treatments, with greatest concentrations observed at 6–7 m from the transducers. Like test periods prior to fill events, mean fish densities were greatly reduced at the depth of the culvert entrance during strobe light treatments as compared with controls. Two sample paired *t*-test results indicated significantly lower (α = 0.05) fish

Table 4. Mean fish densities among control and test treatments before and during daytime fill events with *P* values from two-tailed paired *t*-test results; *N* = 32.

Range	Mean fish number / m^3 / 10 min			Mean fish number / m^3 / fill		
(m)	Control	Test	*P*	Control	Test	*P*
2 to 3	24.2	13.1	0.240	11.7	12.4	0.817
3 to 4	18.0	31.2	0.151	18.1	15.3	0.569
4 to 5	8.7	44.9	0.018	18.7	23.4	0.515
5 to 6	9.3	55.4	0.003	17.1	44.0	0.063
6 to 7	10.7	26.6	0.069	19.5	63.5	0.022
7 to 8	14.7	8.2	0.398	23.9	38.7	0.114
8 to 9	20.5	2.4	0.069	38.5	12.2	0.003
9 to 10	15.0	0.7	0.044	37.8	2.2	0.001
10 to 11	24.7	0.7	0.046	27.1	0.7	0.001
11 to12	39.1	0.2	0.056	15.4	0.6	0.002
12 to 13	46.3	1.3	0.002	9.6	1.2	0.005
All strata	231.2	184.6	0.580	237.4	214.1	0.676

Table 5. Mean fish densities among control and test treatments before and during nighttime fill events with *P* values from two-tailed paired *t*-test results; *N* = 30.

Range	Mean fish number / m^3 / 10 min			Mean fish number / m^3 / fill		
(m)	Control	Test	*P*	Control	Test	*P*
2 to 3	3.6	2.6	0.454	6.9	12.8	0.113
3 to 4	3.4	2.4	0.483	8.4	14.9	0.105
4 to 5	1.1	1.5	0.568	6.4	7.8	0.391
5 to 6	0.4	0.5	0.647	5.6	8.5	0.133
6 to 7	0.2	0.5	0.219	3.6	4.5	0.371
7 to 8	0.1	0.5	0.151	1.7	3.3	0.104
8 to 9	0.2	2.6	0.184	1.1	1.5	0.334
9 to 10	0.8	1.2	0.474	0.6	1.1	0.109
10 to 11	1.4	1.2	0.599	0.5	0.9	0.040
11 to12	3.6	0.5	0.010	1.2	1.0	0.553
12 to 13	4.9	1.0	0.003	3.3	2.4	0.105
All strata	19.9	14.5	0.273	39.2	58.7	0.075

densities from 8 to 13 m in range from the transducers during strobe on versus control treatments (Table 4). Mean fish densities increased during strobe light treatments at the 6–7 m range strata, but overall mean densities for the entire sample volume did not differ. On average, at-risk fish (proportion of fish within 5 m of the floor) decreased from 54% during control treatments to 7.8% during test treatments.

Compared with daytime distributions during control treatment fill events, proportionally fewer fish were observed at the depth of the culvert during nighttime fill events (Table 5). Vertical distribution patterns were similar among both control and test treatments with most fish concentrated within 2–6 m from the transducers. However, fish densities 10–11 m from the transducers decreased significantly ($\alpha = 0.05$) during strobe light treatments as compared with control conditions. Nighttime fish densities during fill events for both treatments combined were reduced by 78% as compared with daytime fish densities (compare Tables 4 and 5).

Discussion

Estimates of fish distribution during control treatments (strobe lights off) before and during daytime fill events clearly indicate that fish near the north

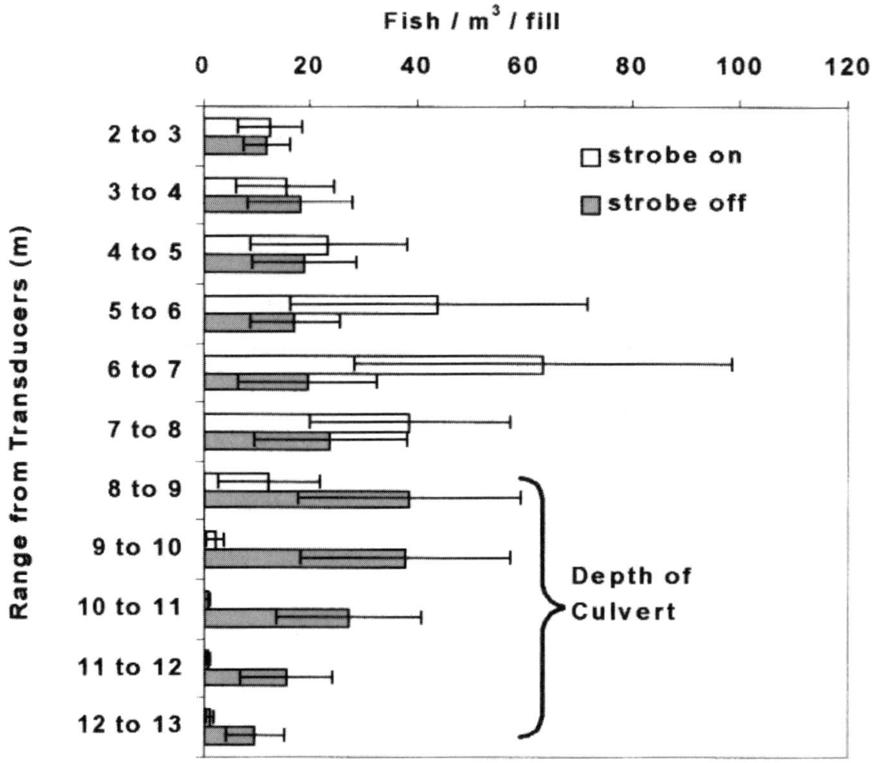

Figure 4. Mean fish densities among control and test treatments during daytime fill events. Error bars indicate 95% confidence limits about the mean.

wall concentrated within 5 m of the floor (Figures 3 and 4). Biosonics (1997) also reported concentrations of fish in this location at these depths during the day based on mobile hydroacoustic surveys. Fish near the north wall and within 5 m of the floor were most vulnerable to the entrainment flows associated with fill events and were thus at risk of injury incurred during passage through the barnacle-lined conduit system.

Fish distribution patterns were altered greatly by strobe light treatments before and during daytime fill events. In both cases, strobe lights shifted the distribution of fish vertically, with greatest concentrations observed above the elevation of the culvert entrance (Figures 3 and 4). Although the zone of influence of entraining flows is unknown, we believe that fish distributed above the elevation of the culvert entrance were much less vulnerable to entrainment into the filling culvert. Strobe lights were shown to effectively clear the immediate area in front of the culvert entrance of fish, reducing the proportion of at-risk fish to less than 3% of total fish density during prefill tests and less than 8% of total fish density during fill event tests. However, one limitation of our study was that we did not estimate entrainment of fish into the culvert. We believe that upward displacement of fish would greatly reduce entrainment, but this could not be determined.

Although we detected significant reductions in fish densities during nighttime strobe light tests at a few range strata, we think there is little relevance to these observed differences because the overall nighttime densities were so low and the densities that were detected were likely non-salmonid resident fish. Hydroacoustic estimates of fish densities near the entrance to the large lock collected in 1996 consistently showed lower fish abundance at night versus day (Dillon and Goetz 1999). Observation of smolt passage over the spillway in 1995 found negligible passage during nighttime hours (Goetz et al. 2001, this volume).

The test location at the Locks offered ideal conditions for field testing of strobe lights due to its manageable scale and low turbidity. Additionally, the 1998 out-migration was estimated at 3 million sockeye salmon smolts, which provided a large number of fish to displace. Purse-seine samples inside the large lock chamber, which were dominated by juvenile salmon, indicated that the majority of the light-displaced fish near the culvert entrance were juvenile salmon. Since the catch data showed greater proportions of sockeye relative to other salmon species during the course of the tests, it is reasonable that specifically, juvenile sockeye avoided the strobe lights and thus likely avoided entrainment and its associated consequences.

Conclusions and Recommendations

Given the encouraging results presented here, we strongly recommend further evaluations of strobe lights for redistributing juvenile salmon away from filling culvert entrances at the Chittenden Locks Project. Future investigations should include, but not be limited to the following:

1. Simultaneously test strobe light effectiveness at both the north and south wall culvert entrances. Fish that are redistributed away from one culvert may be entrained into the opposite culvert.
2. Monitor fish densities with split-beam hydroacoustics during strobe light evaluations. If fish densities are small enough for single target tracking, split-beam data acquisition would allow for calculation of trace direction across the beam and estimation of fish size. Trace direction yields calculation of entrainment estimates and target size information allows for thresholding of smaller or larger fish. Additionally, fish size estimation would allow for calculation of more accurate scaling factors for echo integration processing during periods when fish densities are too great to single target track.
3. In conjunction with split-beam hydroacoustics, monitor species composition with purse seining to verify equal effectiveness among species.
4. Conduct hypothesis testing of strobe light effectiveness as a function of both flash rate and light intensity (i.e., does effectiveness differ among flash rates of 300 and 450 flashes/min?). Compare effectiveness among years to determine optimal system configuration.
5. Automate transducer deployment, data collection, and strobe light control.
6. Exhaustively map the strobe light field in three dimensions to better understand the size of the affected area.
7. Employ an acoustic doppler velocimeter (ADV) at successive distances from the culvert entrance for estimating fill velocities around the culvert. These measurements should be made over a range of tidal elevations.

Strobe lights vertically displaced juvenile salmon near a filling culvert at the Locks, but can pulsating light redistribute juvenile salmon as effectively at other, more complex fish passage facilities? Fish must have an alternative passage route to avoid lights and time to react and swim away before being entrained. This makes in-turbine deployments at hydropower plants unsatisfactory. However, the implications of successful vertical displacement of out-migrant salmon are especially important given the advent of surface collection technology at dams on the Columbia and Snake Rivers. We acknowledge that the area in front of the filling culvert at Chittenden Locks is quite dissimilar to forebay conditions at main-stem Columbia and Snake River Projects, especially in terms of current velocity and water clarity. Although we recorded maximum instantaneous velocities during fill events that exceeded 1.5 m/s, we recognize this differs considerably from the sustained flow velocities that occur near intake structures of hydropower facilities. Columbia River water has been reported to be more turbid than our measurements at the Locks by a factor of 10 (Johnson et al. 1998). Despite these differences in site conditions, we believe the results from this study strongly support consideration of testing strobe lights to enhance surface collection on the Columbia and Snake River System.

Acknowledgments

The tests described and the resulting data presented herein, unless otherwise noted, were obtained from research for Seattle District of the USACE by the U.S. Army Engineer Research and Development Center, Waterways Experiment Station, 3909 Halls Ferry Road, Vicksburg, Mississippi 39180-6199. Permission was granted by the Chief of Engineers to publish this information. Funding for this study was provided by Seattle District USACE through two authorities, Section 22 Planning Assistance to the States and Section 1135 Environmental Restoration. King County was the primary local sponsor under both programs. J. Lombard, King County Land and Water Resources, provided the major support.

Many people contributed to this study, especially the personnel at the Hiram M. Chittenden Locks. J. Post (Maintenance Supervisor) and W. Livermore (Assistant Supervisor) provided logistics and troubleshooting. Lockmasters R. Donelon, J. Dodge, L. Meyer, and S. Nitta provided false lockages during strobe light tests. Mechanics S. Hansen and J. Kragstaedt aided in deployment of velocity meter. S. Denny supported placement of the floor-mounted strobe heads. M. Weiland (AScI) and M. Burczynski (DynTel) assisted in deploying strobe lights and cables, and Mr. Weiland helped develop and implement transducer deployments. Biosonics, Inc. provided hydroacoustic and computer expertise, especially C. Sullivan, L. Harkness, J. Dawson, and J. Hedgepeth. D. Seiler and L. Kishimoto, Washington Department of Fish and Wildlife, provided purse-seine data on species composition. C. Schilt provided comments that greatly improved this manuscript.

References

Amaral, S. V., F. C. Winchell, B. C. McMahon, and E. P. Taft. 1998a. Evaluation of behavioral guidance technologies for diverting chinook salmon smolts at the Roza Dam Screening Facility. Report number 218-98/E873F, prepared for the Yakima/Klickitat Fisheries Project, Washington Department of Fish and Wildlife, Yakama Indian Nation, Bonneville Power Administration by Alden Research Laboratory, Holden, Massachusetts.

Amaral, S. V., G. N. Garnett, and N. Taft. 1998b. Review of downstream fish passage and protection technology evaluations and effectiveness. Technical Report TR-111517. Electric Power Research Institute, Palo Alto, California.

Anderson, J. J., K. J. Puckett, and R. S. Nemeth. 1988. Studies on the effect of behavior on fish guidance efficiency at Rocky Reach Dam: avoidance to strobe light and other stimuli. Report Number FRI-UW-8801. Fisheries Research Institute, University of Washington, Seattle.

Bernier, K. R. 1995. Report on effectiveness of the permanent downstream passage system for Atlantic salmon at Weldon Dam. Mattaceunk Project, Federal Energy Regulation Commission Report Number 2520.

Biosonics, Inc. 1997. Evaluation of salmonid smolt passage into the Hiram M. Chittenden Locks, Spring, 1997. Report to U.S. Army Corps of Engineers, Seattle District, Seattle.

Dillon, J., and F. Goetz. 1999. Manual analysis of fish presence and abundance at a large lock using fixed hydroacoustics techniques. Report to the Lake Washington Ecological Studies Group, Seattle District, U.S. Army Corps of Engineers, Seattle.

Fresh, K. L. 1994. Lake Washington fisheries: a historical perspective. Journal of the North American Lake Management Society 9:52–55.

Goetz, F. A., J. J. Dawson, T. Shaw, and J. Dillon. 2001. Evaluation of low-frequency underwater sound emitters for guiding salmon, and steelhead smolts away from a navigation lock at the Hiram M. Chittenden Locks, Seattle, Washington. Pages 91–104 *in* Charles Coutant, editor. Behavioral technologies for fish guidance. American Fisheries Society, Symposium 26, Bethesda, Maryland.

Johnson, P. N., G. R. Ploskey, T. J. Carlson, J. M. Nestler, and J. Evans. 1998. Feasibility study for evaluating redistribution of juvenile salmon in the McNary Bypass Channel using behavioral technologies: 1996. Technical Report EL-98-15, U.S. Army Engineers, Waterways Experiment Station, Vicksburg, Mississippi.

Johnson, P. N., and G. R. Ploskey. 1998. Behavioral technologies for bypass channels. Phase 2: Evaluation of infrasound and strobe lights for redistributing migrant salmonid smolts in the McNary juvenile bypass. Technical Report EL-98-17, U.S. Army Engineers, Waterways Experiment Station, Vicksburg, Mississippi.

MacLennan, D. N. 1990. Acoustical measurement of fish abundance. Journal of the Acoustical Society of America 87:1–15.

Maiolie, M. A., B. Harryman, and B. Ament. 2001. Response to free-ranging kokanee to strobe lights. Pages 27–35 *in* Charles Coutant, editor. Behavioral technologies for fish guidance. American Fisheries Society, Symposium 26, Bethesda, Maryland.

McKinley, R. S., and P. H. Patrick. 1988. Use of behavioral stimuli to divert sockeye salmon smolts at the Seton Hydroelectric Station, British Columbia. Proceedings: Fish Protection at Steam and Hydro Plants. EPRI CS/EA/AP-5663-SR. Electric Power Research Institute, Palo Alto, California.

Mueller, R. P., D. A. Neitzel, and B. G. Amidan. 2001. Evaluation of infrasound, and strobe lights to elicit avoidance behavior in juvenile salmon, and char. Pages 79–89 *in* Charles Coutant, editor. Behavioral technologies for fish guidance. American Fisheries Society, Symposium 26, Bethesda, Maryland.

Nemeth, R. S., and J. J. Anderson. 1992. Response of juvenile coho and chinook salmon to strobe and mercury lights. North American Journal of Fisheries Management 12:684–692.

Ploskey, G. R., and P. N. Johnson. 2001. Effectiveness of strobe lights, and two sound devices for eliciting avoidance by juvenile salmon. Pages 37–55 *in* Charles Coutant, editor. Behavioral technologies for fish guidance. American Fisheries Society, Symposium 26, Bethesda, Maryland.

Puckett K. J., and J. J. Anderson. 1987. Behavioral responses of juvenile salmonids to strobe and mercury lights. Report Number FRI-UW-8717, Fisheries Research Institute, University of Washington, Seattle.

U.S. Army Corps of Engineers. 1998. Coho salmon adult return rate: preliminary information on baseline survival. In Section IIE, Appendix F, Part 1, Howard Hanson Dam Additional Water Storage Project Feasibility Report. Seattle District, Seattle.

Waller, T. N., T. L. Fagerburg, and D. C. McVan. 1999. Prototype evaluation of Lake Washington Ship Canal Navigation Lock, Seattle, Washington. Draft Technical Report Prepared for U.S. Army Engineer District, Seattle.

Washington Department of Fish and Wildlife. 1993. 1992 Washington state salmon and steelhead stock inventory. Olympia, Washington.

American Fisheries Society Symposium 26:27–35, 2001

Response of Free-Ranging Kokanee to Strobe Lights

Melo A. Maiolie, Bill Harryman, and Bill Ament
Idaho Department of Fish and Game, Post Office Box 806, Bayview, Idaho, 83803, USA

Abstract.—We tested the response of kokanee *Oncorhynchus nerka* to strobe lights. Testing was conducted on wild, free-ranging fish in their natural environment (i.e., the pelagic region of two large Idaho lakes). Split-beam hydroacoustics were used to record the distance kokanee moved away from the lights as well as the density of kokanee in the area near the lights. In control tests, where the strobe lights were lowered into the lake but kept turned off, kokanee remained within a few meters of the lights. Once the lights began flashing, the distance between kokanee and the light source increased and was highly significant when compared with the control transects ($P < 0.001$). Kokanee moved an average of 30–136 m away from the lights in waters with Secchi transparencies from 2.8 to 17.5 m. Kokanee densities near the lights were significantly lower ($P < 0.001$) when the lights were turned on than in control samples with no lights flashing and densities dropped between 72% and 100%. Flash rates of 300, 360, and 450 flashes/min elicited strong avoidance responses from the fish. Kokanee remained at least 24 m away from the lights during our longest test of 5 h 50 min. Kokanee appeared to be responding to flashes that were less than 0.00016 lux (about 2.96×10^{-6} micromoles·m^{-2}·s^{-1}) above background lighting.

Entrainment losses of fish through large hydroelectric facilities is an all too familiar problem. Kokanee *Oncorhynchus nerka* seem particularly susceptible to entrainment. High losses of kokanee have been documented at Libby Dam in Montana (Skaar et al. 1996), Banks Lake in Washington (Stober et al. 1979), and Dworshak Reservoir in Idaho (Maiolie and Elam 1998). At Dworshak Reservoir, it was estimated that 1.4 million kokanee (95% of the population) were lost through the dam in a period of 5 months (Maiolie and Elam 1998). These losses of kokanee were severe enough that they greatly affected the reservoir's fishery in subsequent years.

Physically screening Dworshak Dam to prevent fish losses would be very difficult because of the high flows (800 m^3/s) and the large amounts of debris within the reservoir. Our hope was that strobe lights could be used as a behavioral barrier at the water intakes to prevent fish losses while avoiding the problems associated with a physical screen. Strobe lights appeared to have potential for affecting the behavior of a number of fish species including coho salmon *Oncorhynchus kisutch* and chinook salmon *O. tshawytscha* (Nemeth and Anderson 1992), coho salmon (Fields and Finger 1956), American shad *Alosa sapidissima* (Taft 1990), alewife *A. pseudharengus*, rainbow smelt *Osmerus mordax*, gizzard shad *Dorosoma cepedianum*, white perch *Morone americana*, spot *Leiostomus xanthurus*, and Atlantic menhaden *Brevoortia tyrannus* (Patrick and Christie 1985). We therefore tested the response of kokanee to strobe lights to determine if this technology has a high potential as a behavioral deterrent capable of reducing entrainment losses at Dworshak Dam.

Description of Study Areas

Testing was conducted in Spirit Lake and Lake Pend Oreille where kokanee *Oncorhynchus nerka* densities were higher than in Dworshak Reservoir. All testing was conducted in the pelagic area of these lakes on wild, free-ranging fish. Initial testing was conducted in Spirit Lake, which is smaller and relatively shallow. Promising results there were followed by experiments in Lake Pend Oreille, which is deeper, more representative of Dworshak Reservoir, and without the possible influence of light reflecting off the lake bottom.

Spirit Lake is a natural lake located in the northern panhandle of Idaho near the town of Spirit Lake (Figure 1). It is approximately 526 ha and contains a large population of kokanee estimated to be nearly 1,000 fish/ha based on midwater trawling. Water depth in the lake where strobe light testing occurred was 20–25 m. The lake's Secchi transparency varied from 3.7 to 4.7 m at the time and location of testing.

Lake Pend Oreille is also located in the northern panhandle of Idaho between the towns of Sandpoint and Bayview (Figure 1). It is a natural

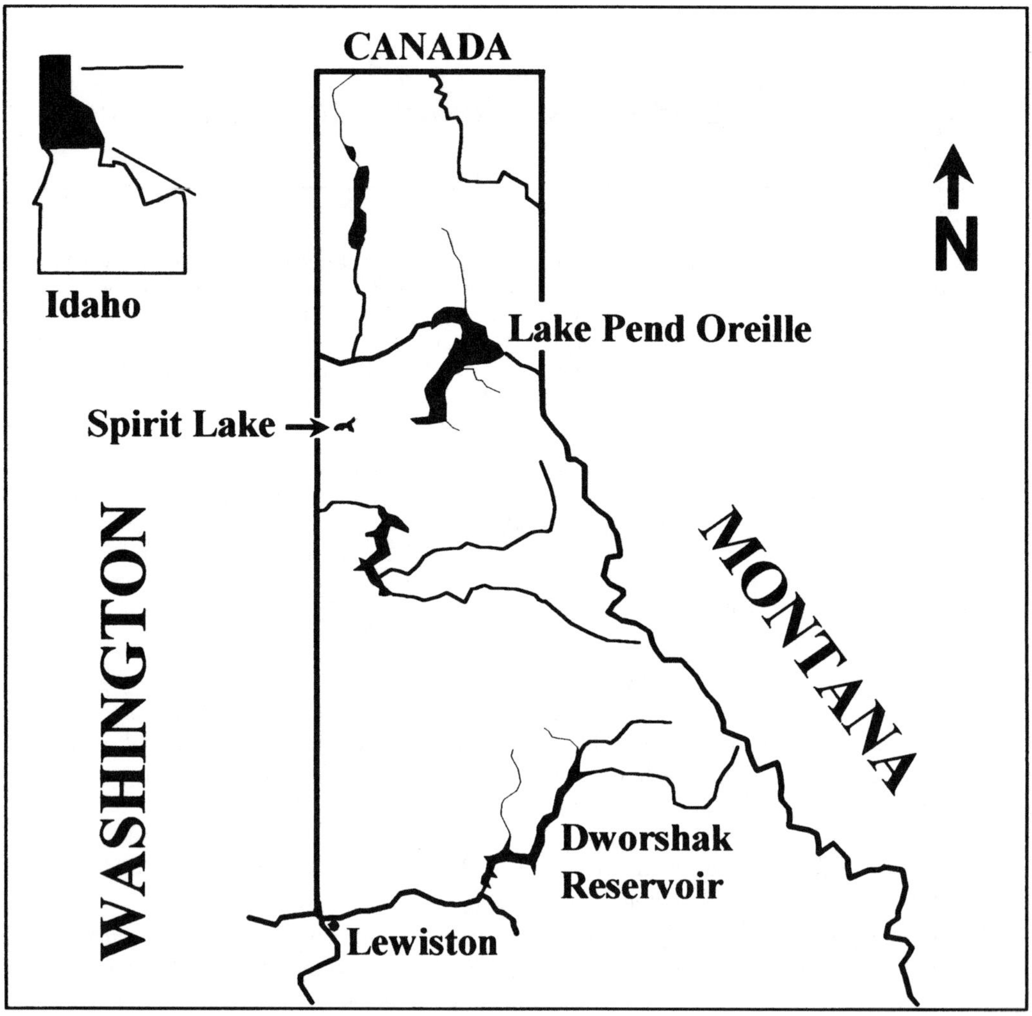

Figure 1. Map of northern Idaho showing the location of the three bodies of water discussed in this study.

lake of about 38,000 ha with a maximum depth of 350 m. Kokanee densities based on trawling were approximately 200 fish/ha, although concentrations of kokanee were found where densities exceeded 1,000 fish/ha. These areas with higher density were used during our testing. All tests were conducted in the south end of the lake where depths exceeded 100 m. Secchi transparency in Lake Pend Oreille varied from 2.8 m to 17.5 m at the time and location of testing.

We felt confident that nearly all pelagic fish in these lakes were kokanee based on midwater trawling at night from 1987 to 1994. Trawl catches in the pelagic zones of both lakes were composed of more than 99% kokanee. These kokanee were the late spawning strain similar to Lake Whatcom kokanee based on their allozyme variability (Winans et al. 1996).

Methods

Strobe lights used in these experiments were built by Flash Technologies, Franklin, Tennessee. Four flash heads were supported in the water by a vertical steel cable with the lights pointed horizontally at 90° angles. We could not synchronize the flash heads so that they flashed at the same time. A 9 kg concrete weight held the lights down (Figure 2). Power was supplied by a 5,000 W, portable generator. During testing, we set the strobe lights to flash rates of 300, 360, or 450 flashes/min. Lights were manufactured so that the brightness of the flash was slightly reduced as the flash rate increased.

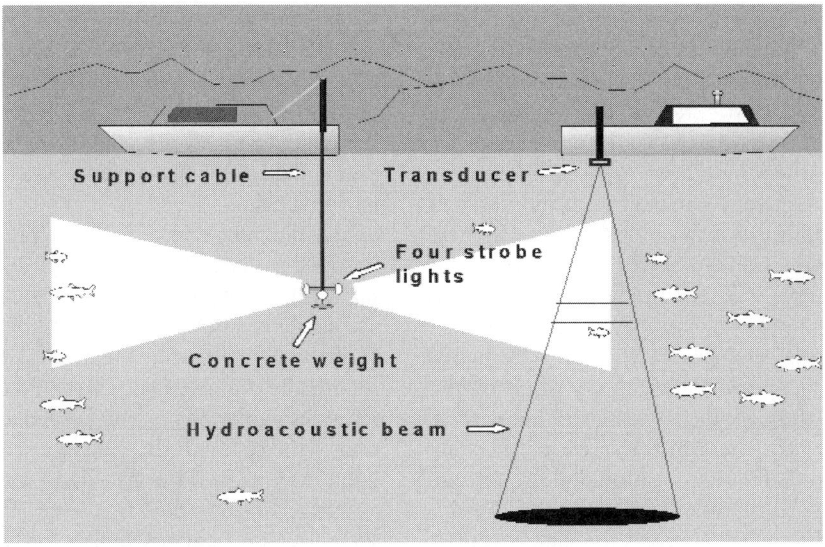

Figure 2. Arrangement of boats in the strobe light tests. One boat raised and lowered the lights while a second one conducted hydroacoustic surveys.

This was done to prolong the life expectancy of the flash heads, but it meant that two variables (flash rate and brightness) changed during our testing. The brightness of the strobe lights was measured using an International Light Radiometer, model IL1700. We used this light meter to integrate the amount of light received for one minute, and then we divided the total by the number of flashes to obtain the brightness of light per flash. Light measurements were made at all three flash rates in waters of 9.5 and 6.0 m Secchi transparencies. Readings from the radiometer were read directly in lux. To convert this to micromoles ($\mu mol \cdot m^{-2} \cdot s^{-1}$), we multiplied the lux value times 0.0185 (Thimijan and Heins 1983). This conversion assumes the spectral output of the strobe lights was similar to sunlight and is therefore an approximation.

One boat, which was equipped with hydraulic winches, raised and lowered the lights. This boat was either allowed to drift, or was anchored during testing. A second boat was equipped with hydroacoustic gear. This boat conducted surveys past the first boat to determine the distribution of kokanee in the vicinity of the lights, and to estimate fish densities in the area of the lights (Figure 2). We used a Simrad EY500 split-beam scientific echosounder with a 120 kHz transducer to conduct the hydroacoustic surveys. Boat speed was 1–1.5 m/s and all surveys were conducted at night. The echosounder was set to ping at 0.5–1.0 s intervals, depending on water depth, with a pulse width of 0.3 milliseconds. We calibrated the echosounder at the beginning of each year using a 23 mm copper calibration sphere with a target strength of –40.4 dB (decibels) (at 23°C). We checked the calibration of the echosounder prior to each test and adjusted the transducer gains if needed. Fish density estimates were calculated using EP-500 software, version 4.5 or 5.2. All fish with echoes stronger than –60 dB (approximately 20 mm total length, Love 1971) were included in the density estimates. Densities were based on echo integration techniques to account for fish within schools that could not be distinguished as single targets.

We estimated the density of fish within 30 m on either side of the lights and 5 m above and below the lights in most tests. The exception to this was the October 29, 1998 testing on Lake Pend Oreille where fish densities were analyzed within 70 m of the lights because of their wider dispersion on this date. We also estimated the distance from the lights to the first group of fish. Distances to the left and right of the lights on each transect were averaged and recorded as the distance for that transect. We used paired Student's *t*-tests with separate variances (Systat version 5.03 software) to determine significance of changes between the control and test samples.

A common protocol was followed in tests at both lakes. First, control information was collected at each site. The lights were lowered to the desired depth, the generator was started, but the lights remained off. The hydroacoustic boat then conducted surveys past the strobe lights to record kokanee density and distribution data. We used the boat's radar to keep the transect straight and to ensure the transect length was at least 200 m on each side of the strobe lights. One to five surveys were run at different angles past the strobe lights. All of the control transects were averaged and used as a single sample since they were collected at the same location.

Testing was conducted with the strobe lights set at 300, 360, or 450 flashes/min. Hydroacoustic surveys were then conducted past the lights in a fashion similar to the controls. The boats were then moved to a different location and the testing was repeated with controls and test samples. Again, all survey transects conducted at the same location were averaged and considered to be a single sample.

All testing was conducted at night. Attempts to test the lights during the day failed because kokanee formed tight schools during the day and were seen only infrequently on the echograms even in our control samples.

During the strobe light testing, we recorded several additional variables to determine their effect on the outcome of our test. These included water clarity (measured by Secchi transparency), water depth, and date.

Experiments were conducted on Spirit Lake on May 29 and June 17, 1997. Testing was then moved to Lake Pend Oreille on June 18, 1997, February 25, 1998 and October 29, 1998 (Table 1). By testing on dates throughout the year, we were able to explore seasonal effects of kokanee responses and the effects of changing Secchi transparencies. All tests conducted in Lake Pend Oreille were in water of more than 100 m depth. This eliminated the effect of light reflecting off of the lake bottom that could have influenced the outcome of the tests in Spirit Lake.

A different type of test was conducted to see if kokanee became habituated to the lights during the night. The boat with the strobe lights was anchored at a single location on Spirit Lake on July 30, 1997. Lights were lowered to the 12 m depth which was the middle of the kokanee layer. Four control transects were surveyed past the lights during the first 20 min of the test with the lights turned off. Then the lights were turned on at a flash rate of 360 flashes/min. Over the next 5 h 50 min, 16 transects were surveyed past the lights; four at the start of flashing, two each hour during the night, and four at daybreak. A separate analysis to look at the response of larger kokanee was conducted on this test. In addition to analyzing the echograms for all fish greater than –60 dB (approximately 20 mm), we also analyzed them for fish greater than –48 dB (approximately 100 mm) (Love 1971).

We analyzed the data to determine the effect of the lake's transparency after completion of these tests. The distance fish moved away from the strobe lights on each night of testing was correlated to the Secchi transparency for the lake at that time. A linear relationship was assumed and a line fit by the least-squares method. We then calculated a correlation coefficient (r) for the data set (Kachigan 1986).

Table 1. Results of testing strobe lights in Spirit Lake and Lake Pend Oreille, Idaho. *P*-Values are given for paired *t*-tests between control and test groups.

Lake	Date	Secchi depth (m)	Flash rate (flashes/min)	Mean distance to first group of fish (m)			Reduction in fish density within 30 m
				control	test	*P*-Value	
Spirit Lake	5/29/97	3.7	450	7	39	P=0.036	83%, P=0.023
Spirit Lake	6/17/97	–	300	7	30	P=0.006	80%, P=0.071
Spirit Lake[a]	7/30/97	4.7	360	1	40	P<0.001	94%, P=0.002
Lake Pend Oreille	6/18/97	2.75	300	4	45	P<0.001	84%, P=0.098
Lake Pend Oreille	6/18/97	2.75	450	6	45	P=0.009	72%, P=0.084
Lake Pend Oreille	2/25/98	17.5	360	14	120	P=0.026	98%, P=0.021
Lake Pend Oreille	10/29/98	9.9	300, 360 and 450[b]	0	136	P=0.004	100%, P=0.015[c]

[a]Testing on July 30, was the habituation experiment described in the methods section.
[b]Analysis of the three flash rates were combined.
[c]Kokanee density estimates were analyzed within 70 m of the lights.

Results

All tests showed a marked reduction in the number of fish near operating strobe lights, compared with controls (Figure 3). We found no response from kokanee during our control (lights off) sampling. The lowering of strobe lights, running the generator, movement of people on the boats, use of boat lights, and repeated hydroacoustic surveys above the fish, did not affect the distribution or density of kokanee in the vicinity of the lights (Figure 3). Densities of fish near the lights and the densities of fish more than 200 m away appeared very similar with no apparent change in density or depths.

Kokanee showed a marked change in their distribution almost immediately after the lights were turned on in each test (Figure 3). Kokanee appeared to quickly leave the area around the lights resulting in lower densities. Changes in the distance from the lights to the first group of fish, as well as the reduction in density of fish within 30 m of the lights, were both highly significant ($P < 0.001$) when analyzed across all tests. During the tests, fish moved 30–136 m away from the strobe lights. Densities of fish within 30 m of the lights were reduced from 72% to 100% less than control samples. Even on individual nights, the distances fish moved away from the strobe lights was significant at the 95% confidence level in all tests (Table 1). The reduction in fish densities, on individual nights, was significant at the 95% confidence level in 4 out of 7 tests (Table 1).

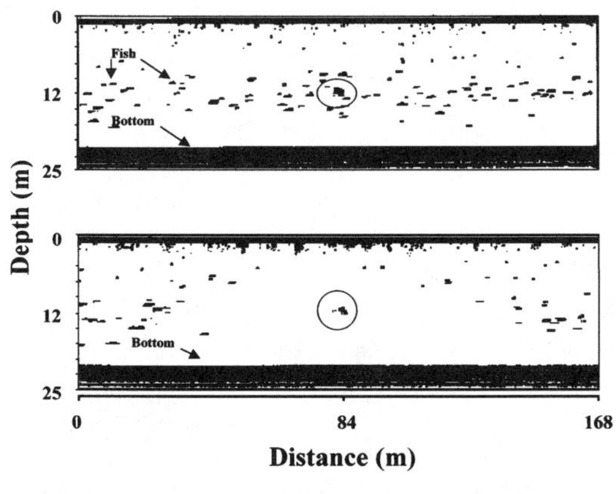

Figure 3. Example echograms of strobe light testing in Spirit Lake, May 1997. Lights were turned off in the top figure (control sample), and were flashing at 450 flashes/min in the lower figure. Strobe lights are circled in the figure.

The first test, conducted in Spirit Lake on May 29, 1997, tested the response of kokanee to a flash rate of 450 flashes/min, at each of three sites ($N = 3$). Secchi transparency was 3.7 m, water depth was about 22 m, and the strobe lights were 12 m deep. In control samples the mean distance to the fish was 7 m and the mean density was 814 fish/ha (Table 1). With lights flashing, the distance to the first group of fish increased significantly to a mean of 39 m ($P = 0.036$, df = 2). Fish densities within 30 m of the lights dropped significantly to 138 fish/ha, on average ($P = 0.023$, df = 2).

Strobe light testing continued on June 17, 1997, in Spirit Lake. Water depth was about 20 m, and the strobe lights were placed 12 m deep. The mean density of fish was reduced from 1200 fish/ha in control samples ($N = 3$) to 235 fish/ha when the strobe lights were turned on with a flash rate of 300 flashes/min ($N = 3$, $P = 0.071$). This was an 80% reduction in fish densities within 30 m of the lights. Fish were repelled from the lights an average distance of 30 m which was a significant increase from the 7 m seen in control samples ($P = 0.006$).

Fish stayed away from the lights for the entire night during the habituation test conducted in Spirit Lake on July 30, 1997 (Figure 4). The boat holding the lights was anchored at a single location for the entire night. In four control transects, with the lights turned off, fish averaged 1.4 m away from the lights, and densities within 30 m of the lights averaged 1,551 fish/ha. Density of larger fish greater than –48 dB (approximately 100 mm total length, Love 1971) was 1,160 fish/ha within 30 m of the lights in these control samples.

Horizontal distance to the first group of fish increased to 52 m once the lights began flashing (Figure 4). Density of all fish within 30 m of the lights dropped to 343 fish/ha and density of larger fish (over –48 dB) decreased to 0 fish/ha. This initial movement of fish away from the lights was somewhat farther than later samples. Between the first and fifth hours, fish stayed an average of 40 m away from the lights without any evidence of habituation. The last four transects, collected after the fifth hour, were made as dawn was beginning to break and background light levels were increasing. The mean distance to the first group of fish declined to 31 m for the last four samples. Even in the last four samples (5 h 30 min to 5 h 50 min) kokanee greater than –60 dB remained an average of 31 m away from the lights (Figure 4), with fish densities of 126 fish/ha within 30 m of the lights. For the

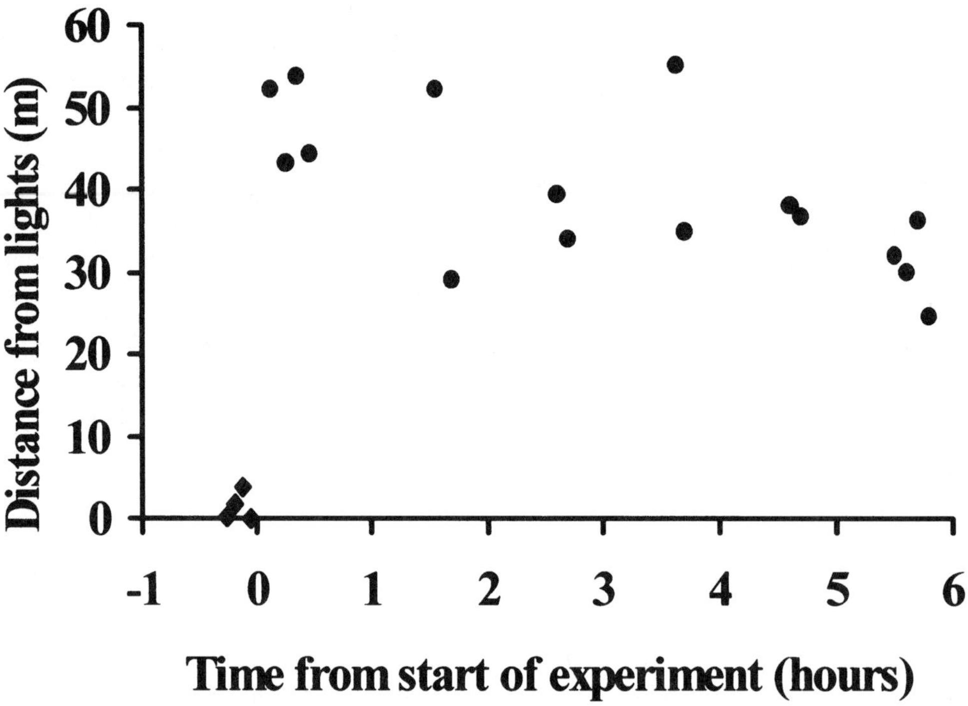

Figure 4. Distances fish moved away from strobe lights during a 5 h 50 min test. All fish larger than −60 dB were included. Diamonds represent control samples collected before the lights began flashing at time zero.

entire test, fish were repelled an average of 40 m and densities within 30 m of the lights declined 94% for all fish and 95% for fish larger than −48 dB.

Results in Lake Pend Oreille, having deep water without bottom reflectance, were similar to results in Spirit Lake (Table 1). We found during our testing on Lake Pend Oreille, June 18, 1997, that kokanee remained close to the strobe lights in the control samples (lights off). In control samples paired with the 300 flash rate tests, the mean distance to the first group of fish was 4 m and densities within 30 m of the lights averaged 751 fish/ha. When the lights were flashing at a rate of 300 flashes/min, the mean distance to the fish increased to 45 m ($P < 0.001$), and the mean density dropped to 120 fish/ha ($N = 4$, $P = 0.098$). In control samples paired with the 450 flash rate test, the mean distance to the first group of fish was 6 m, and densities within 30 m of the lights averaged 548 fish/ha. When the lights were flashing at a rate of 450 flashes/min, the mean distance to the fish increased to 45 m ($P = 0.009$), and the mean density dropped to 154 fish/ha ($N = 3$, $P = 0.084$). Secchi transparency during this test was 2.75 m.

Fish were repelled even more effectively in winter. We tested a flash rate of 360 flashes/min in Lake Pend Oreille on February 25, 1998. Water had a Secchi transparency of 17.5 m. Kokanee were an average of 14 m away from the lights in control samples, but moved to an average of 120 m away from the lights when they were flashing ($N = 3$, $P = 0.026$). Density of fish within 30 m of the lights dropped from 372 fish/ha in control samples to 4 fish/ha once the lights were turned on ($P = 0.021$) (Table 1).

Autumn test results were similar to those in winter. Our last testing was conducted on Lake Pend Oreille on October 29, 1998. Secchi transparency was 9.9 m. The average distance from the light to the first group of fish was 0 m in the control samples since fish were found directly above and below the lights in all control samples ($N = 4$). Once the lights were turned on to flash rates of 300, 360 and 450 flashes/min, fish moved away from the lights a mean distance of 136 m ($N = 4$, $P = 0.004$). Density of fish within 70 m of the lights dropped from 816 fish/ha to 0 fish/ha once the lights began flashing ($P = 0.015$).

Prior to this test, brightness measurements were recorded for the strobe lights in water of 9.5 m Secchi transparency. Flash heads produced a mean brightness of 59 lux (1.09 $\mu mol \cdot m^{-2} \cdot s^{-1}$) and 56 lux (1.04 $\mu mol \cdot m^{-2} \cdot s^{-1}$) at about 2 cm from the lights at a flash rates of 360 and 450 flashes/min, respectively. Brightness declined logarithmically with distance from the light. At 44 m from the flash head, brightness had decreased to a mean of 0.00016 lux (2.96×10^{-6} $\mu mol \cdot m^{-2} \cdot s^{-1}$) at the 360 flashes/min rate and 0.00019 lux (3.52×10^{-6} $\mu mol \cdot m^{-2} \cdot s^{-1}$) at the 450 flashes/min rate. We felt this was the lower limit of what we could measure since readings became more variable and tended to blend into small changes in background lighting.

Our strobe light testing was conducted in water with Secchi transparencies of 2.8–17.5 m. Distance that fish moved away from the lights was positively correlated to the lake's transparency (r^2 = 0.69, N = 6, Figure 5).

Discussion

We believe that these tests show conclusively that kokanee can be dispersed away from a selected area by use of strobe lights. This technology should, therefore, have a high potential for successful reduction in numbers of fish near turbine intakes of dams like Dworshak, where additional testing is warranted.

Kokanee responded in both deep (>100 m) and shallow (<25 m) water tests. Thus, the reflection of light off of the lake bottom did not appear to be the trigger for avoidance responses of kokanee. We conclude that the lights would work equally as well if they are positioned over shallow or deep water. Little difference was seen between flash rates of 300, 360, and 450 flashes/min; however, this was not extensively tested.

Considerable, and statistically significant, reductions in kokanee densities were found in winter as well as spring testing (Table 1). This is important since high entrainment losses at Dworshak

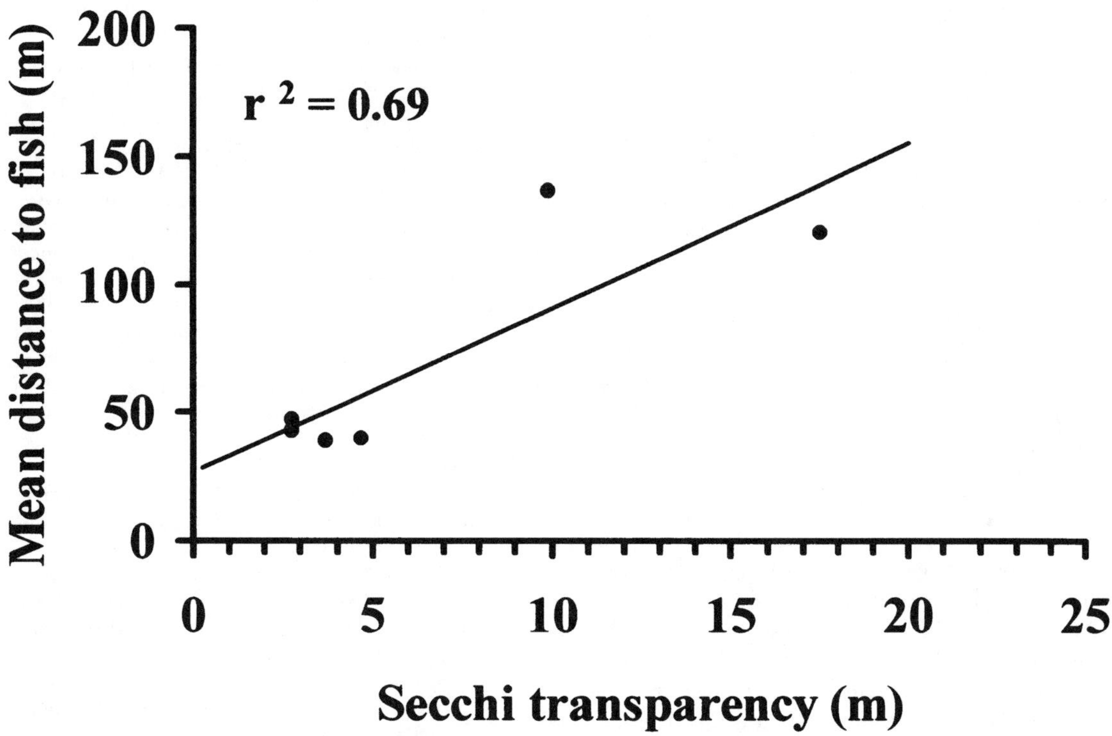

Figure 5. Distances fish moved away from the strobe lights in water of different Secchi transparency. Each point represents the mean of a night of testing.

Reservoir often occur during winter. Strobe lights proved to be even more effective during winter most likely because of greater water transparency which explained nearly 70% of the variation in the distance kokanee moved away from the lights (Figure 5).

Kokanee appeared to be responding to light levels well below 0.00016 lux (2.96×10^{-6} $\mu mol \cdot m^{-2} \cdot s^{-1}$) in our tests. This was the brightness at a distance of 44 m from the lights, whereas kokanee avoided the area within 135 m of the lights in tests at similar water clarity. Brightness at 135 m from the lights was below our ability to measure.

The response of kokanee to strobe lights in our study was much more pronounced than the response of many other fish species tested with behavioral deterrent devices. Low frequency sound evoked spontaneous avoidance responses from Atlantic salmon *Salmo salar* if they were within a distance of 2 m (Knudsen et al. 1992). Carlson (1994) summarized an extensive literature base and concluded that salmonids' response to sound occurred at less than a few meters in range. The extent of influence of a single pneumatic popper appeared to be approximately 10 m in repelling adult alewife *Alosa pseudoharengus* (Haymes and Patrick 1986). Other studies that tested strobe lights were conducted in raceways or enclosures where the full effect of distance could not be realized (Nemeth and Anderson 1992; Patrick 1982; Patrick and Christie 1985). Strobe lights were shown to reduce total fish entrainment at the Milliken Station at Cayuga Lake, New York (Homa et al. 1994). Results on individual species, however, were mixed. Some species or life stages were repelled while others were attracted at certain times of the year. Kokanee were not present in their study. Our tests on free-ranging fish allowed us to see that kokanee would react negatively to strobe lights at distances up to 136 m under the right conditions.

This study had several limitations, however. All testing was conducted at night. We suspect that daytime avoidance of the lights would be less since background levels of light would be much higher. Our tests were conducted in the relatively calm waters of lakes. As such they may not be indicative of kokanee responses in moving water near the face of a large dam. Our longest test was 6 h. Installation on dams may require the lights to be on for much longer periods. Habituation to the lights for periods greater than 6 h has not been tested. Lastly, all testing was conducted in water with Secchi transparencies of 2.8–17.5 m. Trends in our data suggest the effectiveness of the lights would be reduced in more turbid water (Figure 5).

Kokanee were lost in very high numbers from Dworshak Reservoir during some years (Maiolie and Elam 1998). Our hope was to find a way to minimize these losses in the future. Strobe lights appeared to show promise for repelling kokanee in our off-site tests. All age classes responded during each season of the year. Further testing, preferably on-site at Dworshak Dam, appears warranted and is recommended.

Acknowledgments

The authors would especially like to thank S. Elam, E. O'Brien, and M. Duclos for assisting with field work and strobe light testing. We would also like to thank C. Craig and the Bonneville Power Administration for funding and contracting this project.

References

Carlson, T. J. 1994. Use of sound for fish protection at power production facilities: a historical perspective of the state of the art. Battelle Pacific Northwest Laboratories, Final Report, Prepared for Bonneville Power Administration, Project Number 92–071, Portland, Oregon.

Fields, P. E., and G. L. Finger. 1956. The effectiveness of constant and intermittently flashing light barriers in guiding young silver salmon. University of Washington College of Fisheries Technical Report 22.

Haymes, G. T., and P. H. Patrick. 1986. Exclusion of adult alewife, *Alsoa pseudoharengus*, using low-frequency sound for application at water intakes. Canadian Journal of Fisheries and Aquatic Sciences 43:855–862.

Homa, J., M. Stafford-Glase, and M. E. Conners. 1994. An evaluation of the effectiveness of the strobe light deterrent system at Milliken Station on Cayaga Lake, Tompkins County, New York. Report of Ichthyological Associates Incorporated to New York State Electric and Gas Corporation, Binghamton, New York.

Kachigan, S. K. 1986. Statistical analysis: an interdisciplinary introduction to univariate and multivariate methods. Radius Press, New York.

Knudsen, F. R., P. S. Enger, and O. Sand. 1992. Awareness reactions and avoidance responses to sound in juvenile Atlantic salmon, *Salmo salar* L. Journal of Fish Biology 40:523–534.

Love, R. H. 1971. Dorsal-aspect target strength of an individual fish. Journal of the Acoustic Society of America 49:816–823.

Maiolie, M. A., and S. Elam. 1998. Kokanee entrainment losses at Dworshak Reservoir. Idaho Department of Fish and Game, Annual Progress Report, Prepared for Bonneville Power Administration, Project number 87-99, Portland, Oregon.

Nemeth, R. S., and J. J. Anderson. 1992. Response of juvenile coho and chinook salmon to strobe and mercury vapor lights. North American Journal of Fisheries Management 12:684–692.

Patrick, P. H. 1982. Responses of alewife to flashing light. Ontario Hydro Research Division Report, number 82-305-K.

Patrick, P. H., and A. E. Christie. 1985. Responses of fish to a strobe light/air-bubble barrier. Fisheries Research 3:157–172.

Skaar, D., J. DeShazer, L. Garrow, T. Ostrowski, and B. Thornburg. 1996. Investigation of fish entrainment through Libby Dam, 1990-1994. Montana Department of Fish, Wildlife and Parks, Prepared for Bonneville Power Administration, Project Number 83-467, Portland, Oregon.

Stober, Q. J., R. W. Tyler, C. E. Petrosky, K. P. Johnson, C. F. Cowman, Jr., J. Wilcock, and R. E. Nakatani. 1979. Development and evaluation of a net barrier to reduce entrainment loss of kokanee from Banks Lake. Fisheries Research Institute, Final Report, Prepared for United States Bureau of Reclamation, Contract Number 7-07-10-50023, University of Washington, Seattle.

Taft, E. P. 1990. Fish protection systems for hydro plants. Interim Report GS-6712, Prepared for Electric Power Research Institute, Palo Alto, California.

Thimijan, R. W., and R. D. Heins. 1983. Photometric, radiometric, and quantum light units of measure: a review of procedures for interconversion. HortScience 18:818–822.

Winans, G. A., P. B. Aebersold, and R. S. Waples. 1996. Allozyme variability of *Oncorhynchus nerka* in the Pacific Northwest, with special consideration to populations of Redfish Lake, Idaho. Transactions of the American Fisheries Society 125:645–6663.

American Fisheries Society Symposium 26:37–56, 2001

Effectiveness of Strobe Lights and an Infrasound Device for Eliciting Avoidance by Juvenile Salmon

Gene R. Ploskey[1]
U. S. Army Engineer Research and Development Center, Waterways Experiment Station
Fisheries Engineering Team, Columbia River Basin, Post Office Box 40
North Bonneville, Washington 98639, USA

Peter N. Johnson
AScI Corporation, Fisheries Engineering Team, Columbia River Basin
Post Office Box 40, North Bonneville, Washington 98639, USA

Abstract.—Our primary objectives were to videotape and characterize vertical responses of juvenile salmon to strobe light and infrasound in controlled tests in a net pen that was 4-m long and 1.5-m in diameter. Experiments were conducted near the Hiram M. Chittenden Locks, Seattle, in spring and summer, 1997, for the U.S. Army Engineer District, Portland. Four underwater video cameras mounted on the perimeter of the vertically oriented net pen were used to record responses of schools of juvenile salmon during test and control treatments. When strobe lights were tested within 1 m of the pen, the frequency of avoidance by hatchery-reared juvenile coho *Oncorhynchus kisutch* and chinook *O. tshawytscha* salmon was 80–100%. Fish usually moved toward the top of the pen in response to strobe light directed either horizontally or 45° upward into the bottom section of the pen. In contrast, fish did not respond to strobe light directed straight up through the bottom of the pen. A particle-motion generator consisting of a high-volume pump and rotary valve produced 10–35 Hz infrasounds that failed to elicit avoidance from hatchery yearling coho and subyearling coho and chinook salmon.

Our study of the effectiveness of strobe light and infrasound for stimulating avoidance behavior of juvenile salmon resulted from the need to improve the bypass efficiency of lower Columbia River dams. Success in increasing fish bypass efficiency likely will require modification of the vertical distribution of out-migrating smolts, especially in summer, to improve the performance of surface collectors and in-turbine screens. The deep distribution of outmigrating smolts is suggested by low fish-guidance efficiency of screens when measured by Fyke netting (e.g., Gessel et al. 1989) or by fixed-aspect hydroacoustics (Ploskey and Carlson 1999).

The goal of this study was to evaluate the performance of promising light and sound devices for redistributing juvenile salmon. We were interested in the potential of behavioral devices to redistribute smolts in general and to move them up in the water column, in particular. Testing the potential for redistributing smolts vertically seemed to be a logical first step in exploring potential solutions to the vertical distribution problems at Columbia River dams. Our primary objectives were to videotape and characterize vertical responses of juvenile salmon to strobe light and infrasound in controlled tests in a vertical net pen.

Constant light, strobe light, and sound are complicated physical phenomena that fishery biologists and anglers have long hoped to use to concentrate, redistribute, or guide fish. A strobe light is defined as a device emitting extremely rapid, brief, and brilliant flashes of light. Infrasound refers to sounds less than 50 Hz, water-particle accelerations in the near field local flow of volume-displacement sources that diminish in amplitude in proportion to 1/range[3]. Ultrasound refers to high-frequency pressure waves (>20 kHz) that are above the documented hearing range of humans and most fishes and occur in the far field of small displacement sources. Amplitudes of ultrasound decrease in proportion to 1/range.

Light is an effective stimulus to which juvenile salmon and other fishes respond. However, the nature of the response (avoidance or attraction) may be inconsistent and depends upon the state of acclimation of fish to ambient lighting and the

[1]Current address: Pacific Northwest National Laboratory, Post Office Box 241, North Bonneville, Washington 98639, USA.

intensity of the test light (e.g., Puckett and Anderson 1987; Nemeth and Anderson 1992). Acclimation may explain the many apparently inconsistent responses obtained in tests with constant light. Craddock (1956) reviewed some early, unpublished research on guiding downstream migrating salmon with light in the northwest, and Fields et al. (1958) examined light attraction and repulsion in fast turbid water.

A substantial volume of literature documents fish avoidance of strobe lights, as described in a review by Winchell et al. (1994). For example, strobe lights were 56% effective in diverting downstream migrating sockeye salmon *Oncorhynchus nerka* away from turbine intakes at Seton Hydroelectric Station in British Columbia (McKinley and Patrick 1988). In a flume under static conditions, hatchery reared sub-yearling chinook salmon *O. tshawytscha* and steelhead *O. mykiss* consistently avoided strobe light (Puckett and Anderson 1987; Nemeth 1989; Anderson et al. 1988).

Carlson (1994) conducted an extensive review of the use of sound to protect fish at hydropower facilities, and Nestler et al. (1994) developed a rationale for an acoustic-technology research program. These papers provide a good historical overview of state-of-the-art acoustic technologies. Popper and Carlson (1998) also reviewed the application of sensory signals, particularly sound, for possible use in modifying fish behavior.

Fish are broadly classified as hearing specialists if they have anatomical connections between inner ear and a swim bladder or another pressure transducing device (Popper and Fay 1993). Examples of hearing specialists include catfishes, minnows, and clupeids. Salmon are classified as hearing nonspecialists because their elongated swim bladder is dorsal and posterior in the body cavity and plays no part in hearing (Hawkins and Johnstone 1978).

The most consistently successful applications of sound devices at generating facilities involved the use of ultrasound on alewives *Alosa pseudoharengus* (Dunning et al. 1992; Ross et al. 1993), blueback herring *Alosa aestivalis* (Nestler et al. 1992; Ploskey et al. 1995), and American shad *Alosa sapidissima* (Ploskey et al. 1994). In contrast, reported successes in redistributing fishes with low-frequency, sound-pressure waves (Haymes and Patrick 1986; Loeffelman et al. 1991a, 1991b; Hanson Environmental, Inc. 1995) are rare relative to reported failures (e.g., Moore and Newman 1956; Burner and Moore 1953, 1962; Smith and Anderson 1984; Electric Power Research Institute 1994a,b; Taft et al. 1995; Ploskey et al. 2000).

Recent evidence has shown that juvenile salmon perceive and are highly responsive to infrasound in laboratory-scale experiments (Knudsen et al. 1992, 1994; Taft et al. 1995; Mueller et al. 1998). Apparently, the water-particle-motion component of infrasound is an effective stimulus for eliciting avoidance responses in Atlantic salmon *Salmo salar* smolts (Knudsen et al. 1992; Taft et al. 1995). Infrasound testing of chinook salmon and steelhead smolts longer than 150 mm in tanks at Oregon State University in 1995 provided videotaped evidence of vigorous avoidance of infrasound (Knudsen et al. 1997). Mueller et al. (1998) observed strong avoidance responses of wild and hatchery-reared juvenile chinook salmon less than 70 mm long to infrasound signals.

Methods

Site Description and Experimental Setup

A 7-m long, 2.4-m wide pontoon boat was located in Salmon Bay adjacent to the north side of the wall of the large lock chamber at Hiram M. Chittenden Locks, Seattle, Washington (Figure 1). The boat was oriented with the front end of the pontoons touching the pier so that we had pier access to the boat and could suspend behavioral devices at least 3 m from the pier.

A 4-m long, 1.5-m diameter cylindrical frame was constructed of five aluminum hoops attached to 4-m long pieces of aluminum channel. Four underwater cameras were attached to the hoops at 1-m intervals beginning at one end of the frame. Cameras were aimed into the interior of the frame, which was then covered with 3.2-mm white mesh webbing. For testing with the pen in a vertical position (Figure 2), the top was left open and the webbing extended out of the water between the pontoons of the boat to facilitate introduction and removal of fish. A 1.4-m diameter circular net served as a false bottom for the pen and was raised or lowered with vertical lines to remove fish from the pen after testing. For testing the range of effectiveness of strobe lights, the pen was oriented horizontally with the cameras aimed upward and webbing over the frame was pursed and tied at both ends.

Behavioral Devices

Three strobe lights were aimed in the same direction and bolted on a single mount that was attached

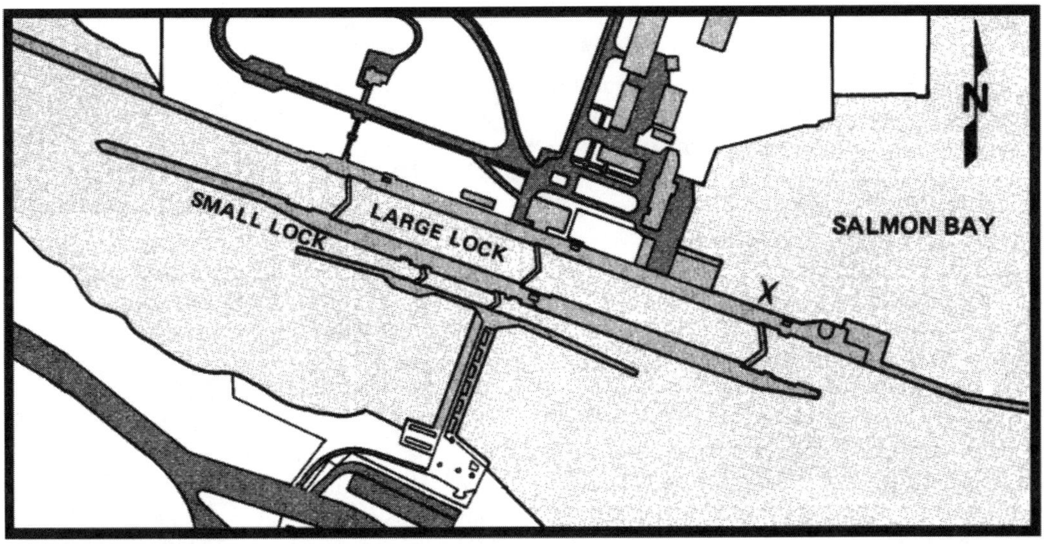

Figure 1. Plan view of the study site indicated with an "X" in Salmon Bay immediately north of the Hiram M. Chittenden Locks, Seattle, Washington.

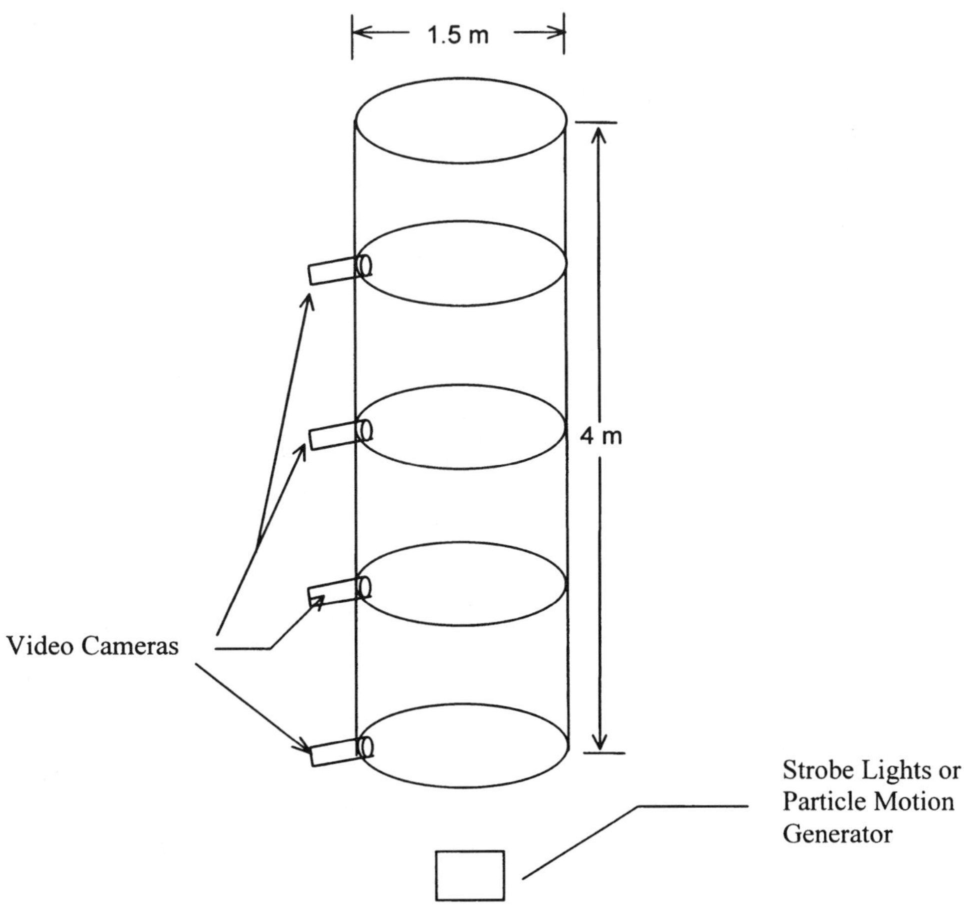

Figure 2. Vertical net pen with cameras positioned at successive 1-m distances from the stimulus source.

to a 6.4-cm-diameter pole for deployment. The flash rate (5 s^{-1}) and power output of strobe lights (400 W) were kept constant throughout the study. In horizontal tests to describe range of effectiveness, lights were located 1–5 m from one end of the pen and aimed at the pen. In vertical tests, lights were located about 1 m from the pen in one of three positions: (1) directly under the pen and aimed upward; (2) 45° below a bottom corner and aimed up into the bottom 1-m section; or (3) adjacent to the bottom 1-m section and aimed horizontally into it.

The particle-motion generator (PMG) made by Alden Research Laboratories (ARL) was housed in a screened chamber that was 2.7 m long, 1.1 m high, and 0.6 m wide. The water pump for the PMG was capable of delivering up to 44.2 L per s through a rotary valve with three 5 × 10 cm rectangular openings. The speed of the valve determined the frequency of water-particle accelerations, while pump rate determined amplitude. The pump required about 7–10 s to reach full pressure and therefore, infrasound amplitudes ramped up at the start of a new test. Changes in amplitude during some prolonged tests with varying pump rates likewise were not immediate. For most tests, the PMG was placed below and within 0.3 m of the bottom of the vertical net pen. It was suspended by straps from winches located on opposite ends of the pontoon boat. Horizontal tests were limited to one school of fish in spring and one in summer to verify that lateral responses of smolts were similar to vertical responses. In these tests, the PMG was located within 0.3 m of one end of the pen.

Measuring Stimuli

We measured the intensity of strobe lights with a LI-COR LI-250 light meter using a LI-195A quantum sensor at ranges from 0.1 to 7.6 m. In spring, we estimated light intensity by subtracting ambient measures of light intensity during the day from measures taken while strobe lights were flashing at 5 times per s. In summer, we measured strobe light intensity directly at night.

The propagating wave of infrasound (10–50 Hz) was measured with a Bruel and Kjaer (B & K) 8103 hydrophone connected to a B & K 2635 charge amplifier and recorded to tape in a Sony digital-audio-tape recorder. Measurements were taken on both ends of the net pen to allow us to determine the range of sound pressure levels that fish experienced. Water-particle accelerations at varying distances from the PMG were measured with a tri-axial accelerometer (three orthogonal model SM-7130 accelerometers from Silicon Microstructures) via a laptop computer with Daq processing software by Iotech, Inc.

Fish Testing Protocols

All test fish were obtained from the Lake Washington watershed. Most, including yearling coho salmon tested in April and sub-yearling chinook and coho salmon tested in June were obtained from the Isaquah Fish Hatchery in Isaquah, Washington. Fish were transported from the hatchery in two 115-L ice chests, each equipped with battery powered aerators. A few schools of sub-yearling coho salmon and one school of sub-yearling chinook salmon were collected in a smolt trap in the adult fish ladder located south of the spillway at Chittenden Locks in June. One large school of coho salmon was obtained from purse-seine samples taken in the large lock by the Washington Department of Fish and Wildlife. Obtaining some fish from the lock area was desirable to provide broader inference from results than could be obtained based upon tests of fish from one fish hatchery. Fish were held in two 1.2 m^3 net pens 40 m from the pontoon barge until they were tested. They were fed hatchery pellets three times a day during the week of testing in early April and two weeks of testing in June.

Test schools consisted of 10–25 juvenile salmon that were introduced to the net pen and allowed to acclimate for 30–60 min before exposure to strobe lights or infrasound. Except for one test school of sub-yearling coho and one of sub-yearling chinook salmon, all specimens appeared to be in good condition and active when placed in the test pen. The last school of fish tested on a given day usually was tested again the next morning. A school of fish was exposed to no more than 3–8 stimulation trials lasting 30–60 s and one final 5- or 10-min trial before being released into Salmon Bay. The short trials were designed to detect startle responses and immediate avoidance, while the final 5–10 min trial was to describe sustained avoidance. Trials were bounded by 30–120 s leading and trailing periods without a stimulus and tests consisting of leaders, trials, and trailers were interspersed with controls. Controls also had a leader, trial, and trailer of similar duration to tests but trials had no stimulus. All tests reported in this paper (Table 1) were conducted during daylight hours between 8:00 a.m. and 6:00 p.m. The number and design of trials was limited by the temporal avail-

Table 1. List of all net pen tests conducted at Hiram M. Chittenden Locks, Seattle, Washington in 1997. Tests are categorized by stimulus and device position or frequency. Schools tested ranged in size from 10 to 25 fish.

Stimulus and position or frequency	Species	Number of schools	Number of trials	Duration (min)
Strobe lights aimed horizontally into	subyearling chinook	4	9 each	0.5
the bottom of the pen	subyearling coho	4	9 each	0.5
Strobe lights aimed 45 degrees up	subyearling coho	4	9 each	0.5
into the bottom of the pen	yearling coho	6	4 each	0.5
		7	1 each	10
10-Hz particle motion generator	yearling coho	3	1 each	2
20-Hz particle motion generator	yearling coho	5	3,4,2,2,2	2
20-Hz particle motion generator	subyearling coho	1	12	1
30-Hz particle motion generator	yearling coho	2	4,1	2
10- to 35-Hz particle motion generator	subyearling coho	2	1 each	constant
10- to 35-Hz particle motion generator	subyearling chinook	1	1	constant

ability of fish of different species and age groups. Results of testing of the effectiveness of a 300 and 400 Hz blend of sounds is described in another publication by Ploskey et al. (2000).

Acquiring and Processing Video Data

The four monochrome video cameras (Sony SSC-M350), fitted with 105-degree wide-angle lenses and encased in underwater housings, were sampled sequentially with a Robot MV94e multiplexer at about seven frames per s each. Multiplexed images were recorded on tape in a Sony EV-C200 Hi 8 video-camera recorder and were replayed through the multiplexer to obtain a composite view from every camera.

The field of view of each camera was divided into two parts so that fish counts could be assigned to eight physical locations in the net pen. The midpoint of each of the eight locations was a known distance from the stimulus source. Video processing involved playing a videotape, stopping the tape at 5-s intervals, counting fish in each of eight locations, and recording the sum of numbers by location. We multiplied the number of fish per location by midpoint distance from the behavioral device and averaged weighted counts to obtain a mean location for the school at 5-s intervals. We did not attempt to track the swim path of individual fish because fields of view of adjacent cameras did not overlap at short ranges. Consequently, we could not always identify an individual fish as it passed from out of one camera's field of view and into another. Processing intervals were slightly longer than 5 s for strobe-light tests recorded in late June because we inadvertently set the record settings on the VCR to time lapse mode. This mistake prevented us from displaying a synchronous four-camera image through the multiplexer. Consequently, asynchronous recordings had to be processed separately.

Evaluating Effectiveness

A common behavior of juvenile salmonids in a vertical net pen involved repeated up and down movements in the water column, which we referred to as pacing. Pacing made difficult any assessment of effects of stimuli during brief exposures because apparent avoidance could be the result of the starting position of a school. Pacing was more common for freshly introduced fish than for fish that had been in the pen more than 30 min, but some schools paced more than others did. When fish were pacing, we had to evaluate effects based upon changes in behavior during exposure relative to behavior during a brief period immediately before exposure and a variable length period after exposure. When avoidance was less obvious, we compared the frequency of apparent avoidance responses in test and control schools. We relied heavily on plots of swimming patterns to assess changes in behavior and displacement. In many tests, smolts appeared to prefer the upper part of the pen and results of those trials were ambiguous because fish could not move further away from the stimulus.

When fish were not pacing, we looked for movement up and away from the stimulus. Under these conditions, a comparison of the end position relative to the preexposure position was sufficient to assess avoidance. When avoidance was consistently evident relative to control swim patterns, we used *t*-tests to compare mean vertical displacements of test and control schools. We used different schools of fish as replicate samples and tested schools exposed once separately from schools exposed two, three, or four times to avoid repeated measures testing. We also tested to see if mean displacements differed among schools that were exposed varying numbers of times.

We could not replicate all tests sufficiently for different types of stimuli and species and age classes of fish to provide high statistical power for detecting differences. However, our purpose was to identify strong and consistent stimuli for eliciting desired avoidance responses. Therefore, we were satisfied to have 80% power to detect large differences (>50%) in displacement or frequency of occurrence at a 5% level of significance for one-tailed tests. All tests were based upon α = 0.05. A device eliciting only minor or infrequent responses would not be an acceptable tool for redistributing fish at dams on the lower Columbia River.

Results

Strobe Light Tests

A strobe light flash greater than or equal to 1 μmol/s × m^2 (1 μE sec^{-1} m^{-2}) was detectable by a quantum sensor at ranges of about 5 m in April and June. Loss of light to absorption and scattering was similar in both months, although our measurements were more variable in April than in June (Figure 3). In April, two daytime light measurements were used to

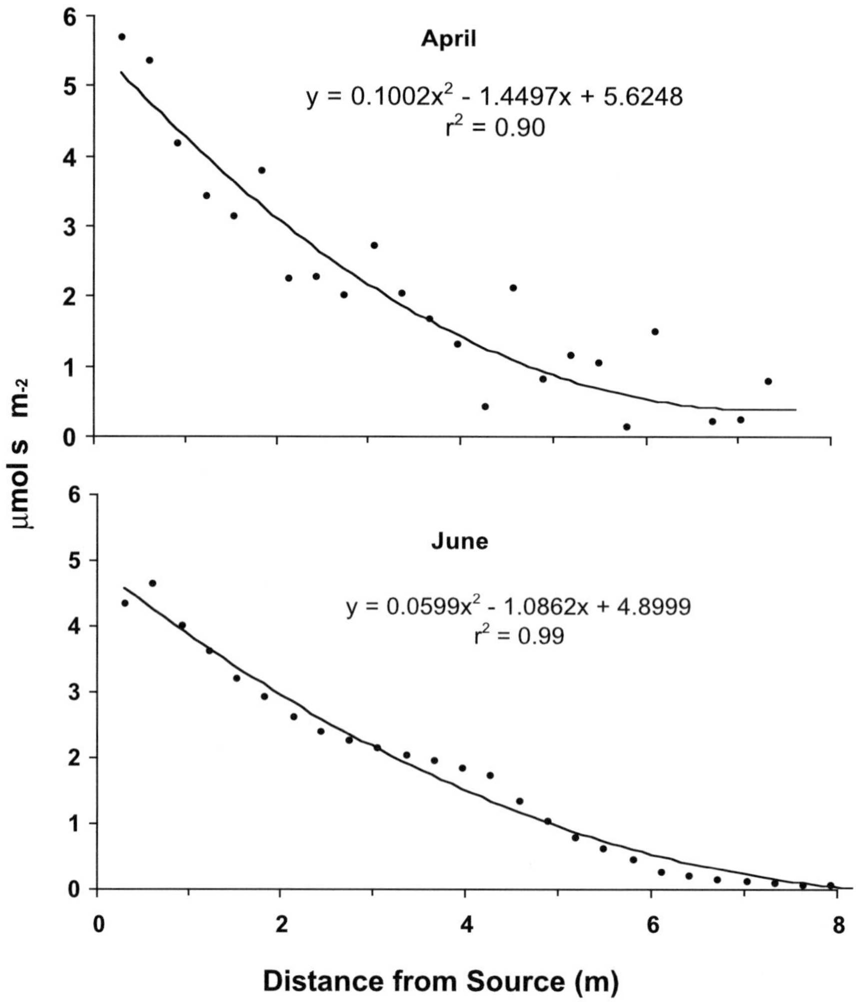

Figure 3. Measurements of light levels as a function of distance from three 400-W strobe lights in April and June 1997.

estimate strobe-light intensity, whereas one night-time measurement was used in June. Water clarity was similar during all strobe-light testing (about 1 Nephelometric Turbidity Unit). The frequency of avoidance by smolt schools ranged from 80 to 100% in vertical tests at ranges less than or equal to 4.5 m to 45% in horizontal tests where the distance from lights to most fish was 3.5–6.5 m. Avoidance frequency was only 19% when the end of the horizontal net pen was 3.5 m from strobe lights and fish usually were 4.5–6.5 m from the lights when trials began. The frequency of avoidance of the light source by control schools was 17%. The greatest distance that yearling coho salmon moved away from strobe lights in horizontal tests was about 7 m, usually when clouds obscured direct sunlight.

In vertical net-pen tests in spring, we began with 30-s reactance tests of yearling coho salmon to strobe lights. Yearling coho salmon schools often exhibited a preference for the bottom half of the net pen. We made qualitative comparisons of the responsiveness of several schools to strobe light entering the pen from different angles. Strobe light entering the bottom 1-m of the pen from the side or from 45° below elicited vertical avoidance responses. However, strobe lights aimed through the vertical pen from directly below (i.e., five trials on each of two different schools) elicited no response. All results described below for yearling coho salmon are from tests of the horizontal or 45° aiming angles.

The vertical movement of yearling coho salmon away from strobe lights directed laterally or 45° upward into the bottom 1 m of the pen was significantly greater than responses of the same fish during control tests without strobe lights. Schools moved significantly higher in the water column during strobe light treatments than during control treatments on each of four successive exposures (Table 2). Average vertical displacement of yearling coho salmon did not differ significantly among the four exposures ($P = 0.52$) per school, and we did not observe a consistent change in the rapidity of response with repeated exposure.

Results of 10-min trials were more informative than results of 30-s trials, although we did not know this until all testing was completed and videotapes had been processed. With 10-min trials, we were able to record the entire avoidance response and continued avoidance, but we could not conduct as many trials on each school without potentially overexposing fish. All schools of yearling coho salmon tested exhibited similar behavior. In contrast to control schools, schools exposed to strobe light usually moved quickly, more than or equal to 3 m away from the lights and then gradually to the uppermost section of the pen until strobe lights were turned off (Figure 4). Mean distance from lights decreased slightly after the exposure ended.

Despite testing problems, patterns of avoidance were apparent for both sub-yearling chinook and sub-yearling coho salmon. Tests on sub-yearling chinook salmon in summer were limited to strobe lights aimed horizontally into the bottom of the pen because we ran out of fish before we could test the 45°-aiming angle. We were able to test both aiming angles for sub-yearling coho salmon. For both species, schools located near the top of the net pen usually stayed there throughout strobe-light tests. However, we dropped those observations from our evaluation of the frequency of different behaviors because the fish could not move up further and exhibit avoidance.

Relative to the frequency of behavioral responses in control tests, both species were much more likely to move up away from the stimulus source and less likely to move toward the source when exposed to strobe lights (Table 3). During

Table 2. Results of paired *t*-tests on the null hypothesis that the maximum upward displacement (m) of the center of yearling coho schools did not differ between treatments.

Schools exposed	Strobe on mean	Strobe off mean	Number of schools (*N*)	df	*t*	*P*
First time	3.53	2.22	6	5	2.53	0.0150
Second time	3.34	1.85	6	5	3.99	0.0013
Third time	3.25	1.44	6	5	2.51	0.0024
Fourth time	3.24	2.14	6	5	2.51	0.0155
Pooled	3.34	1.91	24	23	6.54	<0.0001

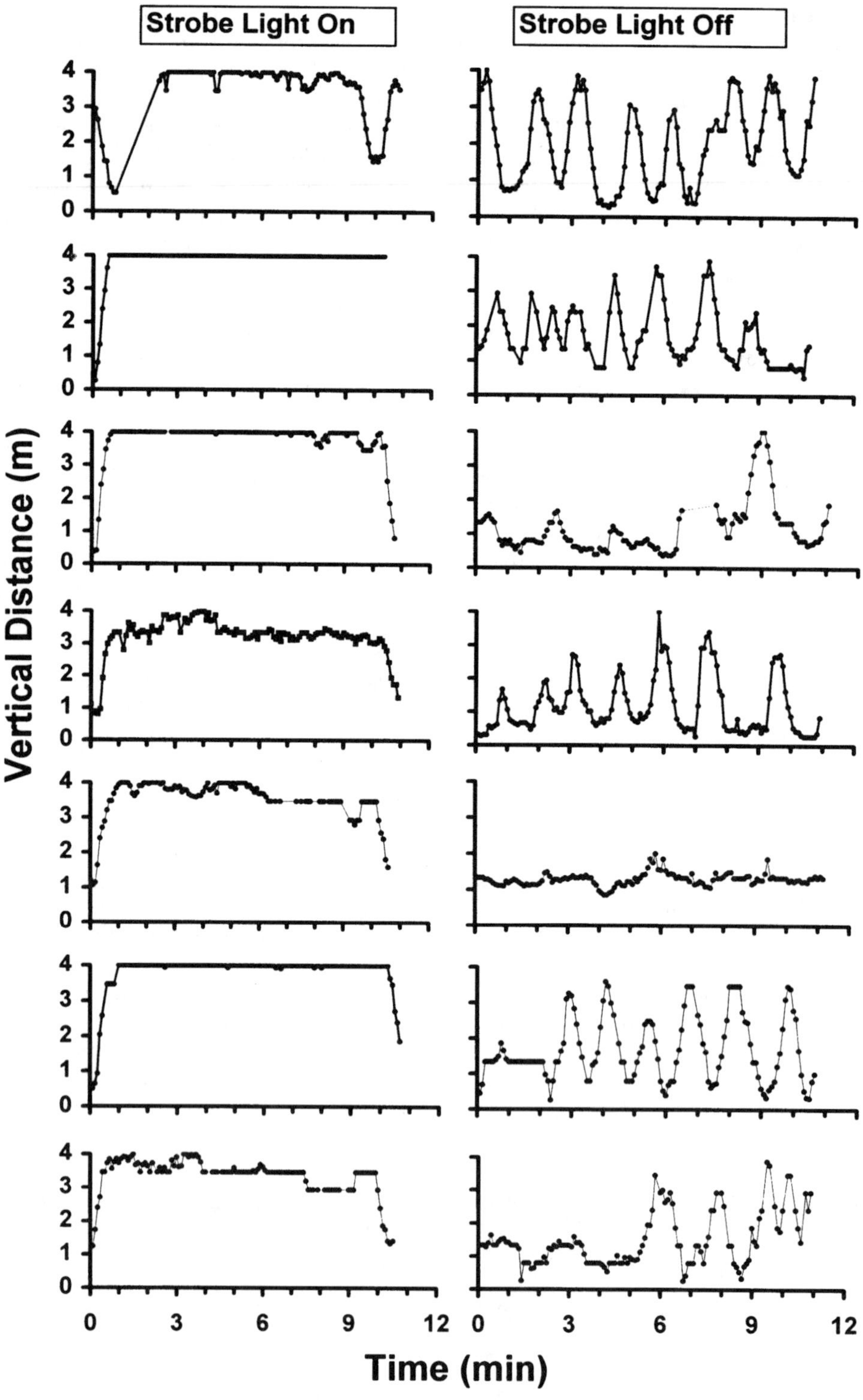

Figure 4. Vertical distance of the center of seven different schools of yearling coho salmon above strobe lights that were flashing 5 times/s from 0 to 10 min (left) and the same distances for schools during control treatments without strobe light (right).

Table 3. Frequency of four responses of sub-yearling chinook and coho salmon during those strobe-light tests and control tests where fish could avoid the stimulus (i.e., fish were not already at the top of the pen). Strobe lights were aimed into the bottom section of the net pen either horizontally or upward at 45 degrees.

Behavioral response	Species: test type:	Chinook control	Chinook strobe	Coho control	Coho strobe	Coho strobe
				Number of trials		
		5	13	14	11	10
				Percent of trials		
Upward away from stimulus		20.0	92.3	28.6	100.0	80.0
Downward toward stimulus		40.0	0.0	28.6	0.0	0.0
Limited vertical movement		20.0	7.7	35.7	0.0	20.0
Pacing up and down		20.0	0.0	7.1	0.0	0.0

control treatments, 40% of sub-yearling chinook and 29% of sub-yearling coho salmon schools moved down toward the bottom of the pen, a behavior that was not observed during strobe-light treatments. Sub-yearlings exposed to a strobe light directed horizontally into the bottom section of the pen also were less likely than controls to stay at one depth or to pace up and down the pen. Sub-yearling chinook salmon schools less than 2.5 m from lights (7 of 13 trials) responded more rapidly than schools 2.5–3.5 m from lights (5 of 13 trials). Two schools of sub-yearling coho salmon exposed to strobe lights located at 45° and 0.5 m below the pen failed to respond to strobe lights.

Particle Motion Generator

Frequency spectra of most PMG signals also had high amplitude harmonics of the fundamental frequency (Figure 5). The gradient of 20-Hz water-particle accelerations ranged from about 0.01 m/s^2 at 3.5 m from the source up to 0.175 m/s^2 at 0.9 m.

April testing of infrasound frequencies from 10 to 30 Hz with the PMG usually failed to elicit consistent upward avoidance from yearling coho salmon schools (Figures 6 and 7). We observed no evidence of avoidance in three trials at 10 Hz. At 20 Hz, 10 of 13 trials showed no significant avoidance. The proportion of 20-Hz trials resulting in some slight avoidance (23%) was similar to the proportion of control trials that also showed slight avoidance (20%) when no sound was transmitted. Four of five 30 Hz trials failed to elicit avoidance, and therefore the proportion that appeared to avoid the stimulus (20%) was the same as for control treatments.

Initial 20-Hz trials in summer on sub-yearling coho salmon produced few avoidance responses, similar to what we observed in spring for yearling coho salmon. Rather than repeatedly testing one specific frequency at a time, we decided to expose fish to continuous infrasound but with varying frequency and amplitudes through time. The hope was to identify some promising combination of frequency and pump rate before returning to repetitive testing. We observed nothing consistent or obvious enough to warrant continued testing for sub-yearling coho salmon (Figure 8 and 9). Even when apparent avoidance was observed as frequencies were switched, fish returned to less than 1 m of the device within 1–2 min while the signal was being transmitted. Throughout the first 10 Hz exposure, sub-yearling coho salmon stayed within 0.8–1 m of the PMG.

Schools of sub-yearling chinook salmon generally were higher in the water column than sub-yearling coho salmon when tested. However, we observed them moving toward and within 1.2 m of the device on three occasions at 10 Hz and three at 35 Hz (Figure 10). The school also once moved down within 1.2 m of the device at 20 Hz and again at 25 Hz. From 7–24 min into the test, sub-yearling chinook salmon stayed an average distance of 2 m above the PMG. We observed no consistent movement away from the PMG and toward the top of the pen, although smolts could have moved up 1.5–2 m.

Discussion

Strobe Lights

The presence of ambient light clearly affected our ability to define the maximum range of effectiveness of strobe lights. During daylight testing with the net pen oriented horizontally, smolts exhibited avoidance behavior at ranges less than or equal to 4.5 m. Avoidance frequency has been observed to increase at intensities ranging from 1 to 4 μmol/s ×

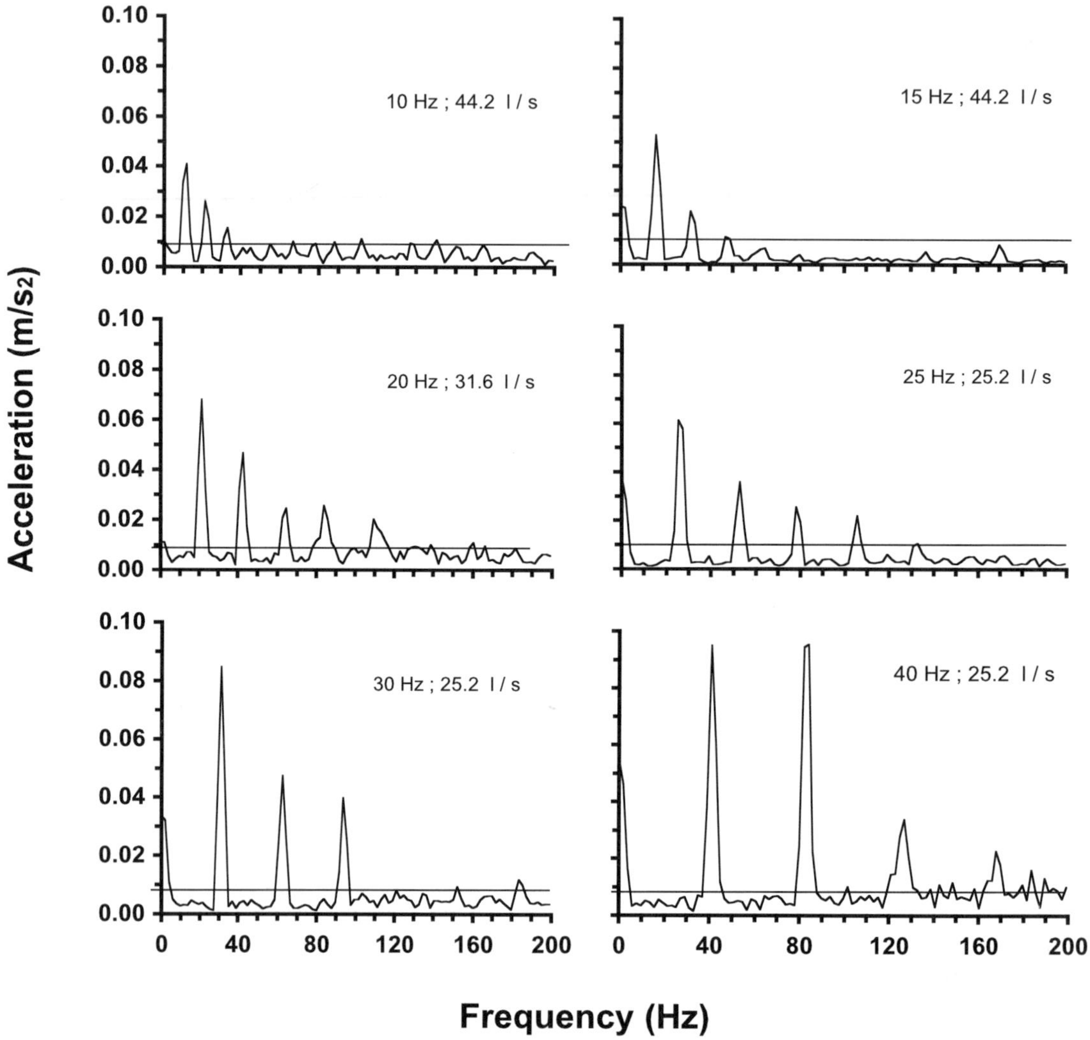

Figure 5. Frequency spectra for six infrasound signals generated by the particle motion generator. Numbers after frequencies indicate pump rates. Horizontal lines indicate the response threshold identified by Knudsen et al. (1994) for Atlantic salmon.

m^2 and decrease at lower and higher intensities, although intensities ranging from 0.1 to 5 μmol/s × m^2 generally were avoided (Anderson et al. 1988). We measured intensities of about 4 μmol/s × m^2 at a range of about 1 m, 1.2–1.4 μmol/s × m^2 at a range of 4.5 m, and less than or equal to 0.1 μmol/s × m^2 at a range of greater than or equal to 6 m. The frequency of avoidance by juvenile salmonids was 80–100% within the length of the 4-m long net pen when strobe lights were located within 0.5 m of the bottom of the pen. However, avoidance frequency for yearling coho salmon decreased to about 45% in horizontal tests when the pen was moved 2.5 m away from lights and the initial range of most fish from lights was 3.5–6.5 m. Avoidance frequency decreased to about 19% and was similar to that of control schools when the range of most fish from strobe lights was 4.5–6.5 m when tests began. Apparent differences in the minimum effective power level of strobe lights in our tests and those of Anderson et al. (1988) likely were a function of ambient light intensity, as all of our tests were conducted during the day. In addition, our horizontal tests were affected more by daylight than our vertical tests because the entire length of the net pen was within 1 m of the surface of the

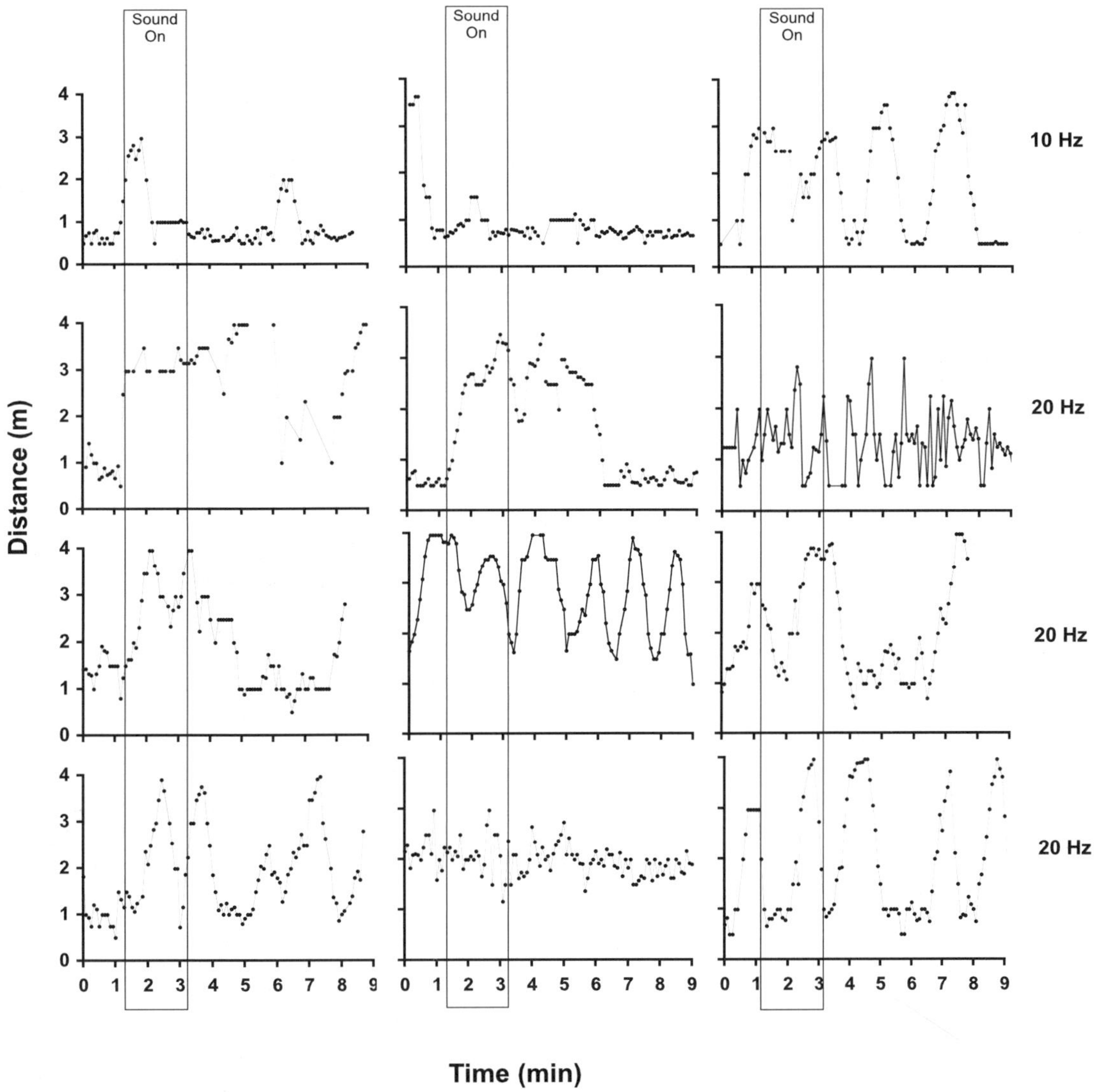

Figure 6. Plots of vertical positions of yearling coho salmon schools during 9-min tests of the particle motion generator. Infrasound was generated only during the 2-min period highlighted by the vertical "sound on" bars at frequencies of 10 and 20 Hz and a pump rate of 31.6 l s^{-1}.

water when the pen was horizontal. The white webbing of the net pen also reflected light and likely reduced effective range.

The literature suggests that the effectiveness of strobe lights, including range, would be greater at night than during the day. Nemeth and Anderson (1992) reported that responses depended upon the light intensity to which fish were acclimated. When adapted to higher light levels, fish exhibited weaker responses to light than when they were adapted to lower light levels. Therefore, smolts acclimated to darkness and encountering a strobe light would be expected to react to lower light intensities and at greater ranges than smolts acclimated to daylight. Anderson et al. (1988) found the strongest avoidance at 11:00 p.m. Our tests probably provide conservative estimates of strobe light effectiveness because they were conducted during daylight hours. The reflectance of the net-pen webbing also would result in underestimates of effective range.

Our 30-s reactance tests in April and June were adequate to detect only the initial avoidance reactions of smolts. Longer 10-min tests in spring

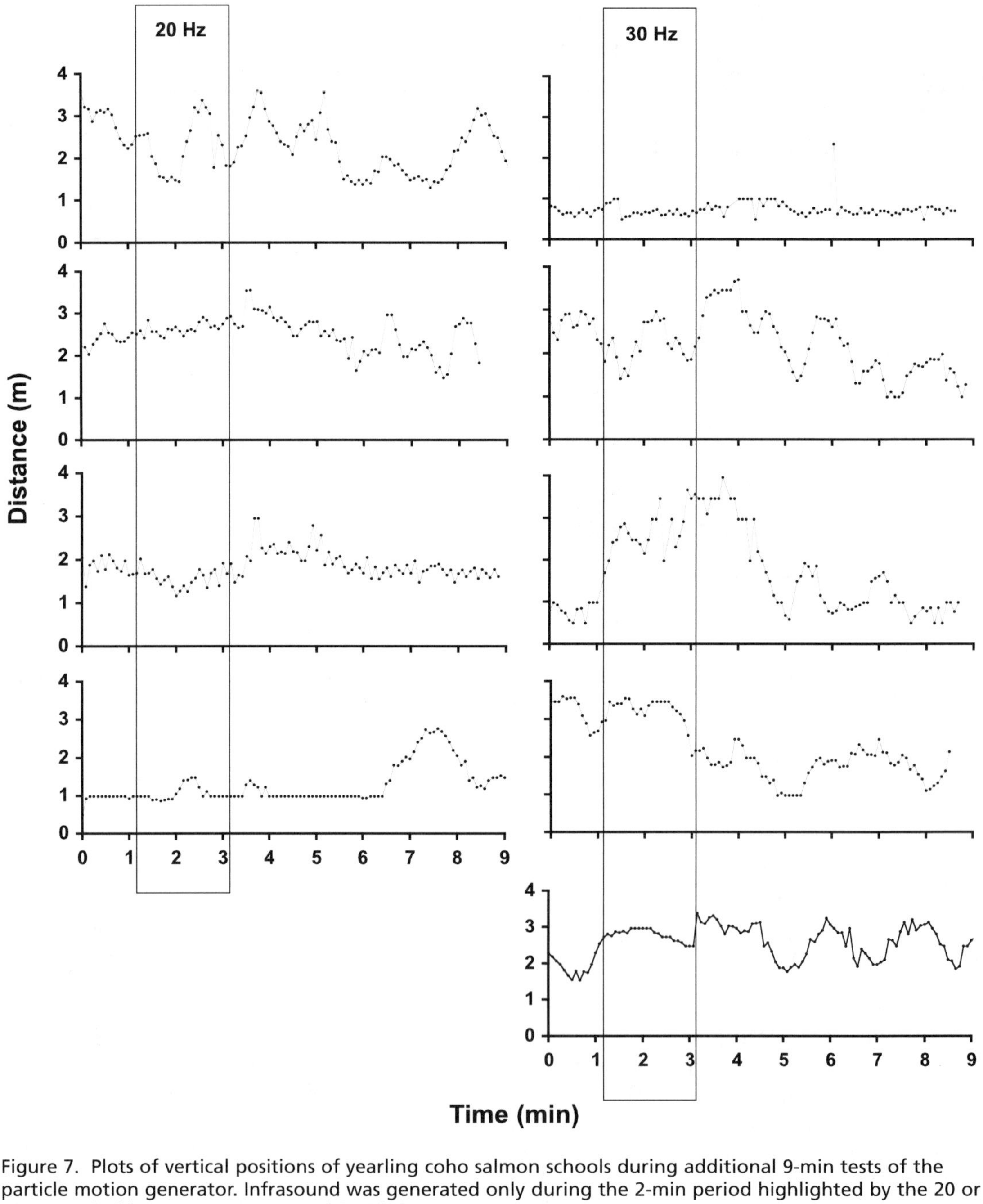

Figure 7. Plots of vertical positions of yearling coho salmon schools during additional 9-min tests of the particle motion generator. Infrasound was generated only during the 2-min period highlighted by the 20 or 30 Hz vertical bars at a pump rate of 31.6 l s^{-1}.

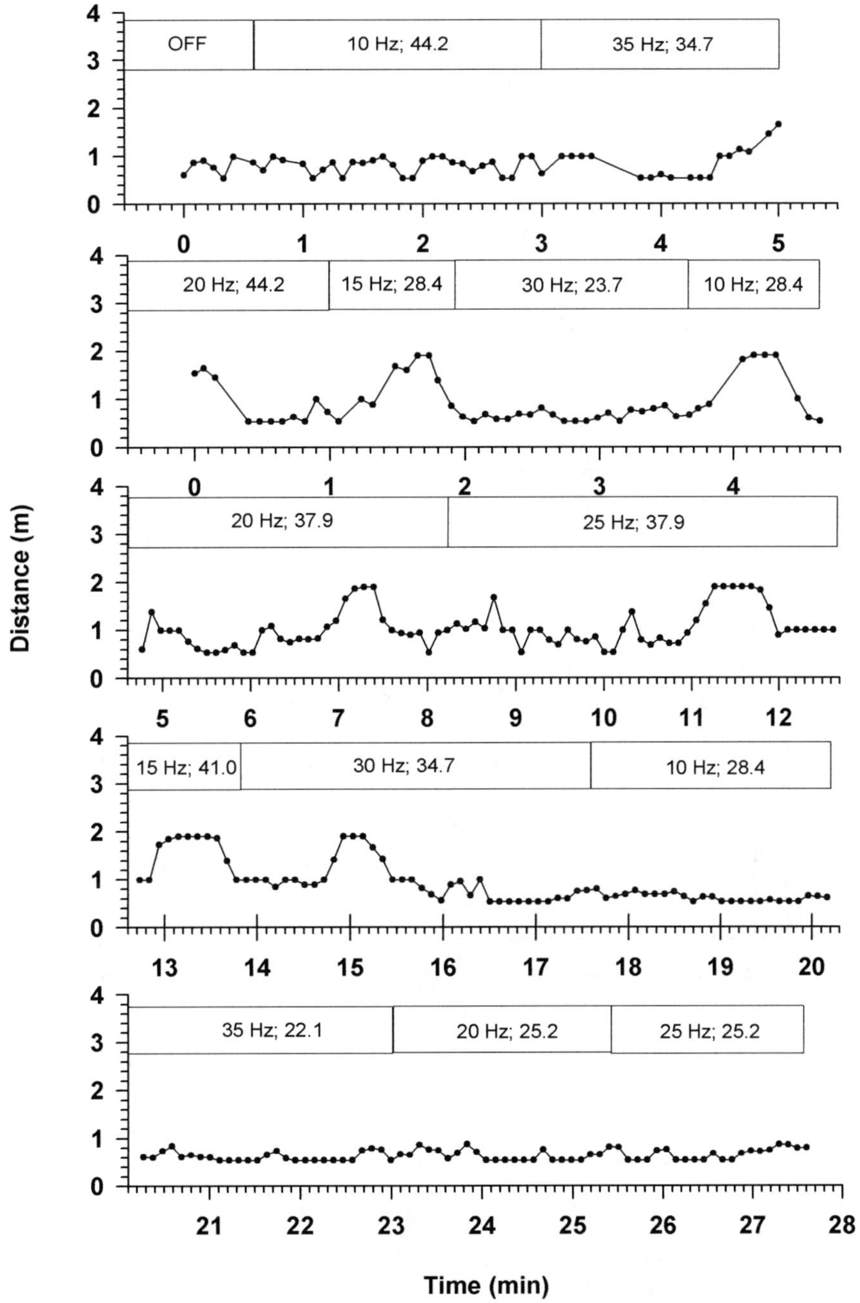

Figure 8. Plots of vertical positions of a sub-yearling coho salmon school during one 5-min exposure to two infrasound frequencies and one 28-min exposure to frequencies ranging from 10 to 35 Hz. Numbers after frequencies are pump rates (l s^{-1}).

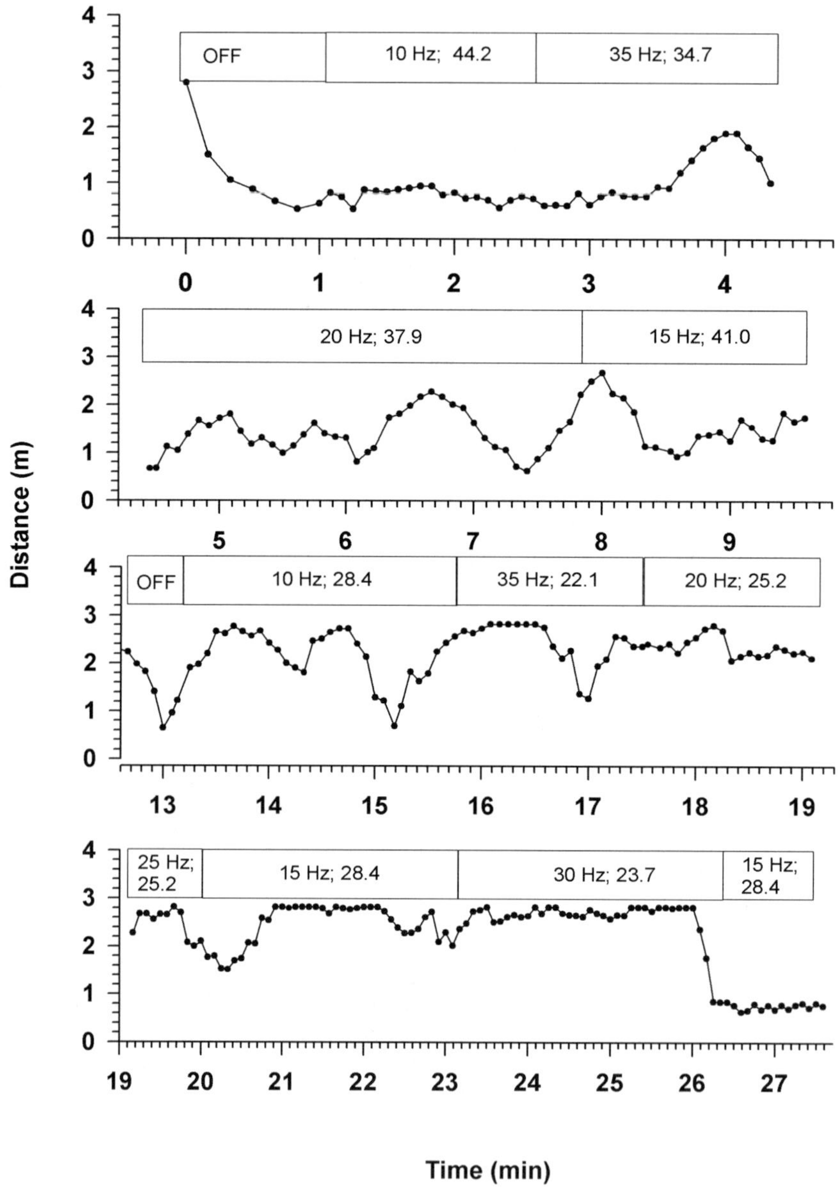

Figure 9. Plots of vertical positions of a sub-yearling coho salmon school exposed to a variety of infrasound frequencies. Numbers after frequencies are pump rates (l s^{-1}).

revealed greater displacement than 30 s tests and prolonged avoidance of the stimulus (Figure 4) relative to the behavior of control fish. Nevertheless, the frequency of false positive responses in 30-s control tests (17%) was much lower than the 80–100% frequency of avoidance observed for yearling coho and sub-yearling coho and chinook salmon schools exposed to strobe lights. Unfortunately, we were unaware of the value of longer duration tests until all tapes had been processed at the end of summer. During testing, we emphasized short duration trials to obtain as many trials as possible within the time available for testing. In retrospect, the minimum trial duration should have been 3 min to make certain that responses to lights were complete before the exposure ended. In five of seven 9-min trials on yearling coho salmon, maximum avoidance reactions were completed within

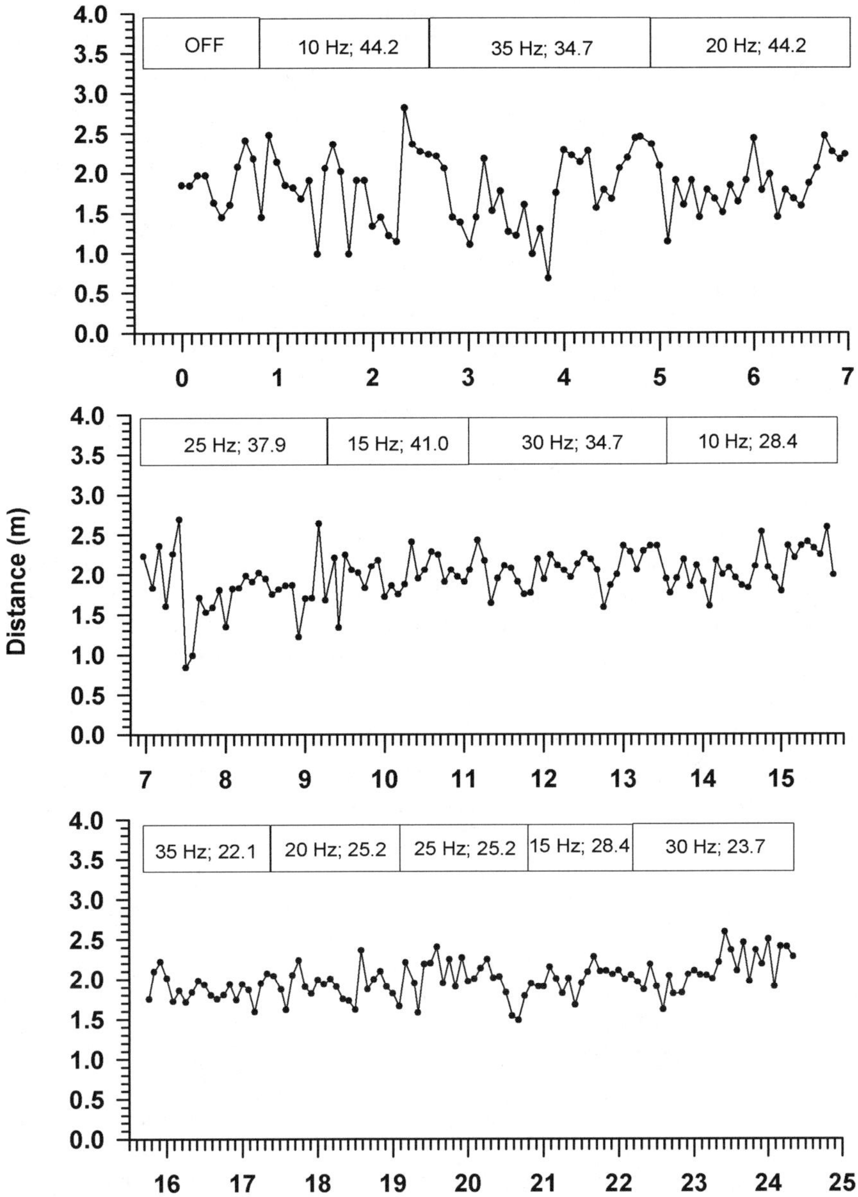

Figure 10. Plots of vertical positions of a sub-yearling chinook salmon school exposed to a variety of infrasound frequencies. Numbers after frequencies are pump rates (l s^{-1}).

15–60 s, but in two trials, the maximum upward displacement took as long as 3 min. In one trial, the school actually moved toward lights during the first min and then moved to the top of the pen within the next min and stayed at the top the remaining 6.5 min.

Observations with yearling coho salmon indicated that horizontal and 45°-upward aiming of strobe lights into the bottom 1 m of the net pen were very effective for eliciting upward avoidance, whereas lights aimed directly upward were not effective. Differences in effectiveness may be related to the angle of incidence of light to laterally positioned eyes or to the contrast between the source of light and the escape path, or both. Light flashed into the bottom of the pen from the side or 45° below the pen struck juvenile salmonids from the side and did not illuminate the escape path in the upper 3/4 of the pen. In contrast, light flashed up through the pen struck smolts from below where they may not be as sensitive. Vertical flashes also lit up the entire pen as light reflected off the sides all the way to the surface of the water. Smolts may need to perceive the stimulus and a contrast between its source and an escape path. A light stimulus from an adjacent source may provide directional information that is lacking when light from below is perceived simultaneously by both eyes. Thus, the position and orientation of strobe lights relative to smolts' angle of encounter and the desired direction of guidance may be important considerations for successful applications.

Only 2 of 16 schools of yearling coho salmon tested did not respond to strobe lights and we speculate that those schools were stressed by handling before they were tested. In one case, smolts were held at high densities in tubs for about 3 h during purse-seining operations before we could test them. The same school of purse-seined fish that did not respond on the day of capture responded vigorously the next day. The other nonresponsive school was the last to be tested and may have been stressed by repeated fishing of holding pens.

We did not observe obvious increases in the rate or distance of avoidance with up to five repeated exposures. However, our tests were not designed to test reaction time or measure escape velocity, as fish were varying distances from strobe lights when trials began. Anderson et al. (1988) found that juvenile rainbow trout in a raceway learned to avoid strobe lights more quickly within 15 repeated exposures. However, Anderson et al. (1988) herded fish under strobe lights and so all fish had the same initial position during each trial. Both the initial distances of fish from lights and fish orientation likely would affect reaction time and escape speed.

Would an upward displacement of 4.5 m per exposure be useful for increasing the effectiveness of a surface collector or in-turbine screens? The answer likely would depend upon the number of strobe lights, how they are deployed, and physical conditions at the site. Experiences at Rocky Reach Dam on the Columbia River (Anderson et al. 1988) suggest that strobe lights should be deployed upstream of turbine or surface collector trash racks so that these structures do not introduce conflicting stimuli such as vibration or turbulence. Also, smolts inside a turbine intake are entrained in rapid flow and would not be successful in avoiding strobe lights. The rate of flow upstream of turbine intakes should be slow enough for smolts to hold in response to a stimulus, (i.e., not more than 0.61 m/s). From our observations, 4-m vertical avoidance usually requires less than 40 s. We recommend future testing of strobe lights by deploying in an area upstream of trash racks where smolts have turned to face upstream and are backing tail-first downstream (Coutant and Whitney 2000). The best deployment likely would be to mount strobe lights upstream of piers on either side of an intake and aim them inward across the opening laterally or up at 45° toward the centerline of the intake. These orientations would take advantage of smolts' likely upstream orientation and increase the probability of smolts sensing and responding to the stimulus. Lights should not be aimed directly upward but might be aimed slightly upstream to prevent smolts from turning and swimming with the flow into the intake. Using a triangular frame, the deepest lights could be located farthest upstream with successive lights downstream located at shallower depths to provide repeated stimulation to increase range of guidance.

Particle-Motion Generator

The ineffectiveness of the ARL particle-motion generator in eliciting avoidance may have been a function of a ramped startup and the higher frequency harmonics and poor sinuosity in the fundamental waveform, although this hypothesis remains to be tested. We observed differences in the waveform of infrasounds produced by the piston device tested in 1995 by Knudsen et al. (1997) and pump and rotary-valve device tested in this study (Figure 11). The waveform of the PMG was less sinusoidal and had more high frequency sound riding on the 11-Hz infrasound signal than did the piston device test by Knudsen et al. (1997). The 10-s ramping of the infrasound signal from the PMG also may be less effective at eliciting the initial startle and subsequent escape response than a piston device that comes on instantly. Initial startle and escape responses result from input from several sensory systems including the octavolateralis, lateral line, and visual system (Popper and Carlson 1998).

We expected PMG frequencies ranging from 20 to 30 Hz to elicit moderate to strong avoidance responses, as they did from juvenile Atlantic salmon (Taft et al. 1995). However, the frequency of avoidance responses was similar to those observed in control treatments. An 11-Hz signal

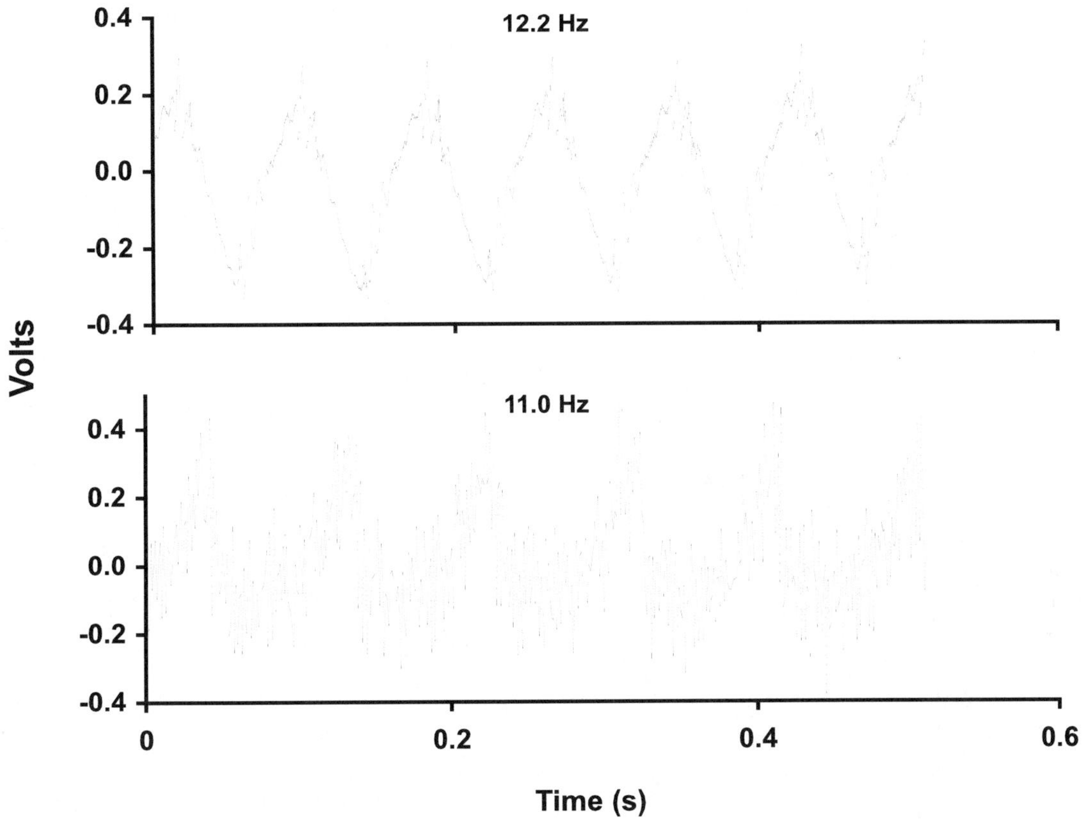

Figure 11. Waveforms of the piston infrasound device (Knudsen et al. 1997; top) and the particle motion generator tested in this study (bottom).

produced by the ARL device in our study clearly had more high frequency sound riding on the fundamental frequency than did the 12.2 Hz signal generated by the piston device tested by Knudsen et al. (1997). The fundamental 12.2-Hz frequency of this piston device was the predominant feature of the signal that elicited strong avoidance responses from a variety of juvenile salmonids in the northwest (Mueller et al. 1998; Mueller et al. 1999; Knudsen et al. 1997). The waveform of the PMG also ramped up more slowly than it fell off which resulted in a slight skew in the waveform, although this likely is a characteristic of the pump device that is always present. The fundamental frequency may have been clearer when the PMG was used on Atlantic salmon than when we tested the device in 1997 (Neal Brown, personal communication). We tried unsuccessfully to muffle higher frequencies from the PMG in summer, and spectral analysis revealed that most of the higher frequencies were strong harmonics of the fundamental and were less than 200 Hz and within the hearing range of juvenile salmonids. Amplitudes of the PMG at 20 Hz were more than adequate to elicit avoidance from fish located within 3.5 m of the source, as compared with a threshold found by Knudsen et al. (1994) of 0.01 m/s^2 at 10 Hz.

We speculate that the hatchery fish we tested may have been less responsive because they were raised in a noisy environment and had not learned to associate infrasound with predation. However, the same fish that failed to respond to PMG infrasounds responded vigorously to strobe lights and shadows of birds that sometimes flew over the test site. Fish also exhibited nondirectional startle responses when exposed to 150- and 180-Hz sounds (Ploskey et al. 2000). Knudsen et al. (1992) observed differences in responses of hatchery and wild juvenile Atlantic salmon to 10 Hz infrasounds but not to the extent that wild fish responded and hatchery fish did not respond. The avoidance response of wild fish was to flee to the deepest part of the pool, whereas hatchery fish only swam away from the source.

Given the weight, size, cost, and limited range of effectiveness of infrasound devices, strobe lights would seem to be the preferred choice for additional testing on the Columbia and Snake Rivers. Strobe lights likely would have a greater range of effectiveness than a piston infrasound device (ca. 1–2 m) unless turbidity is exceptionally high. This is not to say that infrasound should be discarded as a potential tool for guiding smolts at smaller scales. However, additional basic research on infrasound generation and measurement will be required to fully develop its potential as a smolt guidance tool. Apparently, subtle differences in infrasound signals may affect the effectiveness of infrasound devices.

Based upon results of 1997 testing, we recommend future investigations on the effectiveness of strobe lights in open water with flows greater than 0.61 m/s. We have learned a lot in net-pen testing, but the use of a net pen affects results in subtle ways that prevented us from extrapolating performance in a pen to performance in a forebay. For example, the webbing of the net pen reflects light, and we probably underestimated ranges of effectiveness when we moved the pen away from the light source. Also, fish located near the distal end of a net pen may actually move toward a stimulus when startled because they have nowhere else to go.

Acknowledgment

The tests described and the resulting data presented herein, unless otherwise noted, were obtained from research for Portland District of the U.S. Army Corps of Engineers by the U. S. Army Engineer Research and Development Center, Waterways Experiment Station, 3909 Halls Ferry Road, Vicksburg, Mississippi 39180-6199. Permission was granted by the Chief of Engineers to publish this information. Portland District Planning Group provided funding and oversight for this study. Valuable advice on study design and reviews of the original draft report were provided by Thomas Carlson, Pacific Northwest National Laboratory, Richland, Washington, John Nestler, U. S. Army Engineer Research and Development Center, Waterways Experiment Station, Vicksburg, Mississippi, and Arthur Popper, University of Maryland, Rockville, Maryland. Some of the strobe-light results first appeared in the HydroVision '98 Conference Papers, published by HCI Publications, Web Site: www.hcipub.com, and were used with permission.

References

Anderson, J. J., K. J. Puckett, and R. S. Nemeth. 1988. Studies on the effect of behavior on fish guidance efficiency at Rocky Reach Dam: avoidance to strobe light and other stimuli. Report FRI-UW-8801, Fisheries Research Institute, University of Washington, Seattle.

Burner, C. J., and H. L. Moore. 1953. Attempts to guide small fish with underwater sound. Special Science Report on Fisheries Number 111, U.S. Fish and Wildlife Service, Washington, D.C.

Burner, C. J., and H. L. Moore. 1962. Attempts to guide small fish with underwater sound. Special Science Report on Fisheries Number 403, U.S. Fish and Wildlife Service, Washington, D.C.

Carlson, T. J. 1994. Use of sound for protection at power production facilities: a historical perspective of the state of the art. Phase I Final Report under Task Order 49432: Evaluation of the use of sound to modify the behavior of fish. Report DOE/BP-62611–4. U.S. Department of Energy, Bonneville Power Administration, Portland, Oregon.

Coutant, C. C., and R. R. Whitney. 2000. Fish behavior in relation to passage through hydropower turbines: a review. Transactions of the American Fisheries Society 129:351–380.

Craddock, D. R. 1956. A review of studies in guiding downstream migrating salmon with light. Pages 1–13 *in* Progress Report on Fisheries Engineering Research Program, North Pacific Division, U.S. Army Corps of Engineers, Portland, Oregon.

Dunning, D. J., Q. E. Ross, P. Geoghegan, J. Reichle, J. K. Menezes, and J. K. Watson. 1992. Alewives in a cage avoid high-frequency sound. North American Journal of Fisheries Management 12:407–416.

Electric Power Research Institute. 1994*a*. Fish protection/passage technologies evaluated by EPRI and guidelines for their application. Report EPRI TR-104120. Electric Power Research Institute, Palo Alto, California.

Electric Power Research Institute. 1994b. Research update of fish protection technologies for water intakes. Report EPRI TR-104122. Electric Power Research Institute, Palo Alto, California.

Fields, P. E., A. K. Murray, D. E. Johnson, and G. L. Finger. 1958. Guiding migrant salmon by light repulsion and attraction in fast and turbid water. College of Fisheries, University of Washington, Seattle.

Gessel, M. H., B. H. Monk, D. A. Brege, and J. G. Williams. 1989. Fish guidance efficiency studies at Bonneville Dam first and second powerhouses-1988. Annual Report of Research, Coastal Zone and Estuarine Studies Division, National Marine Fisheries Service, Seattle.

Hanson Environmental, Inc. 1995. Georgiana Slough acoustic barrier applied research project: results of

1994 phase II field tests. San Luis and Delta-Mendota Water Authority, Department of Water Resources, Sacramento, California.

Hawkins, A. D., and A. D. F. Johnstone. 1978. The hearing of the Atlantic Salmon, *Salmo salar*. Journal of Fish Biology 13(6):655–674.

Haymes, G. T., and P. H. Patrick. 1986. Exclusion of adult alewives, *Alosa psedoharengus*, using low-frequency sound for application at water intakes. Canadian Journal of Fisheries and Aquatic Sciences 43:855–862.

Knudsen, F. R., P. S. Enger, and O. Sand. 1992. Awareness reactions and avoidance responses to sound in juvenile Atlantic salmon, *Salmo salar*. Journal of Fish Biology 40:523–534.

Knudsen, F. R., P. S. Enger, and O. Sand. 1994. Avoidance responses of low frequency sound in downstream migrating Atlantic salmon smolt, *Salmo salar*. Journal of Fish Biology 45:227–233.

Knudsen, F. R., C. B. Schreck, S. M. Knapp, P. S. Enger, and O. Sand. 1997. Infrasound produces flight and avoidance responses in Pacific juvenile salmonids. Journal of Fish Biology 51:824–829.

Loeffelman, P. H., D. A. Klinect, and J. H. VanHassel. 1991*a*. Fish protection at water intakes using a new signal development process and sound system. Pages 355–365 *in* WaterPower '91. Proceedings of the International Conference on Hydropower, Volume I. American Society of Civil Engineers, Denver.

Loeffelman, P. H., J. H. Van Hassel, and D. A. Klinect. 1991b. Using sound to divert fish from turbine intakes. Hydro Review 10(6):30–43.

McKinley, R. S., and P. H. Patrick. 1988. Use of behavioral stimuli to divert sockeye salmon smolts at the Seton Hydro-electric Station, British Columbia. Pages 1–7 in Chapter 4 of Proceedings: Fish Protection at Steam and Hydroelectric Plants. Report EPRI CS/EA/AP-5663-SR, Electric Power Research Institute, Palo Alto, California.

Moore, H. L., and H. W. Newman. 1956. Effects of sound waves on young salmon. Special Science Report on Fisheries Number 172. U.S. Fish and Wildlife Service, Washington, D.C.

Mueller, R. P., D. A. Neitzel, W. V. Mavros, and T. J. Carlson. 1998. Evaluation of low and high frequency sound for enhancing the capacity of fish screening facilities to protect outmigrating salmonids. Report DOE/BP-62611-13. U.S. Department of Energy, Bonneville Power Administration, Portland, Oregon.

Mueller, R. P., D. A. Neitzel., and B. G. Amidan. 1999. Evaluation of Infrasound and Strobe Lights to Elicit Avoidance Behavior in Juvenile Salmon and Char. Report DOE/BP-00029-6. U.S. Department of Energy, Bonneville Power Administration, Portland, Oregon.

Nemeth, R. S. 1989. The photo-behavior responses of juvenile chinook and coho salmon to strobe and mercury lights. Master of Science Thesis, University of Washington, Seattle.

Nemeth, R. S., and J. J. Anderson. 1992. Response of juvenile coho and chinook salmon to strobe and mercury lights. North American Journal of Fisheries Management 12(4):684–692.

Nestler, J, M., G. R. Ploskey, J. Pickens, J. Menezes, and C. Schilt. 1992. Responses of blueback herring to high-frequency sound with implications for reducing entrainment at hydropower dams. North American Journal of Fisheries Management 12:667–683.

Nestler, J. J. George, F. Ahmad, T. Carlson. 1994. Developing acoustic technologies for improving fish passage and protection in the Columbia River Basin: Program Rationale. Technical Report EL-95–8, U.S. Army Corps of Engineers, Waterways Experiment Station, Vicksburg, Mississippi.

Ploskey, G. R., and T. J. Carlson. 1999. Comparison of hydroacoustic and net estimates of fish guidance efficiency of an extended submersible bar screen at John Day Dam. North American Journal of Fisheries Management 19:1066–1079.

Ploskey, G. R., J. L. Pickens, and G. N. Weeks. 1994. Behavioral responses of anadromous blueback herring and American shad to high frequency sound, flow, and a wing wall in St. Stephen Tailrace, South Carolina. Technical Report EL-94-7, U.S. Army Corps of Engineers, Waterways Experiment Station, Vicksburg, Mississippi.

Ploskey, G. R., J. M. Nestler, G. N. Weeks, and C. Schilt. 1995. Evaluation of an integrated fish-protection system. Pages 162–171 *in* J. J. Casidy, editor. Waterpower '95, Proceedings of the International Conference on Hydropower, San Francisco, California, American Society of Civil Engineers, New York.

Ploskey, G. R., P. N. Johnson, and T. J. Carlson. 2000. Evaluation of a low-frequency, sound-pressure system for guiding juvenile salmon away from turbines at Bonneville Dam, Columbia River. North American Journal of Fisheries Management 20:951–967.

Popper, A. N., and R. R. Fay. 1993. Sound detection and processing by teleost fishes: A critical review. Journal of the Acoustical Society of America 53:1515–1529.

Popper, A. N., and T. J. Carlson. 1998. Application of sound and other stimuli to control fish behavior. Transactions of the American Fisheries Society 127:673–707.

Puckett K. J., and J. J. Anderson. 1987. Behavioral responses of juvenile salmonids to strobe and mercury lights. Report FRI-UW-8717. Fisheries Research Institute, University of Washington, Seattle.

Ross, Q. E., D. J. Dunning, R. Thorne, J. K. Menezes, G. W. Tiller, and J. K. Watson. 1993. Response of alewives to high-frequency sound at a power plant intake on Lake Ontario. North American Journal of Fisheries Management 13:291–303.

Smith, E. J., and J. K. Anderson. 1984. Attempts to alleviate fish losses from Allegheny Reservoir, Pennsylvania and New York, using acoustics. North American Journal of Fisheries Management 4:300–307.

Taft, E. P., F. C. Winchell, S. V. Amaral, N. A. Brown, J. P. Ronafalvy, and M. W. Haberland. 1995. Recent advances in sonic fish deterrence. Page 1724-1733 *in* J. J. Casidy (Editor) Waterpower '95, Proceedings of the International Conference on Hydropower, San Francisco, California. American Society of Civil Engineers, New York.

Winchell, F. C., S. V. Amaral, and E. P. Taft. 1994. Research update on fish protection technologies for water intakes. Report TR-104122, Electric Power Research Institute, Palo Alto, California.

American Fisheries Society Symposium 26:57–77, 2001

Turbulent Attraction Flows for Guiding Juvenile Salmonids at Dams[1]

Charles C. Coutant
Environmental Sciences Division, Oak Ridge National Laboratory
Post Office Box 2008, Oak Ridge, Tennessee 37831-6036, USA

Abstract.—Induction of mild turbulence and increased water velocity in slowly moving water in dam forebays is proposed as a way to attract and guide downstream-migrating juvenile salmonids to dam bypasses. Attraction flows have been used successfully for decades at the downstream ends of fish ladders to attract adult salmonids to fish ladder entrances. Similar hydraulic concepts for simulating a turbulent stream have generally not been implemented to divert juvenile migrants into dam bypasses and away from damaging turbine intakes. Turbulent flow appears to be a normal feature of water movement in natural rivers that guides and assists juvenile salmon migration. It is hypothesized that enhancement of turbulent flow in selected zones of quiescent dam forebays would spatially expand the opportunities for fish to discover entrances to fish bypasses (such as surface-flow bypasses). A "trail of turbulence" could lead fish to surface bypass entrances. Enhanced turbulence at bypass entrances and within bypass channels could also improve the effectiveness of juvenile salmonid guidance and retention in the bypasses. Turbulence could be enhanced by passive (using existing flow momentum or head directed by vanes) and active (pumped water jets or propellers) methods, which can be tailored to specific situations. Initial tests have confirmed the validity of the concept.

Passage of downstream-migrating juvenile salmonids through dams without death or injury is one of the major challenges for providing multiple uses of mainstem regulated rivers such as the Columbia and Snake (OTA 1995). Pacific salmon *Oncorhynchus spp*. and steelhead *O. mykiss* in the Columbia River basin that spawn in the upper mainstem or headwater tributaries must pass through up to nine dams for completion of their life cycle. Although detailed bioengineering research in the early days of dam building (1930s and 1940s) developed successful combinations of physical structures and hydraulics to attract and pass adults upstream through fish ladders (Collins et al. 1962; Weaver 1963; Bell 1990; Clay 1994), similar success with guiding and bypassing juveniles in dam forebays before they enter turbine intakes has not been achieved (Whitney et al. 1997; Coutant and Whitney 2000; Johnson et al. 2000). Delays of migration in the slowly moving waters of dam forebays appear to be a main factor in prolonging migration times of some juvenile salmonids in the impounded lower Snake River (Venditti et al. 1997). Dam breaching is being considered seriously to avoid migration delays at dams (Harza 1996).

An important feature of fish ladders for adults is an attraction flow at the base of a dam (Clay 1994). Using pumped water or water from the reservoir, hydraulic jets create zones of increased velocity and turbulence that simulate large volumes of water emerging from the base of the ladder, even though the main ladder flow is small (Figure 1). These attraction flows mimic the turbulence and water movement of the natal river and entice the adult salmonids to enter the ladder. The ladders themselves are a series of simulated pools, waterfalls, and high-velocity chutes that maintain highly turbulent flow. These common hydraulic features of numerous ladder designs encourage most adult salmonids to continue onward through the ladder.

In contrast, bypass systems for juvenile salmonids migrating downstream have not employed attraction flows other than the inflow currents that are drawn by hydraulic head to enter turbines, spillways, trash sluices, or other passage routes. Early bypass devices used surface spills or variable-depth submerged orifices such as the "Merwin Trap" (Stockley 1959) or pumps such as floating migrant collectors at Baker River, Wash-

[1]An earlier version of this paper was published as Oak Ridge National Laboratory Report Number ORNL/TM-13608. The author may be reached at 1-865-576-6830 or coutantcc@ornl.gov.

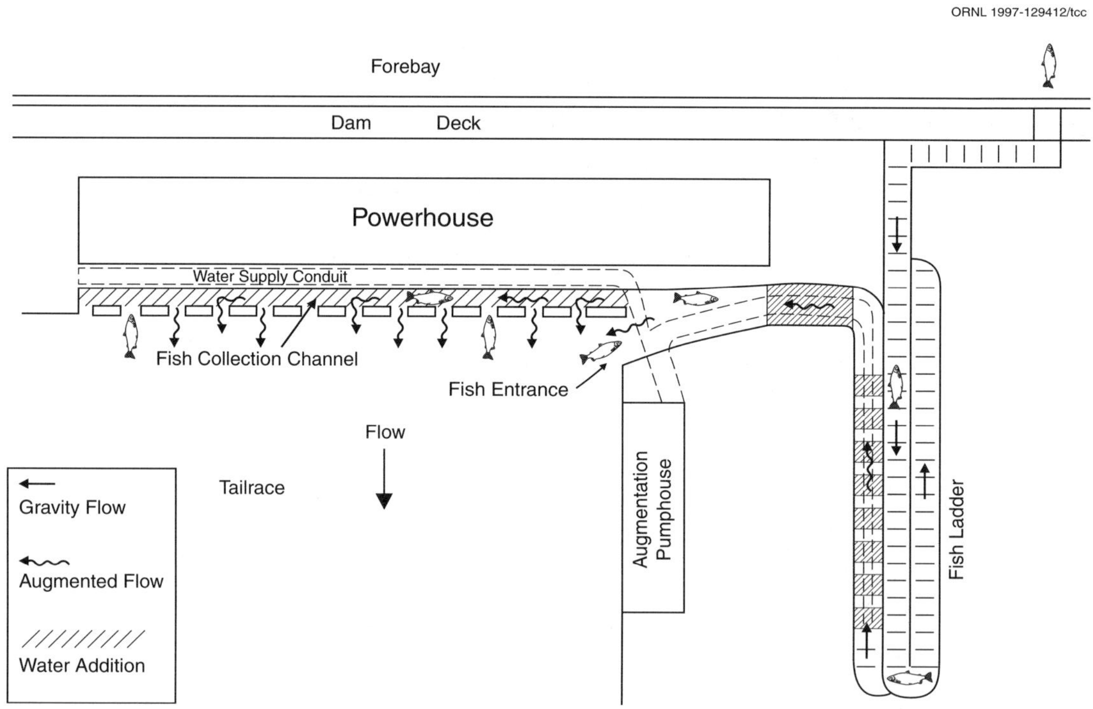

Figure 1. The attraction flow system for adult salmon migrants at the downstream entrance to the south fish ladder at Ice Harbor Dam, Snake River. Attraction water is pumped from the tailwater into the lowest segment of the fish ladder and the collection channel above the turbine discharges (hatched zones). The water that attracts adult salmon into and up the ladder is a mixture of water derived by gravity from the ladder and the pumped additions. (Adapted from blueprints supplied by D. Hurson, U.S. Army Corps of Engineers, Portland, Oregon).

ington, dams (Wayne 1961; Quistorff 1966) to create hydraulic head differences and thus generate attraction flows for outmigrants that pulled the fish into a bypass portal or enclosure. Each device apparently tried to simulate the volume and velocity of a natural stream without concern for natural turbulence. Early studies of turbulent water jets to deflect juveniles to bypasses at water intakes (Bates and VanDerWalker 1964) were not further applied to fish guidance despite positive results.

The most common "bypass" system today in the Columbia River basin uses submerged traveling screens to remove juvenile salmonids from the deep turbine intakes (Mighetto and Ebel 1994). These bypasses operate counter to the fish's natural tendency to remain near the water surface, and probably account for much of the delay in migration at the forebays. Fish near the surface that are drawn by bulk water flow into deeper water at a turbine intake swim vertically in an attempt to counteract the downward flow (Coutant and Whitney 2000; Johnson et al. 2000). To more closely match surface orientation of migrants, bypasses with surface entrances are being tested (Sweeney et al. 1997; Johnson et al. 1997a, 1997b; Johnson et al. 2000).

To facilitate new "surface-flow bypasses" or "surface collectors" that make use of the fish's surface orientation, considerable effort has focused on defining the hydraulic features of the regions where flow moves toward a particular dam portal (the portal's "flownet"; Johnson et al. 1997a). These flownets are generated by the differences in hydraulic head between the reservoir and the portal. The water is pulled toward the portal with gentle acceleration and smooth flowlines (Figure 2). Fish are presumed to detect the flownet and follow flowlines to the portal. Dam operations (e.g., turbine operations or spillway use) have been modified experimentally to alter the relative positions of the flownets leading toward turbines and alternative fish passage routes. Different bypass entrance sizes and shapes, including baffles, have been examined for the flownets they create, which affect the discovery and use of the bypass by fish.

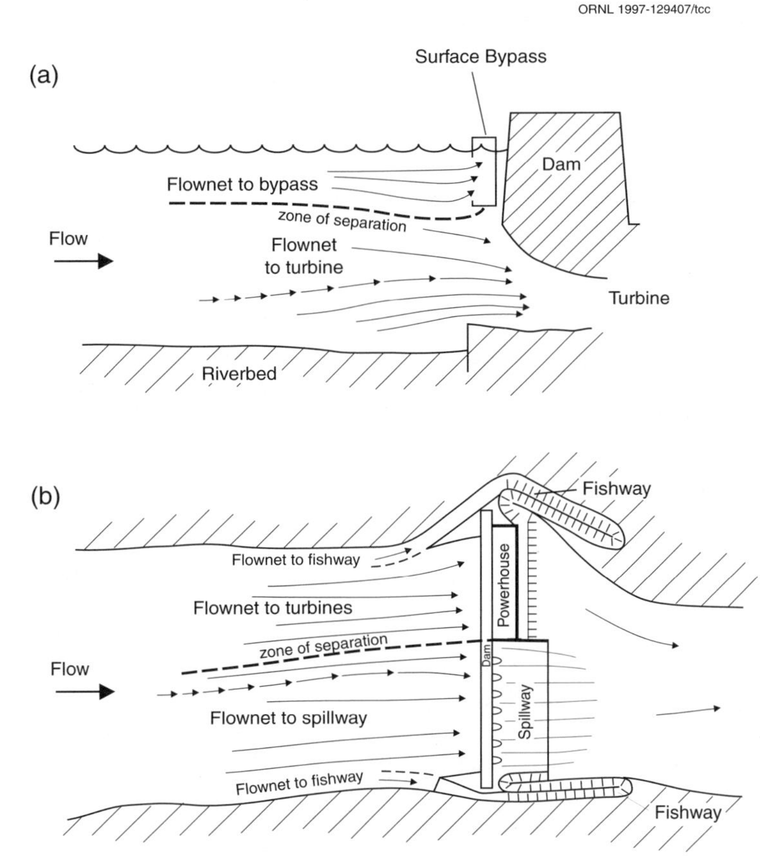

Figure 2. Typical vertical (a) and horizontal (b) flownets in the forebay of a mainstem dam. Flownets are the flow streamlines drawn to a particular dam portal. One streamline in (a) and (b) illustrates the increasing velocity (longer arrows) as the portal is approached. Zones of separation (heavy dashed lines) divide the trajectories of particles heading toward different portals. After Johnson et al. 1997a.

Flownets are passive and generally not very turbulent. They arise gradually in the forebays of dams where impounded waters are quiescent compared with an unimpounded river (even though bulk flow may still be recognizable). Flow velocities typically increase slowly in a linear direction as the dam portal is approached (Figure 2). Turbulence within the flownet has not been the focus of study as attention has been directed toward defining average velocity fields in the direction of the portals and the location of the "zone of separation" between two flownets, one going toward the bypass and the other toward the turbines (Johnson et al. 1997a).

Although the flownets near bypass entrances appear to attract migrants, their effectiveness diminishes with increasing distance from the portal. Johnson et al. (1997a) distinguished three zones in forebays of Columbia River basin mainstem dams: a near field within about 10 m of the portal, an intermediate field (10–30 m), and a far field (>30 m). Bulk flow dominates the far field, flownets separate toward various portals in the intermediate field, and specific portal conditions dominate in the near field. Rainey (1997) noted that there is little opportunity for fish to discover the entrance when they are outside the near field of a bypass flowing at about 2 m/s. Johnson et al. (2000) found that about half of the smolts migrating within 10 m of a bypass entrance actually entered it. Mechanisms for expanding the "opportunity for discovery" are a high priority for surface bypass development (Johnson et al. 1997a; Rainey 1997).

Flows within bypass channels near their entrances have generally not been engineered to keep juveniles in the bypasses once they have entered, as has been done for adult fish ladders.

Studies of several prototype surface flow bypasses have shown that a large proportion of the juveniles that enter bypass structures reverse course and pass through dams by another route (Pevan et al. 1996; Johnson et al. 1997a; Stevenson et al. 1997).

The purpose of this paper is to suggest that a missing component for guiding juvenile salmonids toward and through bypass systems at dams is river-like turbulence that would act as an attraction flow. The turbulence component may be especially important for surface flow bypasses, that otherwise seek to match the natural preferences of juvenile salmonids for surface waters and orientation with the direction of flow. It would be in accord with developing more "normative" fish passage conditions that match our understanding of the functional features of a natural system (R. Williams et al. 1996, 2000; Coutant 1997). Further, this paper suggests that levels of mild turbulence scaled appropriately for guiding fish could be created near bypass entrances by either passive or active methods. Passive methods would insert baffles or vanes in flownets to make use of existing water momentum to alter turbulence; active methods would use pumped water jets (as used in attraction flows at fish ladders) or propellers at distances in flownets where momentum is weak. The paper presents a general hypothesis that could be tested at fish passage prototypes and experimental bioengineering facilities.

Turbulence in Natural Rivers

Natural rivers, even large ones, are turbulent. Even a casual observer is aware of eddies at rocks or bridge pilings that continue for many meters downstream, "boils" where water surges to the surface creating horizontally spreading surface patches, and "rips" that separate fast-moving waters of the channel from slower-moving water at the edges or at mouths of side channels or sloughs. Flow over bottom sills or boulder fields generates standing waves and chutes of faster-moving water. Less apparent are waves propagated downstream by changes in flow volume or other disturbances, which interact with topographic features and add to turbulence. This is the natural environment of a downstream-migrating juvenile salmon. Many of these features are treated in detail in textbooks of applied hydrology and fluid dynamics and are summarized here.

Fluid dynamics of water flowing in rivers (open channels) is a field of physical sciences that matured 15–20 years ago (Liggett 1994) but that has not been brought to bear adequately on questions of juvenile salmon migrations or fish guidance. Study of the relationships between biology and fluid flow has generally been concerned with static (attached) life in moving fluids, such as the shapes of organisms in flowing waters (Vogel 1994), or design of swimming animals for propulsion, lift, and minimizing drag (Webb 1994; Triantafyllou and Triantafyllou 1995). Yet mobile, migrating salmonids undoubtedly interact with the complex fluid dynamics of riverine environments. The basic mathematical expressions of features of water flow, elevation, and velocity in rivers can be found in textbooks such as Chow (1964), Cunge et al. (1980), Abbott and Basco (1989), and Chaudhry (1993). Recent work has combined this basic understanding into computational fluid dynamics (CFD) models useful for designing engineering works for fish passage (e.g., Sinha et al. 1998). Turbulence has received renewed attention as a likely driver of stream channel dynamics as well as biology (Nikora and Smart 1997).

Several features of fluid dynamics of open channel flow (that is, flow in channels with solid sides and bottoms and a water surface exposed to air) may influence juvenile salmon migration. Water rarely moves as uniform flow along a streamline. Flow is most uniform when the depth and width are constant along the direction of flow and the bottom is smooth, a situation unlikely to occur in natural rivers frequented by salmon. Similarly, natural streams rarely have steady flow, in which the velocity at a point does not change with time. The normal pattern for a stream, viewed at the scale of a 50- to 200-mm fish, is to have velocities and directions of flow (vectors) that change in often complex ways, whether viewed while moving along a longitudinal reach or from one stationary point over time. With changes in velocity and direction, usually as vortices, come changes in water pressure. These complex changes in velocity, direction, and pressure are collectively considered turbulence and have several causes in a river.

Streams of water have different turbulent structures depending upon the interactions of flowing fluid and solid objects (e.g., Laufer 1951, 1954). Although the field of fluid dynamics has been greatly concerned with generation of turbulence at fluid-solid interfaces (for resolving such questions as drag in mobile vehicles), the environmental interest for fish migrating in rivers is with so-called free turbulent flows. Free turbulent flows are characterized by the phenomenon of intermittency, first noted in a jet by Corrsin (1943) and later more clearly recognized and studied in detail in the wake of a cylinder by Townsend (1949) and in two-dimensional channel flow by Laufer (1951). It is now known that intermittency is present in flows like wakes, jets, and those near a free surface of a boundary layer, and that the intermittency is a prominent feature of turbulent structure (Laufer 1954).

One large feature of riverine turbulence is the turbulent burst. A turbulent burst is the high-speed, turbulent ejection of fluid and suspended solids away from the sediment bed, often after encountering a streambed obstruction (Leeder 1983). Over distances of 4–5 times the water depth, accelerated flow events occur that act alternately toward and away from the bed (Figure 3). The rising burst of flow propagates downstream in the water column as a 3-dimensional flow feature and is seen at the water's surface as a roughly circular upwelling or "boil." A view of the surface of a swiftly moving river such as the unimpounded Columbia at Hanford, Washington, is of a patchwork of these boils. Water velocities in the leading edge of boils exceeds that of the general surrounding water. Bursts have been studied for their importance in moving sediment, but not fish. For every zone of upwelling there is a corresponding zone of downwelling, which impinges on the bottom and can move sediment. Carling (1992) has suggested that riverine sediment transport is related to the inherent turbulent structure of rivers. Unsteady bedload transport in the sea also is driven by "bursting-type" cycles (Thoren et al. 1989). These velocity bursts are a function of flow rate, water depth, and bottom roughness. Reduction of flow velocity and increase in depth (as in impoundments) would greatly dissipate, and likely terminate such turbulent structures. It seems reasonable that salmonids emigrating in rivers would have evolved to make use of the zones of accelerated velocity in these turbulent bursts to assist them in downstream movement.

Vortices in wakes are another feature of turbulent flow, in which the propagating mechanism for the 3-dimensional flow feature usually acts initially in the horizontal plane rather than the vertical (as in bursts). Rows of vortices are shed behind solid bodies and trail behind in a wake, a feature called a "Von Kármán trail" (Figure 4). If the body is in midstream, there is a wake of roughly parallel vortices, forming first on one side of the body and then the other. Each vortex rotates horizontally in the opposite direction of the preceding and succeeding ones. If the body is a projection from shore, the vortices trail in single file in what is often referred to as a shoreline "rip." In either case, water velocities on the outside of the wake of vortices are more rapid than the general (average) water flow.

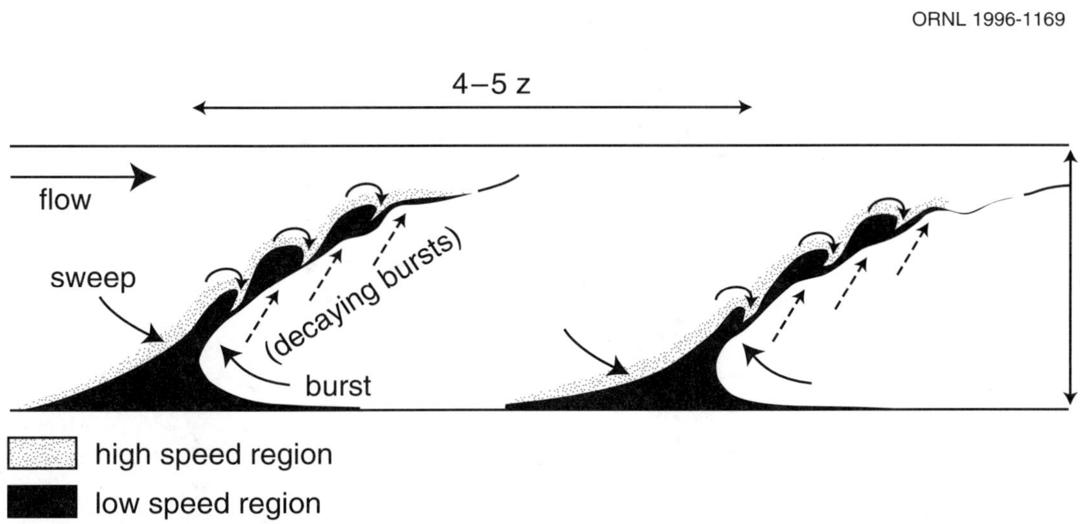

Figure 3. Anatomy of a typical turbulent burst, showing high-speed turbulent ejection of fluid and suspended particles away from the bed of a river. Zones of increased water velocity occur at the leading edges of bursts as they rise toward the water surface at a rate faster than general water movement. After Leeder 1983.

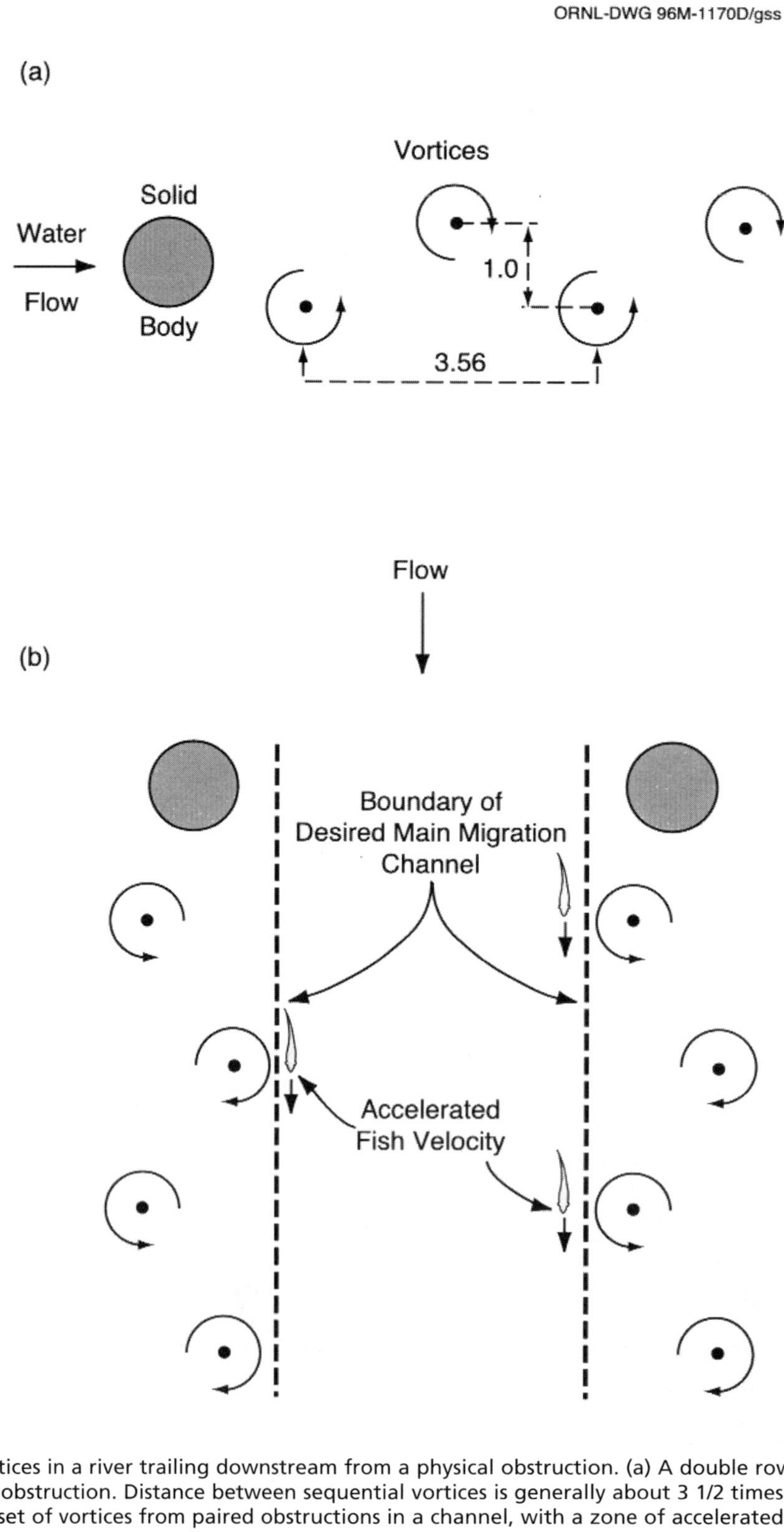

Figure 4. Vortices in a river trailing downstream from a physical obstruction. (a) A double row of vortices from a single obstruction. Distance between sequential vortices is generally about 3 1/2 times the lateral spread. (b) A set of vortices from paired obstructions in a channel, with a zone of accelerated flow between obstructions.

Migrating juvenile salmonids may have evolved to seek and enter these higher velocities.

Responses of Juvenile Salmonids to Turbulence

It was in this turbulent environment that juvenile salmonids evolved their downstream migrations. Although there has been little direct study of the orientation of juvenile salmonids to the well-understood features of turbulence, it is reasonable to assume that they normally used the enhanced velocities of turbulent microenvironments to assist their downstream migration (Triantafyllou and Triantafyllou 1995; R. Williams et al. 1996, 2000). Every bit of assistance from a downstream vector of turbulence would mean less metabolic energy to be expended in swimming. It would be surprising, indeed, if downstream-migrating fish did not take advantage of opportunities for recapturing kinetic energy from turbulent wakes, waves, and bursts. Triantafyllou and Triantafyllou (1995) concluded that fish instinctively exert precise and effective control of the flow around their bodies to extract energy. In fact, most studies of juvenile salmon orientation during out-migration show them positioned head upstream so as to use their energy most efficiently in maintaining orientation with flow rather than in active downstream swimming (Smith 1982; Cada et al. 1997; Coutant and Whitney 2000). In the physical sciences, the extraction of energy from unsteady flows using a stationary foil is called the Katzmayr effect, after the German engineer who first studied it (Triantafyllou and Triantafyllou 1995).

Turbulence has generally not been considered as an attractive feature for guidance of downstream migrants for several reasons. Turbulence developed a negative reputation because of the damaging effects of intense turbulence, in which turbulent shear caused by high-pressure water jets caused gross trauma (Groves 1972). Much of the emphasis in developing "fish-friendly" turbines has been to reduce turbulence in the fish's trajectory through the turbine system (Cada et al. 1997). Even when turbulence intensity was much lower, early experiments showed fish avoided areas of turbulence near bypass intakes (Brett and Alderdice 1958). Recent studies have found that structures like trash racks at bypass portals can cause fish to be repelled, which is often attributed to "turbulence." Rapid changes in direction and velocity of water movement at louvers have been used to repel juvenile salmonids and guide them to bypasses. The hydraulic principle behind louver fish guidance systems is a rapid change in direction (and velocity) of water as it passes through louvers oriented nearly perpendicular to water flow (review by Taft 1986). Fish detect the changed flow conditions at some distance upstream of the louvers and follow the angled bank of louvers to a bypass system. The many studies that have evaluated guidance efficiency of louver systems did not measure turbulence *per se* but may show the influence of it. For example, Ruggles and Ryan (1964) observed clear behavioral changes in sockeye smolts in their experimental flume when velocities (and we can speculate turbulence, also) were changed. At low velocities (to about 40 cm/s) and presumed low turbulence, smolts actively swam downstream and exhibited an in-and-out exploration of the bypass entrance; at about 75 cm/s smolts moved about rather randomly; at 115 cm/s and probably higher turbulence, smolts were all oriented tail-first moving downstream and were guided to a bypass.

These brief examples indicate that it is important to consider turbulence at the appropriate scale. The two characteristics of turbulence that may be most important for defining biological effects are vortex size in relation to the size of a fish and turbulence intensity (Figure 5). At high turbulence intensities and small vortex diameters, shear effects can be damaging, as shown by Groves (1972). At somewhat lower turbulence intensities, fish may recognize the potential danger, and avoid the turbulent zones. At even lower intensities, there seems to be an attraction, in which there is an actual migration aid or simply a behavioral cue (specifically to flow direction or just to the presence of water movement). At the lowest intensity of turbulence (natural waters are never completely free of turbulence), the fish may not recognize turbulence at all. These relationships remain conceptual, but some day may be defined quantitatively.

Researchers in Russia have investigated the influence of flow turbulence on fish behavior since the late 1970s, but only some of this literature is available in English (Shtaf et al. 1983; Skorobogatov and Pavlov 1991). Much of their emphasis has been on nonmigratory species (roach *Rutilus rutilus*, minnow *Phoxinus phoxinus*, and gudgeon *Gobio gobio*) and their ability to maintain position in turbulent flows rather than their use of turbulence to assist migration. Turbulence was shown experimentally to

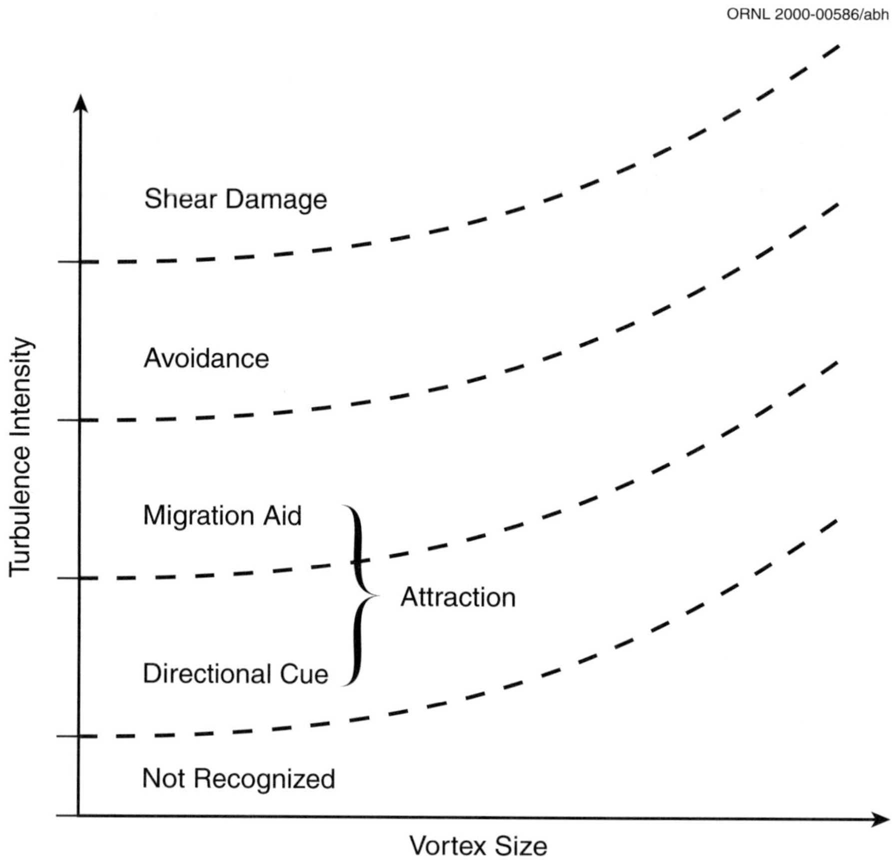

Figure 5. Conceptual view of the scales of turbulence affecting fish, ranging from damaging high turbulence intensities and small vortices to unrecognizably low turbulence intensities. Low turbulence intensities are hypothesized to provide directional cues and migration aids to downstream migrating fish.

lower the critical flow velocity for swimming (or critical swimming speed, the velocity at which a fish is no longer able to maintain itself in a current, and therefore drifts downstream; Pavlov et al. 1994). Whether through fatigue or preference, the fish moved downstream at lower average velocities than in less turbulent flow. A significant result for salmon is the demonstration that the horizontal movement of drifting young fish is facilitated by narrow channels with higher turbulence (Pavlov et al. 1995). Turbulence appeared to assist the fish in their daily cycle of horizontal movements alternating between near-shore (day) and stream-channel (night) occupancy, similar to the normal daily cycle of underyearling chinook salmon *O. tshawytscha* in the Columbia River basin.

Few experiments have been carried out to examine the responses of downstream-migrating juvenile salmonids to 3-dimensional changes in water velocity at distances within 10–20 body lengths. Johnson (1996) reviewed numerous laboratory and field studies of smolt orientation to water velocity, spanning more than 40 years, with the conclusion that evidence for movement to higher velocities was mixed.

A sampling of research suggests that fish can discriminate higher velocities that would be found in small turbulent eddies. Dijkgraaf (1933) showed that blinded fish in midwater could orient to a small jet of water. McKinnon and Hoar (1952) observed that chum salmon *O. keta* and coho salmon *O. kisutch* showed a preference for the stronger of parallel laminar currents. Gregory and Fields (1962) found that juvenile coho and chinook salmon were capable of discriminating very low velocities without visual stimulation (in the order of 0.02 ft per second or 0.4 m/min.). Recently, J. Williams et al. (1996) reported experiments with yearling and underyearling chinook salmon outmigrants in an experimental flume with water moving at 0.7–0.8 m/s. These fish were recorded as they reacted to the presence of a physical barrier inserted in their path (Figure 6). The fish avoided low velocities in front of the barrier and actively swam (oriented head upstream) to position themselves in the high-velocity flow lines above and below the barrier. The fish detected and responded to velocity differences of about 10 cm/s with a detection distance of approximately 10 cm. The detection distance approximated the length of the fish's pressure-sensing lateral line. Anesthetized controls confirmed that the response by alert fish was an active one and not merely a function of fluid dynamics around the barrier. No observations were made in the turbulent zone downstream of the barrier. Quite possibly the mixed results of studies of velocity orientation by smolts are the product of lack of concern for the guiding effects of small-scale turbulence.

At least one field study may have inadvertently shown the beneficial effects of a turbulent attraction flow. A log boom at Lower Granite Dam on the Snake River was shown by Swan et al. (1985) using fyke nets, and Wilson et al. (1991) using radiotelemetry, to guide yearlings of chinook salmon and steelhead toward the dam spillway. The boom was acutely angled from the south shore of the forebay across the powerhouse entrance, terminating at the juncture of powerhouse and spillway. Hydroacoustic surveys of turbine intakes with and without the boom definitely showed a higher abundance of migrants at the turbines nearest the termination of the boom when the boom was in place, especially in daytime (Kuehl 1986). Although the boom provided little physical barrier to movement of fish passing under it in the upper water column, one can speculate that a zone of turbulent water was created along the boom, which may have provided cues for guidance along the boom.

Juvenile salmonids have a highly developed lateral line system for detecting changes in water pressure. More is known about its anatomy than about its function in a natural environment, however (Bleckmann 1986; Popper and Platt 1993). As suggested by the review of R. Williams et al. (1996), this sensory system could be the mechanism whereby downstream migrants facing upstream guide themselves along the microregions of highest velocity in a turbulent hydraulic environment. The fish

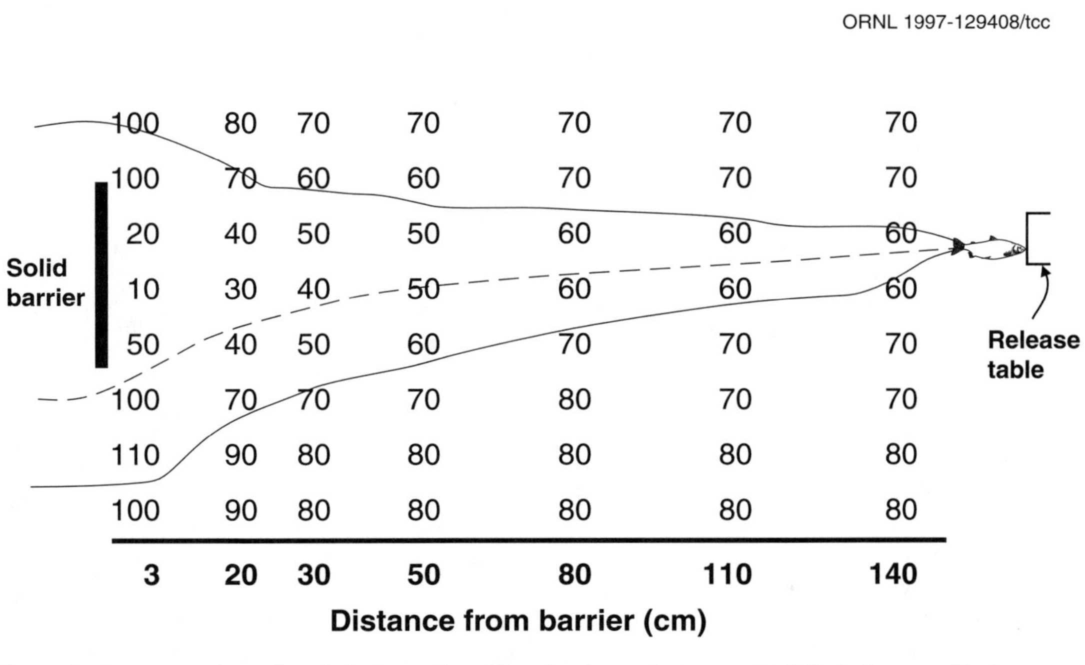

Figure 6. Average trajectories of alert yearling chinook salmon that moved tail-first above and below a solid barrier (solid lines) in an experimental flume, superimposed over average water velocities (cm/s). The trajectory of an anesthetized salmon that moved under the barrier is shown for comparison (dashed). After J. Williams et al. 1996.

might thus easily discover and follow turbulence generated in dam forebays at the entrances to surface bypasses.

It is difficult to extract information about possible fish behavior in turbulent flows from the literature because flow structure generally has been simplified to measures such as mean water column velocity (e.g., Johnson 1996). Highly sensitive acoustic doppler current profiler data that could give information on turbulence have been averaged both in time and dimension (e.g., depth) and high-velocity "errors" discarded to remove all evidence of small-scale variations (turbulence). Not surprisingly, the sanitized velocity data are not well correlated to fish distribution, even at locations such as Wells Dam where there is good fish guidance to bypass entrances and nearly 90% of downstream migrants are passed in approximately 10% of the river flow (Johnson 1996). Realistically, the relationships will not be clarified or quantified until measurements and data analyses appropriate to understanding turbulence at the scale of migrating fish are undertaken (such as Stacey and Monismith 1997). In the meantime, considerations for management can be given on the basis of likely responses based on indirect evidence.

Induction of Turbulence

I propose that the "opportunity for discovery" (Rainey 1997) be enhanced in the intermediate and far fields of dam forebays by creating a "trail of turbulence" leading to bypass entrances. Further, I suggest that such induced, turbulent flows could be useful for guiding downstream migrants in reservoirs wherever the normal, riverine turbulence has been dissipated and fish lose their behavioral cues to which direction is downstream. In essence, the turbulent features of currents in a natural, unimpounded river would be simulated in a relatively narrow path in the quiescent surface waters of the forebay's intermediate and far fields. Fish guidance would be "normative" (Coutant 1997) by making use of the normal orientation mechanisms of river-migrating fish to turbulent flow. Both true behavior (i.e., selection of turbulent zones with accelerated velocity) and physical entrainment (fish being incorporated hydraulically into the zones of higher velocity) would be involved, although it would be difficult to distinguish between them. A trail of turbulence could be engineered in several ways, both active and passive. A linear series of multiple devices, active and passive, would likely be most effective for establishing the "trail" in areas such a dam's forebay.

The idea of guiding downstream migrants by creating river-like currents aimed in the direction of a bypass entrance has some precedent. A search of U.S. patent files uncovered an expired 1984 patent by D. L. Koch (Patent Number 4437431: "Method and apparatus for diversion of downstream migrating anadromous fish") for a surface bypass system in which water jets were proposed to generate "artificial stream conditions within the forebay of dams used for hydroelectric facilities." The jets were to create an artificial stream current in the path of fish in the reservoir, direct the migrating fish in such a manner that they sense the artificial stream current, are attracted to it, and enter it. The jets were directed at a funnel-like bypass entrance in a dam forebay and were to push fish from the forebay into the funnel and then through the bypass piping. The patent encompassed the entire bypass system, and the exact features of the current-inducing jets near the entrance were not described in detail. No functioning prototype of this patented procedure seems to have been reported.

More recently, Truebe and Truebe (1997, 1998) proposed and tested a "current inducer" directed toward a surface bypass. Field testing of a surface bypass for Atlantic salmon juvenile migrants in New Hampshire (Upper Penacook Hydroelectric Project on the Contoocook River) without induced currents indicated that migrants were not drawn to the bypass but followed bulk flow to the turbine intake. The bypass entrance was located where there were few surface currents beyond its immediate vicinity. To rectify this problem, a mechanically generated current was aimed at the bypass entrance and across the top of the turbine flow from a point along the shoreline upstream of the powerhouse. Current was generated by two 2-horsepower, low-speed submersible electric motors driving approximately 1-m-diameter propellers mounted in a screened frame. Propellers were calculated to be more energy efficient than water jets (J. Truebe, personal communication). One propeller was positioned horizontally with its centerline 76 cm below the water surface while the other was tilted upward. Whereas the bypass had been ineffective without the current inducer, tests with the artificial flow yielded bypass efficiencies of 93 and 80%. The emphasis of design and testing was to induce water movement (velocity, momentum) toward the bypass, although photographs indicate that highly turbulent flow was created by the propellers. Both turbulent cues for orientation and water momentum for transport probably contributed to the positive results.

Stimulated by publication of an earlier version of this paper in a limited-distribution laboratory report (Coutant 1998), Darland et al. (2001) tested the concept of guiding juvenile Pacific salmon with an induced, turbulent flow. Currents were induced in the forebay of Cowlitz Falls Dam on the Cowlitz River, Washington, in summer 1999 using a variation of the propeller arrangement used by Truebe and Truebe (1997). Migrating juvenile salmon were guided from the north side of the forebay to the south side, as determined by split-beam hydroacoustics and radio telemetry. An on-off statistical design of test conditions was used to demonstrate effectiveness of the induced currents for guiding fish. Fish followed currents that were characterized by drifting drogues. A second season of testing (in spring and summer 2000) yielded similarly positive results (D. Rondorf, personal communication).

Passive Devices

Control structures placed in a reservoir might use the momentum of existing flow fields to induce turbulence and changes in water velocity that could guide fish and aid their movement downstream and toward bypasses. For example, a pair of simple concrete cylinders placed at the edges of the main channel would induce vortices in the direction of a bypass that could be sufficient to guide fish movements (Figure 4). If such cylinders were placed at intervals along one side of a reservoir, a useful channel velocity might be continued from upstream riverine reaches well into the dam forebay where downstream directionality of fish movement disappears and fish begin a back-and-forth searching behavior (Venditti et al. 1997). Submerged berms and dikes are commonly used to direct river flows. Few navigable rivers are without these structures intended to maintain navigation channels through induction of turbulent scour of bottom sediments.

The submerged vane is a control structure that also has found application in redirecting water velocities to aid sediment movement (Odgaard and Wang 1991a). This structure also may be capable of redirecting velocity patterns to aid juvenile salmon migration into bypasses. As applied to sediment management, the vanes are small, flow-training structures (foils), installed near the riverbed and designed to modify the near-bed flow pattern and redistribute flow across the channel cross section (Figure 7a). They are installed at an angle of 15–25° with the flow and their initial height is 0.2–0.4 times the local water depth at design stage. The vanes generate a secondary circulation of flow not unlike other midstream obstacles, but with additional avenues of control through angle, size, and number of vanes. A single vane generates a vertical vortex of flow that would push surface water to the center of a channel (where fish migrate) and sediment toward the sides. Vanes in groups generate larger, combined vortices (Figure 7b). When aligned on opposite sides of a channel, sets of vanes can constrict the flow to a more defined channel (Odgaard and Wang 1991b; Figure 7c). The functionality of vanes in rivers has been evaluated experimentally (Marelius and Sinha 1998) and numerically (Sinha and Marelius 2000).

Although there have been many applications of submerged vanes to sediment control, there has been no experimentation with their use as fish-assisting devices. One can visualize using both vertical and horizontal vanes suspended in flow fields in many orientations to simulate natural turbulent bursts and vortices directed toward bypass entrances. The vanes should be effective wherever there is sufficient momentum in the bulk flow to power the changes in flow direction, velocity, and turbulence.

Active Devices

Active induction of turbulence may be necessary for effective fish cueing or attraction to dam bypasses where ambient water flow and channel topography are insufficient to generate useful turbulence and there is insufficient momentum to use passive devices. Flownet flows induced in bypass structures are generally insufficient to extend flow cues far enough into the quiescent waters of the reservoir for fish to discover and orient to them.

An active approach was advocated by Koch (patent) and Truebe and Truebe (1997). A small amount of pumped water (relative to the bypass flow) could produce turbulent flow cues through hydraulic mixing of an induced jet with surrounding water (Figure 8). The physics of turbulence in an unconstrained fluid jet is well established (Laurence 1956). A pumped water jet (single or in arrays) is highly controllable in volume and direction, and it can be positioned easily in the water column. However, the jets must be designed so that damaging effects of turbulent shear are avoided (Groves

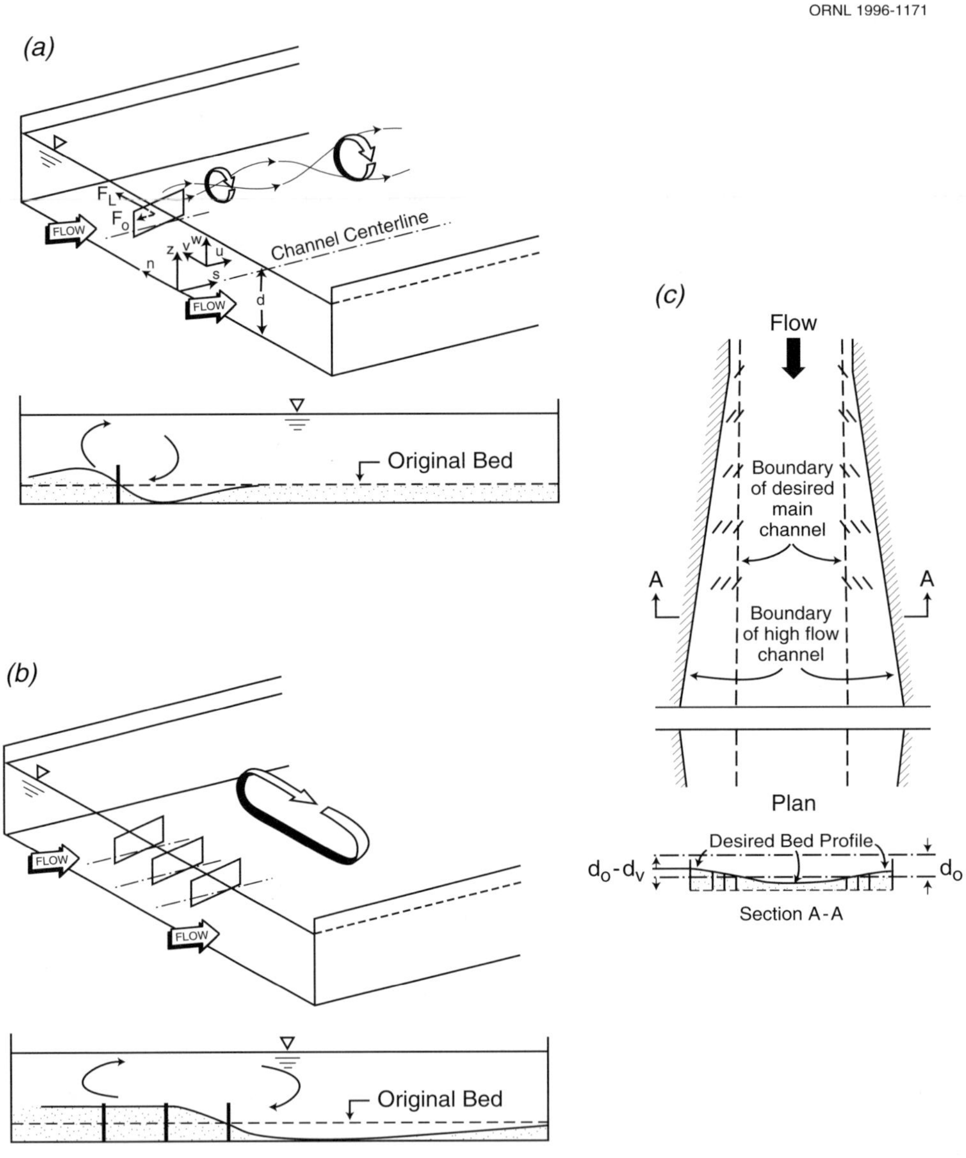

Figure 7. Flow induced by submerged vanes to control sedimentation and clear river channels. (a) Schematic perspective view of flow around a single vane at the side of a channel, showing vertical vortices. The lower panel of (a) shows the circulation pattern in channel cross section. (b) The larger vortex pattern created by a set of closely spaced vanes, shown in perspective and cross section. (c) Multiple vanes installed at the sides of an expanding water body to constrict the higher-velocity channel. After Odgaard and Wang 1991a and 1991b.

1972). It may be preferable to use propellers (as Truebe and Truebe 1997), paddle wheels, or similar mechanical devices. Computer models for the near-field mixing of water jets (such as the CORMIX model developed for mixing of waste discharges) could be used to custom design a series of water jets to form a trail of turbulence through a dam forebay and into a bypass entrance. Considering the high potential for such approaches, high priority should be given to both computational and experimental studies of relative effectiveness and energy efficiency of alternative devices.

Turbulence generators need to produce clearly defined types of turbulence so that fish responses can be compared. An ability to quantitatively define turbulence at scales appropriate for migrating fish and in ways that encompass both vortex size and intensity in three dimensions is essential

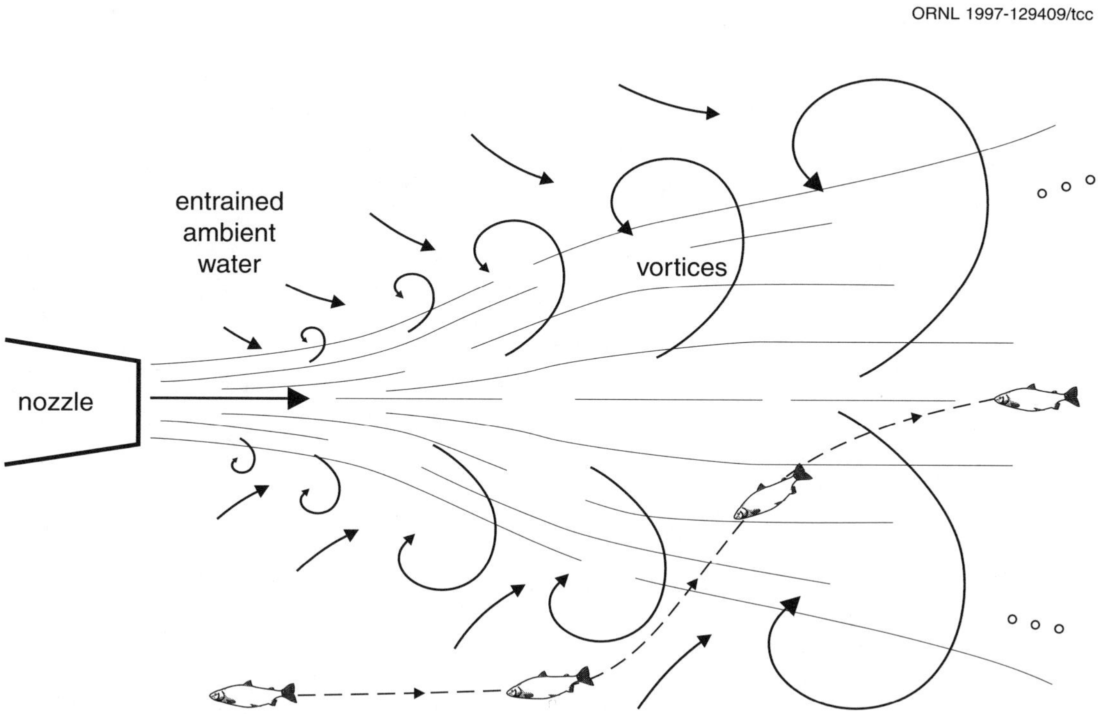

Figure 8. Two-dimensional view of turbulent mixing of a hydraulic jet into a standing or slowly moving fluid. Ambient water is entrained into a series of vortices at the edge of the jet. The momentum of the water at the nozzle is imparted to the sum of jetted and entrained fluid. Downstream-migrating juvenile salmon would discover the jet as they are entrained with surrounding water (dashed trajectory) and likely maneuver to remain in the highest velocity zones as they follow the jet's overall trajectory.

for conducting meaningful guidance experiments and obtaining repeatable results. Behavioral study designs need to be "calibrated" with equivalent hydraulic conditions. Unfortunately, turbulent flows are quite variable and hard to measure. Acoustic and laser doppler velocimeters have greatly advanced measurement techniques (e.g., Nikora and Smart 1997). The calculated "turbulence intensity" is a variable derived from a sequence of near-instantaneous field measurements that may be a useful summary of turbulence (Pavlov et al. 1994). But there is, as yet, no agreed-upon measure of turbulence that has high relevance to fish migration.

Application

Specific designs for locating and installing passive and active devices to create a trail of turbulent flow would be necessary for each reservoir and fish bypass facility, largely because each dam is laid out differently. However, the concept of simulating hydraulic turbulence and momentum in waters through which fish are to be guided should have general application. There appear to be three principal applications: (1) deployment in open waters of a quiescent main reservoir, (2) deployment in the intermediate- and near-field approach to fish bypass entrances (e.g., within 30 m of the bypass entrance), and (3) deployment in the bypasses themselves, to continue the mild turbulence into and through the bypass to prevent fish from sensing a changed environment and swimming back out.

Water nozzles or vanes could be placed underneath a desired surface migration pathway and angled upward, thus closely simulating turbulent bursts (Figure 9). The turbulent flow effects could be concentrated in the near surface waters usually populated by downstream-migrating fish by suspending vanes or jets at middepth. Such a middepth placement in a dam forebay could deflect migrants that might otherwise sink into turbine intakes. An added advantage of upward-angled turbulence and flow induction would be the elevation of somewhat cooler water into a dam forebay where

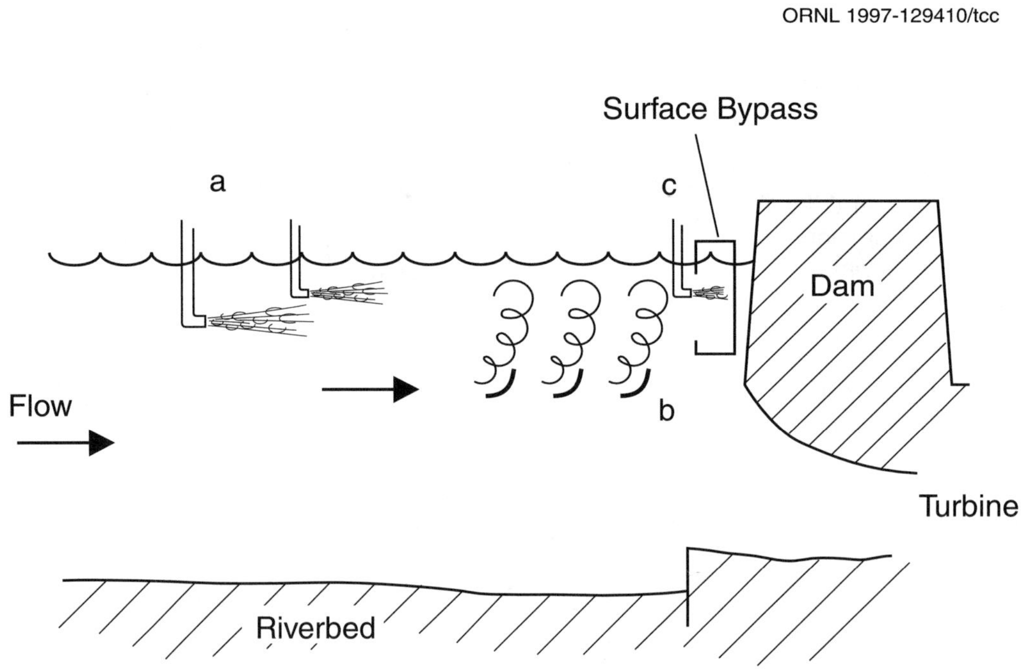

Figure 9. Diagrammatic placement of jets (a, c) and vanes (b) in a vertical section of a dam forebay to generate turbulence in surface waters to aid bypass discovery by juvenile salmon. Jets at (a) provide turbulent orientation in the slowly moving forebay at a distance from the bypass. Vanes at (b) simulate turbulent bursts in upper layers by passively using the increased velocity of water just below the zone of separation and heading toward the turbine. Water jets at (c) direct fish into and in the bypass. Actual placement would depend on physical configuration of the forebay and bypass.

a pool of warm water often accumulates. Warmer surface layers near surface bypasses may otherwise be avoided by migrants in summer.

Design specifications for the hydraulic structures needed to create a useful trail of turbulence at a given facility could be developed several ways. Computational fluid dynamics modeling (computer simulation) could be the most cost effective. However, hydraulic physical models could be used in which scale-model prototypes are constructed and the effects of alternative arrangements would be observed through dye tracers or other visual observations.

Passive and active methods could be employed as part of an integrated system to enhance migration of fish through reservoirs and into fish bypass structures (Figure 10). Iowa vanes or other baffles could be installed at upstream locations to make use of remaining river momentum and impart turbulence and increased microvelocities. Where momentum is too low to be managed effectively, water jets or propellers could be employed to provide turbulent flow zones to guide fish. When sufficient velocity from hydraulic head has been attained in the flownet close to or within a bypass, passive methods may again be suitable.

These measures for using turbulence for fish attraction to bypasses might be integrated with other technologies oriented toward attraction or repulsion of fish at turbine intakes, such as strobe lights (Johnson and Ploskey 1997; Ploskey and Johnson 1997; and papers in this volume) or sound (Carlson and Popper 1997; and papers in this volume). Coutant (2001, this volume) further discusses integrating turbulent flow with other behavioral technologies. Integration of several technologies has been shown to yield high rates of fish diversion at dams even though any one technology alone would not have met management goals (e.g., spill, turbine intake screening, and surface bypass; Johnson et al. 2000). Integration of technologies may be critical to successful diversion in many situations.

This normative fish guidance concept could enhance fish migration at critical low-velocity locations in reservoirs while preserving cultural benefits of hydropower, navigation, and flood control of

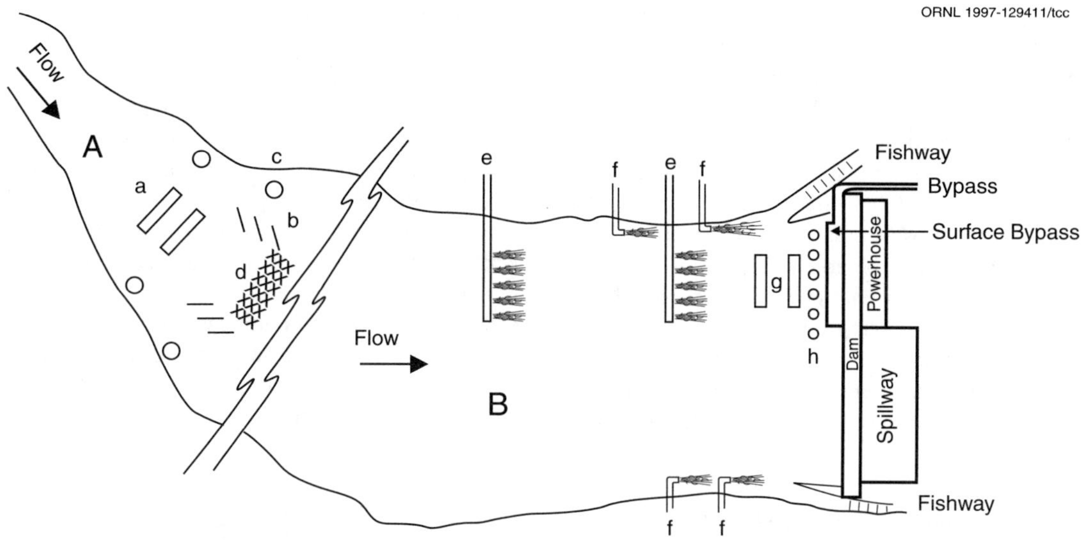

Figure 10. Diagrammatic, plan view of an integrated system of passive and active devices to impart more normative turbulent conditions in a quiescent reservoir and dam forebay. In the upper reservoir (A) where cross-sectional area expands and velocity decreases, the remaining river momentum can be used passively to create surface turbulence for salmon guidance by horizontal (a) and vertical (b) vanes, pilings (c), or submerged berms (d). In a quiescent dam forebay (B), surface turbulence aimed at the bypass can be created actively with banks of jets in diffuser arrays (e) or along the shoreline (f) where adult migration from fish ladders may also be guided. Near the dam, surface turbulence can again be generated passively with horizontal vanes (g) by using the increased velocities as water nears the turbine entrances (as in Figure 9). Fish may be deterred from diving into turbine intakes by banks of flashing strobe lights (h) or other repellants placed below the vertical zone of flow separation.

mainstem dams. Since 1980, taxpayers and electric ratepayers have invested about $3 billion to stem the decline of Pacific salmon in the Columbia River basin, with little success. In 1997, the lowest number of juvenile salmon ever recorded migrated from the Snake River spawning grounds to the ocean. Two scientific reports (NRC 1996; R. Williams et al. 1996) and the Corps of Engineers (Harza 1996) have converged on a key approach to reviving fish runs: replication of natural river habitat that once nurtured salmon. A key feature of the natural migration habitat is the fast, turbulent flow that likely guided and assisted downstream migrations of juveniles.

One alternative for replication of riverine habitat is to breach four dams on the Snake River (and perhaps other dams; Harza 1996), thus returning the river to near-original flow conditions. This approach would likely have large repercussions in the regional economy. It is estimated to reduce average regional hydropower production capacity by approximately 1231 megawatts of renewable energy or near 5% of the region's total generating capacity, which is needed especially on days of peak demand (Fazio 1998). Replacement power would likely come from fossil fuels that generate carbon dioxide and contribute to global warming. It would also reduce river navigation ($500 million/year in grain shipments alone) and irrigation. Dam breaching on the Snake River would cost an estimated $153 million/year for 5 years (minimum), and is under serious consideration by the Corps of Engineers (Harza 1996). If the precedent of dam breaching for fish migrations is adopted in other river basins, the economic loss to hydropower, navigation, and flood control could be large. With international attention on global warming from emissions of greenhouse gases, the loss of renewable hydroelectric power generation could have global consequences. In contrast, the concept of fish attraction flows using induced flow and controlled turbulence provides an alternative way to replicate critical fluid dynamics of rivers in dam forebays and fish bypasses for benefit of migrating juvenile salmonids that could prevent these estimated economic losses.

Another advantage of applying turbulent flow to fish guidance is its likely cost effectiveness. If effective, the costs for installing active and passive devices would not seem to be excessive, particularly when compared with alternatives either in use or contemplated (such as barging fish downstream or dam breaching). Cost effectiveness depends on many factors, however, and we are far from being able to substantiate this initial impression.

There are potential applications of turbulent-flow guidance for cooling-water intakes of thermal-electric power stations, as well as for hydroelectric facilities. Under an October 1995 Consent Decree, the U.S. Environmental Protection Agency currently is engaged in a rulemaking process for implementing Clean Water Act Section 316(b), which applies to cooling water intake structures. The rulemaking is to set guidelines for minimizing the withdrawal of fish from source waterbodies. Fish can be damaged by entrainment through the cooling system (if small) or impingement on intake screens (if large). Behavioral guidance techniques, as well as structural systems, have been tested to divert fish from such intakes (see special issue of Environmental Science & Policy, Vol. 3, Supplement 1, 2000). The search for effective fish diversion approaches for cooling-water intakes is equally as intense, perhaps more so, than it is for hydroelectric facilities. Guidance by induced turbulent flow seems especially relevant to normal fish behavior in flowing water of rivers and tidal estuaries used for cooling.

Needed Work

Neither the basic biology of salmonid migration in turbulent waters nor the potential value of inducing suitable turbulence and flow at bypass entrances or elsewhere in reservoirs is well enough known for attraction flow facilities to be designed and installed everywhere without additional study. Enhanced fish guidance to and within juvenile salmonid bypass systems is expected to be of critical importance for the federal Columbia River basin hydropower system because other approaches that rely just on flow fields induced by the hydraulic head of bypass portals have not proven fully successful (Rainey 1997). Thus, the effort to understand the interaction of biology and hydraulics appears justified. It seems crucial that prototype systems of turbulent attraction flows be evaluated and tested before long-standing decisions are made about the future operation of the Columbia River hydropower system, including dam breaching. The research by Darland et al. (2001) is a welcome beginning.

Contact with researchers in the former Soviet Union who began studies on fish behavior in turbulent waters should be facilitated. Their work may have progressed well beyond what we have available in English. These researchers might be useful consultants for studies and applications in the Columbia River basin and elsewhere. A 1999 Russian book on fish migration through dams of hydropower plants is now being translated by the U.S. Department of Energy's Oak Ridge National Laboratory (Pavlov et al. 1999).

The patterns of turbulence in natural waters during fish downstream migration are poorly understood from the standpoint of fish behavior, despite fairly thorough characterization from the viewpoint of hydraulic engineering. Although engineering understanding may have "matured" a few decades ago (Liggett 1994), the interactive biological and hydraulic aspects could be considered to be in their infancy.

Biological testing of salmon migration behavior in turbulent water should be conducted in experimental flumes with known hydraulics. Recirculating systems such as used by J. Williams et al. (1996) or flow-through designs used successfully for adult fish ladders at the Bonneville Dam engineering test facility and for juvenile Atlantic salmon behavior at the USGS Biological Resources Division Conte facility at Turners Falls, Massachusetts would each have a place in a planned research program. How fish orient to vortices, turbulent bursts, and waves of different physical sizes and energy intensity should be described and quantified. Biological thresholds of detection should be determined. Such a program should be oriented toward the objective of designing trails of turbulence for fish guidance that increase the opportunity for discovery and use of bypass entrances with the least amount of added energy for pumps or propellers.

Advanced, nonintrusive measuring techniques are now available for establishing turbulent flow characteristics near migrating fish. Methods for accurately measuring velocity fluctuations (and consequently turbulence scale and intensity) include laser doppler anemometer (LDA), acoustic doppler velocimeter (ADV), or conventional hot wire anemometer. LDA is most suitable for laboratory applications and is completely non flow-intrusive (but expensive). Several organizations have had a great deal of experience and success in using the ADV for macro-scale velocity measurements near fish bypasses, and this instrument is likely the most practical for field experimental purposes on a finer scale (but it is partially flow intrusive). The third option is a hot wire anemometer which is the least expensive, but flow intrusive. To describe turbulence intensity and scale parameters, velocity fluctuations should be measured at varying locations in the flow field and in proximity to each other (on the size scale of migrating fish). The correlation function can then be determined and finally the turbulence scale is determined based on integration of the correlation function. The best description of these analytical methods are in the classic text by Schlichting (1979). As noted above, the most appropriate descriptors of turbulence for characterizing the responses of juvenile salmonid migrants to it have not been established, however.

On the biological side, innovative techniques for determining the detailed trajectories of fish need to be developed. Radio telemetry used extensively at dam sites (e.g., Johnson et al. 2000) has too poor a spatial resolution to be very useful for detailed trajectories in relation to turbulence structure. Three-dimensional ultrasonic telemetry has better spatial resolution and shows promise for fine-scale fish behavior (K. M. Cash, U.S. Geological Survey, Cook, Washington, personal communication). Stereo film and video techniques such as those used by Boisclair (1992), Krohn and Boisclair (1994), and Trudel and Boisclair (1996) hold promise. Because much salmonid migration occurs at night, techniques using low light may be necessary. Multibeam hydroacoustics has shown a great deal of promise for showing detailed fish motion in relation to water motion (R. Johnson et al. this volume). A particular challenge will be to visualize both fish trajectories and hydraulic patterns of turbulence simultaneously. Perhaps fish trajectories can be determined in flumes with a medium containing neutrally buoyant anisotropic particles for flow visualization, such as used by Matisse and Gorman (1984) and Savas (1985).

Computational fluid dynamics modeling can be used to evaluate alternative methods (passive versus active), designs of hydraulic devices, and layouts in dam forebays. Such mathematical modeling would have to be evaluated for its practicality and utility compared with empirical testing at prototype facilities. Physical models of Columbia River dams are already in place at the U.S. Army Corps of Engineers Waterways Experiment Station in Vicksburg, Mississippi and can be employed when they have unique advantages.

Prototype installation and testing of the concept could be implemented by the Corps of Engineers and the Bonneville Power Administration at any of eight federal Snake and Columbia River hydropower projects. Five nonfederal hydropower projects on the mid-Columbia River owned and operated by public utility districts would also likely test and (if successful) use this technology, under supervision of the Federal Energy Regulatory Commission. Prototype surface flow bypasses are being developed (tested, planned, implemented, or contemplated) at Lower Granite, Ice Harbor, John Day, The Dalles, Bonneville (I & II) (all federal) and Wells, Rocky Reach, and Wanapum nonfederal dams (Johnson et al. 1997b; Johnson et al. 2000).

Smaller dams on tributaries, such as the Cowlitz Falls Dam on the Cowlitz River, are good prototype test sites (Darland et al. 2001). Tests at small dams are logistically less arduous than at mainstem dams. Once the principles have been demonstrated, then larger scale tests at mainstem dams could be conducted more efficiently. Small-scale testing with Atlantic salmon (e.g., Truebe and Truebe 1997, 1998) should continue with emphasis on evaluation of why prototype systems work from the perspective of interactions of migrating fish with a turbulent zone.

The turbulence regime should be quantified in the field wherever fish guidance is tested. The success of guidance at Wells Dam, for example, may derive as much from continuation of riverine turbulent flow near the bypass entrances as to the hydrocombine dam design (bypass located above the turbine) to which success has long been attributed (Johnson 1996; Johnson et al. 1997). Turbulence near structures such as log booms that appear to guide fish (Swan et al. 1985; Wilson et al. 1991) should be measured. These inadvertent trails of turbulence may yield clues to functionality that would be difficult or expensive to replicate experimentally.

The economic benefits of induced attraction flows for juvenile salmon should be evaluated if guidance is proven successful biologically. The monetary costs of installing passive and active turbulent-attraction guidance devices should be compared with alternatives such as transporting fish by barge and dam breaching. Benefits can include more than just improved fish passage. There may be large water conservation benefits where induced attraction flows provide fish passage with small volumes of water, thus preserving bulk flows for hydropower generation or maintaining pool elevations.

Conclusions

Available evidence points to natural use of turbulent flows by downstream-migrating salmon for guidance and assistance in migration. In the hydrosystem, normal downstream migration appears to be most disrupted at locations in reservoirs where turbulent flow is least, especially in large reservoirs and dam forebays. A major problem with developing effective surface bypass systems at dams is insufficient attraction of migrants to, and their retention in, the bypasses. Nonturbulent flownets as a result of hydraulic head differences between reservoir and bypass are insufficient for migrants to detect and follow the flows to the bypasses. Generation of mildly turbulent flows in the direction of downstream flow and dam bypasses (trails of turbulence) would appear to be an effective way to guide migrants. Such flows could be generated either passively (with physical structures such as vanes) or actively (with pumped water jets or propellers). Initial tests in the field have been successful and promising. Further research on this concept and evaluation of the results would appear important for selecting socially acceptable alternatives for operation of the federal Columbia River hydrosystem and other hydropower systems with migrating fish.

Acknowledgments

Background work for this report was supported by three projects under Contract No. DE-AC05–96OR22464 between Lockheed Martin Energy Research Corp. and the U.S. Department of Energy for management and operation of Oak Ridge National Laboratory (ORNL). ORNL is currently managed and operated by UT-Battelle LLC under Contract No. DE-AC05–00OR22725. Funding sources were the U.S. Department of Energy Office of Geothermal Technologies (DOE B&R No. EB4400000), U.S. Department of Energy Bonneville Power Administration (BPA Project No. 8907201), and the Northwest Power Planning Council (DOE Project No. ERD 92–1128). The author thanks M. Bevelhimer, G. Cada, P. Patrick and an anonymous reviewer for valuable comments on a draft. This is Environmental Sciences Division Publication 5077.

References

Abbott, M. B., and D. Basco. 1989. Computational Fluid Mechanics: Introduction for Engineers. Longman, New York.

Bates, D. W., and J. G. VanDerWalker. 1964. Exploratory experiments on the deflection of juvenile salmon by means of water and air jets. Fish Passage Research Program, Review of Progress (U.S. Bureau of Commercial Fisheries) 3(14):1–6.

Bell, M. 1990. Fisheries Handbook of Engineering Requirements and Biological Criteria. Fish Passage Development and Evaluation Program, U.S. Army Corps of Engineers, North Pacific Division, Portland, Oregon.

Bleckmann, H. 1986. Role of the lateral line in fish behavior. Pages 177–204 *in* T. J. Pitcher, editor. The behavior of teleost fishes. Johns Hopkins University Press, Baltimore, Maryland.

Boisclair, D. 1992. An evaluation of the stereocinematographic method to estimate fish swimming speed. Canadian Journal of Fisheries and Aquatic Sciences 49:523–531.

Brett, J. R., and D. Alderdice. 1958. Research on guiding young salmon at two British Columbia field stations. Bulletin No. 117. Fisheries Research Board of Canada, Ottawa.

Cada, G. F., C. C. Coutant, and R. R. Whitney. 1997. Development of biological criteria for the design of advanced hydropower turbines. DOE/ID-10578. U.S. Department of Energy, Idaho Operations Office, Idaho Falls, Idaho.

Carling, P. A. 1992. Instream hydraulics and sediment transport. Pages 101–125 *in* P. Calow and G. E. Petts, editors. The rivers handbook: hydrological and ecological principles. Blackwell, Oxford.

Carlson, T. J., and A. N. Popper. 1997. Using sound to modify fish behavior at power-production and water-control facilities: a workshop. DOE/BP-62611–11. U.S. Department of Energy, Bonneville Power Administration, Portland, Oregon.

Chaudhry, M. 1993. Open Channel Flow. Prentice Hall, New York.

Chow, V. T. 1964. Handbook of Applied Hydrology. McGraw-Hill, New York.

Clay, C. H. 1994. Design of Fishways and Other Fish Facilities, 2nd edition. Lewis Publishers, Boca Raton, Florida.

Collins, G. B., J. R. Gauley, and C. H. Elling. 1962. Ability of salmonids to ascend high fishways. Transactions of the American Fisheries Society 91:1–7.

Corrsin, S. 1943. Investigations in an axially symmetrical heated jet of air. Report WR W-94. National Advisory Committee for Aeronautics, Washington, D.C.

Coutant, C. C. 1997. The normative river: An ecological vision for the recovery of the Columbia River salmon. Pages 50–59 *in* D. J. Mahoney, editor. Waterpower '97, Proceedings of the International Conference on Hydropower. American Society of Civil Engineers, New York.

Coutant, C. C. 1998. Turbulent attraction flows for juvenile salmonid passage at dams. ORNL/TM-13608, Oak Ridge National Laboratory, Oak Ridge, Tennessee.

Coutant, C. C., and R. R. Whitney. 2000. Fish behavior in relation to passage through hydropower turbines: a review. Transactions of the American Fisheries Society 129:351–380.

Cunge, J. A., J. F. M. Holly, and A. Verwey. 1980. Practical Aspects of Computational River Hydraulics. Pittman, New York.

Darland, T. J., S. D. Evans, D. W. Rondorf, G. L. Rutz, D. H. Feil, N. S. Adams, C. F. Merrill, and J. D. Serl. 2001. Test of concept to improve fish guidance using induced water currents at Cowlitz Falls Dam, Washington. Annual Report for 1999. U.S. Geological Survey, Columbia River Research Laboratory, Cook, Washington.

Dijkgraaf, S. 1933. Seitenorgane an Fischen. Zeitschrift fuer Vergleichende Physiologie 20:162–214.

Fazio, J. 1998. Transmission System Impacts of Drawing Down John Day Dam. NPPC 98-3. Northwest Power Planning Council, Portland, Oregon.

Gregory, R. W., and P. E. Fields. 1962. Discrimination of low water velocities by juvenile silver (*Oncorhynchus kisutch*) and chinook salmon (*Oncorhynchus tshawytscha*). Technical Report No. 52 to U.S. Army Corps of Engineers. College of Fisheries, University of Washington, Seattle.

Groves, A. B. 1972. Effects of hydraulic shearing on juvenile salmon. Processed report. Northwest and Alaska Fisheries Center, National Marine Fisheries Service, Seattle.

Harza (Harza Northwest, Inc.). 1996. Salmon decision analysis, lower Snake River feasibility study, final report. Report for U.S. Army Corps of Engineers, Walla Walla District, Walla Walla, Washington.

Johnson, G. E. 1996. Fisheries research on phenomena in the forebay of Wells Dam in spring 1995 related to the surface flow smolt bypass. Final Report. U.S. Army Corps of Engineers, Walla Walla District, Walla Walla, Washington.

Johnson, G. E., N. S. Adams, R. L. Johnson, D. W. Rondorf, D. D. Dauble, and T. Y. Barila. 2000. Evaluation of the prototype surface bypass for salmonid smolts in spring 1996 and 1997 at Lower Granite Dam on the Snake River, Washington. Transactions of the American Fisheries Society 129:381–397.

Johnson, G. E., A. E. Giorgi, and M. W. Erho. 1997a. Critical assessment of surface flow bypass development in the lower Columbia and Snake rivers. Completion Report for U.S. Army Corps of Engineers, Portland and Walla Walla districts, Portland, Oregon.

Johnson, G. E., R. L. Johnson, T. Y. Barila, and D. R. Kenney. 1997b. Fish passage at the prototype surface bypass and collector at Lower Granite Dam. Pages 599–608 *in* D. J. Mahoney, editor. Waterpower '97. Proceedings of the International Conference on Hydropower. American Society of Civil Engineers, New York.

Johnson, P. N., and G. R. Ploskey. 1997. Behavioral technologies for bypass channels. Phase 2: Evaluation of infrasound and strobe lights for redistributing migrant salmon smolts in the McNary juvenile bypass. abstracts for the 1997 annual research review, Anadromous Fish Evaluation Program. U.S. Army Corps of Engineers, North Western Division, Portland, Oregon.

Johnson, R. L., M. A. Simmons, C. S. Simmons, and S. L. Blanton. 2001. A new multibeam sonar technique for evaluating fine-scale fish behavior near hydroelectric dam guidance structures. Pages 161–170 *in* C. C. Coutant, editor. Behavioral technologies for fish guidance. American Fisheries Society, Symposium 26, Bethesda, Maryland.

Krohn, M. M., and D. Boisclair. 1994. Use of a stereo-video system to estimate the energy expenditure of free-swimming fish. Canadian Journal of Fisheries and Aquatic Sciences 51:1119–1127.

Kuehl, S. 1986. Hydroacoustic evaluation of fish collection efficiency at Lower Granite dam in spring, 1985. Prepared for U.S. Army Corps of Engineers, Walla Walla, Washington.

Laufer, J. 1951. Investigation of turbulent flow in a two-dimensional channel. Report No. 1053. National Advisory Committee for Aeronautics, Washington, D.C.

Laufer, J. 1954. The structure of turbulence in fully developed pipe flow. Report No. 1174. National Advisory Committee for Aeronautics, Washington, D.C.

Laurence, J. C. 1956. Intensity, scale, and spectra of turbulence in mixing region of free subsonic jet. Report 1292. National Advisory Committee for Aeronautics, Washington, D.C.

Leeder, M. R. 1983. On the interactions between turbulent flow sediment transport and bedform mechanics in channelized flow. Pages 121–132 *in* J. D. Collinson and J. Lewin, editors. Modern and Ancient Fluvial Systems. Blackwell, Oxford.

Liggett, J. A. 1994. Finite-difference methods for shallow water flow analysis. Pages 33–61 *in* M. H. Chaudhry and L. W. May, editors. Computer Modeling of Free-Surface and Pressurized Flows. Kluwer Academic Publishers, Dordrecht, The Netherlands.

Marelius, F., and S. K. Sinha. 1998. Experimental investigation of flow past submerged vanes. Journal of Hydraulic Engineering 124:542–545.

Matisse, P., and M. Gorman. 1984. Neutrally buoyant anisotropic particles for flow visualization. Physical Fluids 27:759–760.

McKinnon, D., and W. H. Hoar. 1952. Responses of coho and chum salmon fry to current. Journal of the Fisheries Research Board of Canada 10:523–538.

Mighetto, L, and W. J. Ebel. 1994. Saving the salmon: A history of the U.S. Army Corps of Engineers' efforts to protect anadromous fish on the Columbia and Snake rivers. Historical Research, Inc., for U.S. Army Corps of Engineers, North Pacific Division, Portland, Oregon.

NRC (National Research Council). 1996. Upstream. Salmon and society in the Pacific Northwest. National Academy Press, Washington, D.C.

Nikora, V. I., and G. M. Smart. 1997. Turbulence characteristics of New Zealand gravel-bed rivers. Journal of Hydraulic Engineering 123:764–773.

Odgaard, A. J., and Y. Wang. 1991a. Sediment management with submerged vanes. I: Theory. Journal of Hydraulic Engineering 117:267–283.

Odgaard, A. J., and Y. Wang. 1991b. Sediment management with submerged vanes. II: Applications. Journal of Hydraulic Engineering 117:284–302.

OTA (Office of Technology Assessment). 1995. Fish Passage Technologies: Protection at Hydropower Facilities. OTA-ENV-641. U.S. Government Printing Office, Washington, D.C.

Pavlov, D. S., A. I. Lupandin, N. G. Degtyareva, and S. M. Dedov. 1995. Role of turbulence in the distribution of downstream migrating young fishes (early larval stages) in wide and narrow channels. Doklady Biological Sciences 341:211–215 (translated from Doklady Akademii Nauk 341:842–845).

Pavlov, D. S., A. I. Lupandin, and V. V. Kostin. 1999. Fish downstream migration through dams of hydroelectric power plants (in Russian). Russian Academy of Science, Moscow.

Pavlov, D. S., A. I. Lupandin, and M. A. Skorobogatov. 1994. Influence of flow turbulence on critical flow velocity for gudgeon (Gobio gobio). Doklady Biological Sciences 336:215–217 (Translated from Noklady Akedemii Nauk 336:138–141).

Pevan, C. M., T. R. Mosey, and K. B. Truscott. 1996. Biological Evaluation of the Rocky Reach Surface Collector. Chelan County Public Utility District, Wenatchee, Washington.

Ploskey, G. R., and P. N. Johnson. 1997. Development of behavioral technologies to improve the efficiency of surface collection and bypass structures on the lower Columbia River. In abstracts for the 1997 Annual Research Review, Anadromous Fish Evaluation Program. U.S. Army Corps of Engineers, Northwestern Division, Portland, Oregon.

Popper, A. N., and C. Platt. 1993. The inner ear and lateral line. Pages 99–136 *in* D. H. Evans, editor. The physiology of fishes. CRC Press, Boca Raton, Florida.

Quistorff, E. 1966. Floating salmonid smolt collectors at Baker River dams. Fishery Research Papers (Washington Department of Fisheries) 2(4):39–52.

Rainey, S. 1997. "Opportunity for discovery:" An important tool for assessment of surface collection technology potential for improving juvenile salmon passage at Columbia River dams. Proceedings, Fish Passage Workshop (Milwaukee, Wisconsin, May 6–8, 1997). Alden Research Laboratory, Inc., Holden, Massachusetts.

Ruggles, C. P., and P. Ryan. 1964. An investigation on louvers as a method of guiding juvenile Pacific salmon. Canadian Fish Culturist 33:1–68.

Savas, O. 1985. On flow visualization using reflecting flakes. Journal of Fluid Mechanics 152:235–248.

Schlichting, H. 1979. Boundary layer theory (translated by J. Kestin). 7th edition. McGraw Hill, New York.

Shtaf, L. G., D. S. Pavlov, M. A. Skorobogatov, and A. S. Barekyan. 1983. The influence of flow turbulence on fish behavior. Journal of Ichthyology 23:129–140.

Sinha, S. K., and F. Marelius. 2000. Numerical analysis of flow past submerged vanes. Journal of Hydraulic Research 38:1–13.

Sinha, S. K., F. Sotiropoulos, and A. J. Odgaard. 1998. Three-dimensional numerical model for flows through natural rivers. Journal of Hydraulic Engineering 124:1–12.

Skorobogatov, M. A., and S. D. Pavlov. 1991. A study of the orientation of young roach, *Rutilus rutilus*, with respect to current velocity. Journal of Ichthyology 31:142–148.

Smith, L. S. 1982. Decreased swimming performance as a necessary component of the smolt migration in salmon in the Columbia River. Aquaculture 28:153–161.

Stacey, M. T., and S. G. Monismith. 1997. Measuring estuarine turbulence with an ADCP. Pages 155–160 *in* F. M. Holly Jr., and A. Alsaffar, editors. Environmental and coastal hydraulics: protecting the aquatic habitat. Proceedings of the 27th Congress of the International Association for Hydraulic Research. American Society of Civil Engineers, New York.

Stevenson, J. R., A. E. Giorgi, W. R. Koski, K. K. English, and C. A. Grant. 1997. Evaluation of juvenile spring chinook and steelhead migratory patterns at Rocky Reach and Rock Island dams using radio telemetry techniques, 1996. Report for Public Utility District No. 1 of Chelan County, Wenatchee, Washington.

Stockley, C. 1959. Merwin Dam downstream migrant bypass trap experiments for 1957. Unnumbered report. Washington Department of Fisheries, Olympia, Washington.

Swan, G. A., R. F. Krcma, and F. Ossiander. 1985. Development of an improved fingerling protection system at Lower Granite Dam. Final Report. U.S. Army Corps of Engineers and Coastal Zone and Estuarine Studies Division, Northwest and Alaska Fisheries Center, National Marine Fisheries Service, Seattle.

Sweeney, C. E., B. Christman, and B. Gatton. 1997. Update on juvenile fish bypass at Rocky Reach Dam. Pages 589–598 *in* D. J. Mahoney, editor. Waterpower '97, Proceedings of the International Conference on Hydropower. American Society of Civil Engineers, New York.

Taft, E. P. 1986. Assessment of downstream migrant fish protection technologies for hydroelectric application. Report Number AP-4711. Electric Power Research Institute, Palo Alto, California.

Thoren, P. D., J. J. Williams, and A. D. Heathershaw. 1989. In situ acoustic measurements of marine gravel threshold and transport. Sedimentology 36:61–74.

Townsend, A. A. 1949. The fully developed turbulent wake of a circular cylinder. Australian Journal of Scientific Research, Series A 2(4):451–468.

Triantafyllou, M. S., and G. S. Triantafyllou. 1995. An efficient swimming machine. Scientific American 272(3):64–70.

Trudel, M., and D. Boisclair. 1996. Estimation of fish activity costs using underwater video cameras. Journal of Fish Biology 48:40–53.

Truebe, J. P., and E. P. Truebe. 1997. Current Inducer and Downstream Bypass System at Upper Penacook Hydroelectric Plant, July 1997. Prepared by Lakeside Engineering, Inc., Mirror Lake, New Hampshire, for Essex Hydro Associates, Boston.

Truebe, J. P., and E. P. Truebe. 1998. 1998 fish passage report for Upper Penacook Falls. Lakeside Engineering, Inc., Mirror Lake, New Hampshire, for Essex Hydro Associates, Boston.

Venditti, D. A., J. M. Kraut, and D. W. Rondorf. 1997. Behavior of juvenile fall chinook salmon in the forebay of a lower Snake River reservoir. Pages 48–68 *in* D. W. Rondorf and K. F. Tiffan, editors. Identification of the spawning, rearing, and migratory requirements of fall chinook salmon in the Columbia River basin. Annual Report for 1995. DOE/BP-21708-5. U.S. Department of Energy, Bonneville Power Administration, Portland, Oregon.

Vogel, S. 1994. Life in moving fluids. The physical biology of flow. 2nd edition. Princeton University Press, Princeton, New Jersey.

Wayne, W. W. 1961. Fish handling facilities for Baker River Project. Journal of the Power Division, Proceedings of the American Society of Civil Engineers 87:23–54.

Weaver, C. R. 1963. Influence of water velocity upon orientation and performance of adult migrating salmonids. Fishery Bulletin 63:97–121.

Webb, P. W. 1994. The biology of fish swimming. Pages 45–62 *in* L. Maddock, Q. Bone, and J. V. Rayner, editors. Mechanics and Physiology of Animal Swimming. Cambridge University Press, Cambridge.

Whitney, R. R., L. D. Calvin, M. W. Erho, Jr., and C. C. Coutant. 1997. Downstream passage for salmon at hydroelectric projects in the Columbia River basin: development, installation, and evaluation. Report 97–15, Northwest Power Planning Council, Portland, Oregon.

Williams, J. G., M. H. Gessel, B. Sandford, and J. J. Vella. 1996. Evaluation of factors affecting juvenile chinook salmon fish guidance efficiency. Unnumbered report. Coastal Zone and Estuarine Studies Division, Northwest Fisheries Science Center, National Marine Fisheries Service, Seattle.

Williams, R. N., L. D. Calvin, C. C. Coutant, M. W. Erho, Jr., J. A. Lichatowich, W. J. Liss, W. E. McConnaha, P. R. Mundy, J. A. Stanford, R. R. Whitney, D. L. Bottom, and C. A. Frissell. 1996. Return to the river: Restoration of salmonid fishes in the Columbia River ecosystem. Report No. 96–6. Independent Scientific Group, Northwest Power Planning Council, Portland, Oregon.

Williams, R. N., L. D. Calvin, C. C. Coutant, M. W. Erho, Jr., J. A. Lichatowich, W. J. Liss, W. E. McConnaha, P. R. Mundy, J. A. Stanford, R. R. Whitney, D. L. Bottom, and C. A. Frissell. 2000. Return to the river 2000: Restoration of salmonid fishes in the Columbia River ecosystem. Report No. 2000–12. Northwest Power Planning Council, Portland, Oregon.

Wilson, J. W., A. E. Giorgi, and L. C. Stuehrenberg. 1991. A method for estimating spill effectiveness for passing juvenile salmon and its application at Lower Granite Dam on the Snake River. Canadian Journal of Fisheries and Aquatic Sciences 48:1872–1876.

American Fisheries Society Symposium 26:79–89, 2001

Evaluation of Infrasound and Strobe Lights for Eliciting Avoidance Behavior in Juvenile Salmon and Char

Robert P. Mueller, Duane A. Neitzel, and Brett G. Amidan
Pacific Northwest National Laboratory
Post Office Box 999, Richland, Washington 99352, USA

Abstract.—Laboratory tests were conducted using juvenile chinook salmon *Oncorhynchus tshawytscha*, brook trout *Salvelinus fontinalis*, and rainbow trout *O. mykiss* to determine specific behavior responses to infrasound (<20 Hz) and flashing strobe lights. The objective of these tests was to determine if juvenile salmonids could be deterred from entrainment at water diversion structures. Caged fish were acclimated in a static test tank and their behavior was recorded using low light cameras. Species-specific behavior was characterized by measuring movements of the fish within the cage and by observing startle and habituation responses. Wild chinook salmon (40–45 mm TL) and hatchery reared chinook salmon (45–50 mm TL) exhibited avoidance responses when initially exposed to a 10-Hz volume displacement source of infrasound. Rainbow and eastern brook trout (25–100 mm TL) did not respond with avoidance or other behaviors to infrasound. Evidence of habituation to the infrasound source was evident for chinook salmon during repeated exposures. Wild and hatchery chinook displayed a higher proportion of movement during the initial exposures to infrasound when the acclimation period in the test tank was 2–3 h as compared with a 12–15 h acclimation period. A flashing strobe light produced consistent movement in wild chinook salmon (60% of the tests), hatchery reared chinook salmon (50%), and rainbow trout (80%). No measurable responses were observed for brook trout. Results indicate that consistent, repeatable responses can be elicited from some fish using high-intensity strobe lights under a controlled laboratory testing. The species specific behaviors observed in these experiments might be used to predict how fish might react to low-frequency sound and strobe lights in a screening facility.

Salmon outmigrating from tributaries and mainstream rivers in the Pacific Northwest face many obstacles in their journey to the Pacific Ocean. One major obstacle in watersheds with irrigated agriculture is diversion dams and the associated screen and pumping systems. Previous screening evaluations have determined that diversion facilities are effective at protecting juvenile salmonids larger than 40 mm TL in length. However, there is a concern that smaller fry may be vulnerable to passing through or around screens and through gaps in screen seals (Mueller et al. 1995; Neitzel et al. 1990a).

As state and federal agencies change screening criteria to protect smaller fish (e.g., chinook fry), physical barriers may not always be effective. Screen mesh size small enough to protect small fish may be vulnerable to frequent plugging. Small gap tolerances on side and bottom seals may be difficult to install and maintain. However, it may be possible to enhance physical barrier screens with behavioral barriers that cause fish to avoid a hazard. Behavioral barriers may consist of any condition (e.g., sound, light) that moves a fish away from danger to safety. The Columbia Basin Fish and Wildlife Program (NPPC 1984, 1987, 1994) provides oversight and funding to ensure safe passage of juvenile salmonids and other resident species past irrigation diversions and screens. The goal is to ensure that facilities operate in accordance with fish-protective design criteria and to make recommendations based on research, monitoring and evaluations to improve facility effectiveness.

Many laboratory and field studies have been conducted on behavior of fish when exposed to infrasound (Sand and Karlsen 1986; Karlsen 1992). The use of volume displacement sources (VDS) operating in the infrasonic range has gained attention over the past few years as a possible behavioral barrier device. Avoidance behaviors have been documented in laboratory and field studies by Knudsen et al. (1992, 1994, 1997) and Mueller et al. (1998) using piston type VDS. The rationale behind those tests was to determine if sub-yearling salmonids would respond to infrasound with predictable patterns. Those studies concluded that yearling and sub-yearling juvenile Atlantic salmon *Salmo salar*, chinook salmon *Oncorhynchus tshawytscha*, and rainbow trout *O. mykiss* exhibited avoidance responses when exposed to the near field of high particle acceleration infrasound. Knudsen found that smolts (>100 mm TL) responded with an innate avoidance response when the local flow acceleration is greater than 10^{-2} m/s^2 at a frequency of 10 Hz. These tests were replicated using smaller (<50 mm TL) salmonids with similar results (Mueller et al. 1998).

Strobe lights also have been shown to be effective at producing avoidance responses in juvenile salmonids under certain environmental conditions (Ploskey and Johnson 1998; Ploskey et al.1998). Other studies indicate the life stage of fish is important because sensitivity to light changes with fish growth (Anderson et al. 1988; Fernald 1988). We wished to further explore whether sound or light could be used at juvenile fish screening facilities to enhance their effectiveness for protecting small fish from being entrained or impinged. We hypothesized that small fish would respond in a predictable manner and would avoid the source of the stimulus when subjected to infrasound or strobe light. This study was a continuation of studies conducted to evaluate the effectiveness of using infrasound as a behavior deterrent for salmonid fry (Mueller et al. 1998). If the majority of the fish are directed away from becoming entrained, survival of outmigrating salmonids can be increased.

Methods

Testing Apparatus, Fish, and Behavior Devices

Tests were conducted in an outdoor steel reinforced, oval fiberglass fish tank, measuring 3 m wide × 7.3 m long × 1.8 m deep. The top of the tank was covered using a canvas tarp to avoid direct sunlight. Test fish were confined to a net pen constructed of 0.16-cm nylon netting, measuring 1 m wide × 2 m long × 1.5 m deep, attached to a 2.54-cm diameter polyvinyl chloride frame with holes drilled in the bottom and side supports to aid in submergence. Gridlines were placed on the background along the x, y, and z-axis 0.33 m apart to quantify fish movements within the net pen (Figure 1). Three underwater video cameras were

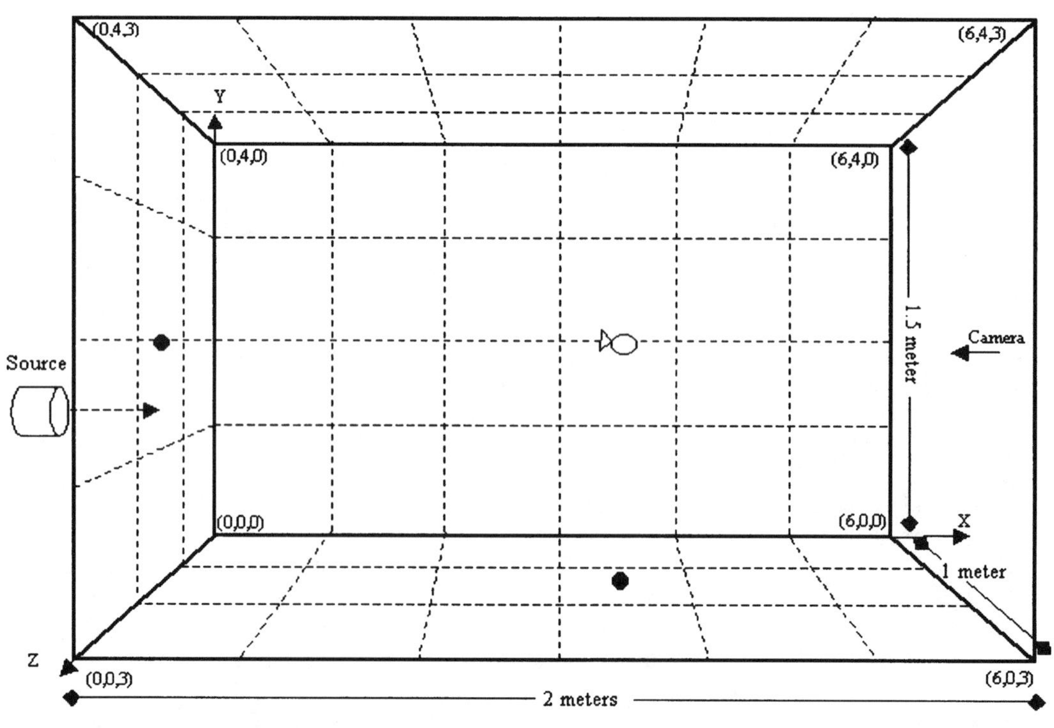

Figure 1. Overhead view of a 3-dimensional depiction of the testing pen (the coordinate system starts in the bottom-back-left corner at [0,0,0] and moves to the top-front-right corner at [6,4,3]).

used to document and record fish movement. Video recording system consisted of high-resolution monochrome Sony cameras, model HVM-352 with a wide-angle lens connected to a Hi 8-mm Sony camcorder, model FX710. Two cameras, with overlapping views, were positioned to record movement in the X, Y plane, and the third was placed at the back of the pen to record movement in the Z plane. Video cameras operated continuously during each test series. The three-video signals were multiplexed and displayed on a single monitor and recorded to Hi 8-mm tape.

Test fish consisted of wild and hatchery reared chinook salmon, hatchery rainbow trout, and hatchery brook trout *Salvelinus fontinalis*. Wild fall chinook salmon fry were seined from the Hanford Reach of the Columbia River and ranged in length from 40 to 45 mm TL. Hatchery fall chinook salmon fry were raised from eggs obtained from the Priest Rapids State Fish Hatchery, Washington, and were 45–50 mm in total length. Rainbow trout (Kamloop) fry were raised from eggs at our aquaculture facility and ranged from 25 to 44 mm TL. Two groups of hatchery-reared brook trout (30–40 mm and 80–100 mm, TL) were used for the infrasound tests. Strobe light tests were conducted with 30–40 mm hatchery brook trout. All test fish were held in raceways or troughs and not fed for up to 10 h before being tested. They were acclimated to chilled well water (14°C) for 3–15 h before being transferred to the test tank.

A high-particle-displacement, local flow infrasound field was generated in the test tank using a VDS built and designed by the University of Oslo, Norway. The components of the VDS included a piston diameter of 21.5 cm and a displacement amplitude (peak to peak) of 4.5 cm. Water particle acceleration was achieved via the movement of two pistons located on opposite ends of the VDS. The VDS was operated within the frequency range 10–14 Hz after an initial motor startup of about 5–15 s. Horizontal particle acceleration curves for the VDS were mapped in 1996 and can be found in Mueller et al. (1998). The VDS was driven by an electric motor requiring 230-V, three-phase power supply connected to a programmable power driver. A more detailed description of the VDS is given in Sand et al. (2000). The VDS was suspended 1.6 m from one end of the tank and positioned 1 m off the bottom (measured to center of pistons). The support structure was isolated from the tank to eliminate any vibration transfer to the tank. One end of the net pen was positioned 0.8 m from the VDS with the piston axis 0.5 m above the bottom of the net pen. Tests with the VDS were conducted from March through July 1998.

The strobe light tests were conducted with Flash Technology AGL 901 Aquatic Guidance Light. The strobe lights were mounted on an aluminum pole and positioned about 0.8 m away from each end of the net pen and angled in a way to create more intense lighting in certain areas of the net pen. Strobe lights operated at 300 flashes per min (fpm) at 400 W. Tests were conducted in April and May 1998. Measurements of light intensity were taken using a LI COR LI-188B photometer with an underwater quantum sensor LI-192S to determine (1) the ambient light levels, (2) if the net pen was influencing the light intensity, and (3) any decrease in light intensity within the net pen. An ambient light level of 10 lux was measured at the 0.5 m below the water surface with the gray tarp covering the tank. No difference in light intensity was observed with and without the net pen in place using clear well water. Measurements of light intensity within the net pen showed an intensity of near 175 lux at 1-m away from the strobe light then a gradual decay rate to 50 lux at 3 m from the strobe light.

Experimental Procedure

The test protocol for the infrasound and strobe lights was similar. Each testing day consisted of three test cycles: morning (between 7:00 and 9:00 a.m.); late morning through early afternoon (11:00 a.m–1:00 p.m.); and late afternoon (2:00–4:00 p.m). The testing scenario included a random selection of 10 stimuli over a 1-h test period. The random selection was determined by segmenting a 1 h test into 20 possible test events (every 3 min). We used a random number generator and selected 10 times for each test in which the stimulus would be activated. A test cycle for the infrasound tests is summarized as follows: (1) *Acclimation period*—Before testing, test fish were held in covered troughs for a minimum of 5 d. Test groups consisting of 20 fish were placed in the pen and remained in the pen for an acclimation period of at least 1-h. The fish tested in the morning had an acclimation period of at least 15-h, because they were placed in the pen the previous day. After the acclimation period, all three cameras began to record. (2) *Control period*—During the time when no stimulus (sound) was being applied, control measurements were taken to record the general movement patterns of the caged fish. Control measurements occurred before the first run, before each of the 10 random events, and after the tenth run, for a total of 11 per test cycle. All the control measurements were taken at the start and the end of a 20-s interval one minute before the stimulus was applied. (3) *Sound on*—The VDS motor was turned on and piston revolutions per minute (rpm) gradually increased for a 5–15-s period. Once maximum rpm were observed, the VDS remained on for 20-s then was turned off. The test group's response to the initial motor noise was separately graded for startle response and movement. The test period was repeated nine more times, for a total of 10 replications.

The strobe light testing protocol consisted of the following: (1) *Control period*—control responses were recorded in the same manner as the sound control period, (2) *First strobe on*—the first strobe was turned on for a 20-s period from a randomly determined end of the pen, (3) *Second strobe on*—the first strobe was turned off, and the opposite strobe was turned on for a 20-s period and was then turned off. The test period was repeated nine more times, for a total of 10 replications. The second strobe light was not acquired until after the first eight tests and therefore, was not used with those tests. The number of tests conducted for each species for the strobe tests were: 5 for eastern brook trout (all during the double strobe tests), 8 for hatchery chinook salmon (1 during the single strobe tests and 8 during the double strobe tests), 11 for the rainbow trout (all during the double strobe tests), and 9 for wild chinook salmon (7 during the single strobe tests and 2 during the double strobe lights).

Sampling Responses and Data Collection

For each test group, we evaluated the response of a school of fish by measuring gross movement of the center of the school before and immediately after the behavior stimuli. Nonschooled fish were excluded from this determination, and a minimum school contained at least eight individuals for inclusion in the analysis. In most cases 12 or more fish were schooled. Tapes were reviewed and analyzed by measuring the school movement and other fish behavior, including startle and avoidance behavior. The location of the center of a school was recorded by using the grid coordinates described previously. Coordinate values were determined by an approximate location to the nearest half grid unit. After each set of "before" and "after" measurements during a treatment (sound on, or strobe on), the displacement of the school away from the VDS was classified by the evaluator into one of the following four categories: N—no response (no greater than 0.15 m), S—slight (movement of 0.15–0.3 m), M—moderate (movement of 0.3–0.6 m), and G—great (movement of 0.6 m or more). The classification grades for each species tested during the infrasound and motor sound testing are found in Table 1. In addition to the movement, responses were classified for startle and possible habituation behaviors. A startle response was characterized as a rapid abrupt movement away from the source and habituation was defined as a decrease of response frequency. For

Table 1. Number of responses by schools of fish in all runs in each of four movement classification grades for infrasound and motor sound tests (and grade totals for each species).

Species	None	Slight 0.15–0.3 m	Moderate 0.3–0.6 m	Great >0.6 m	Number of tests	Number of schools
			Motor response[a]			
Brook trout	134	2	0	0	14	136
Hatchery chinook	209	10	6	1	24	226
Rainbow trout	134	2	0	0	22	136
Wild chinook	213	5	1	0	23	219
			Sound response			
Brook trout	125	6	2	1	14	134
Hatchery chinook	154	42	23	7	24	226
Rainbow trout	209	3	1	0	22	213
Wild chinook	103	50	38	28	23	219

[a]The initial startup of the infrasound device

these categories a minimum of 12 fish in each school had to display the behavior to be included in the analysis.

Results

Infrasound

Avoidance Movement Responses

Test fish were rarely influenced by the sound of the motor starting up and then only to a slight degree (Table 1). Wild and hatchery reared chinook salmon displayed great or moderate movement away from the VDS sound; whereas rainbow trout and brook trout had few responses to the VDS. We ran analysis of variance (ANOVA) using a significance level of $\alpha = 0.05$ and found significant differences ($P = 0.001$) among the species and also among the runs (i.e., percentage of slight movement and the percentage of great or moderate movement).

Responses of chinook salmon decreased as the fish were exposed to the sound (Figure 2), indicating that test groups showed signs of becoming habituated to a series of repeated exposures to the VDS. Chinook salmon showed decreased response frequency generally after the fourth exposure to the VDS. Tests of 40-mm brook trout fry and rainbow trout exhibited no avoidance to the initial motor noise or during the 20 s exposure period. Only slight avoidance (<20%) was observed when tests were conducted with larger (80–100 mm TL) brook trout.

A comparison between short and long acclimation periods indicated that, for wild chinook salmon, the mean proportion of moderate and great movement was 0.48 for tests conducted on the same day ($N = 13$) compared with 0.30 (significantly different; $P = 0.03$) for fish tested the next day ($N = 10$) (based on first six exposures only). For hatchery chinook salmon, the average moderate and great movement was 0.24 for the same day ($N = 12$) compared with 0.11 ($P = 0.5$) for the next-day tests ($N = 8$, first six exposures). No significant differences were observed when comparing acclimation time to fish response for rainbow trout and brook trout.

Startle Movement Responses

Similar to avoidance responses, there was little response to the motor startup, and of the species tested (Figure 3). Chinook salmon (hatchery and wild) had the most startle and flight responses to infrasound. The startle responses were more evident during the initial exposures (near 50%) and were characterized by rapidly moving away from the VDS, usually toward the bottom of the net pen. Startle responses were reduced after the fifth exposure, except for hatchery chinook salmon. Only rainbow trout exhibited "expansion behaviors" characterized by an instantaneous movement in an outward direction by all individuals of the test group, but only during the first few exposures. No startle responses were noted for brook trout.

Vector Analysis

Vector movements were plotted to show the movement during each of these periods and to help compare the direction of movement during a treatment to the normal movement (control

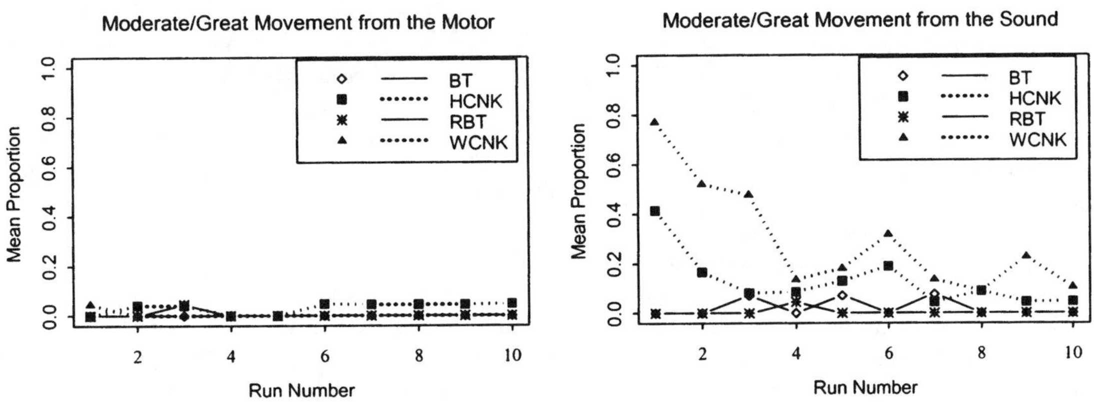

Figure 2. Proportion of tests that showed moderate/great movement for each exposure (run) during motor and sound testing. (BT = brook trout, HCNK = hatchery chinook salmon, RBT = rainbow trout, WCNK = wild chinook salmon).

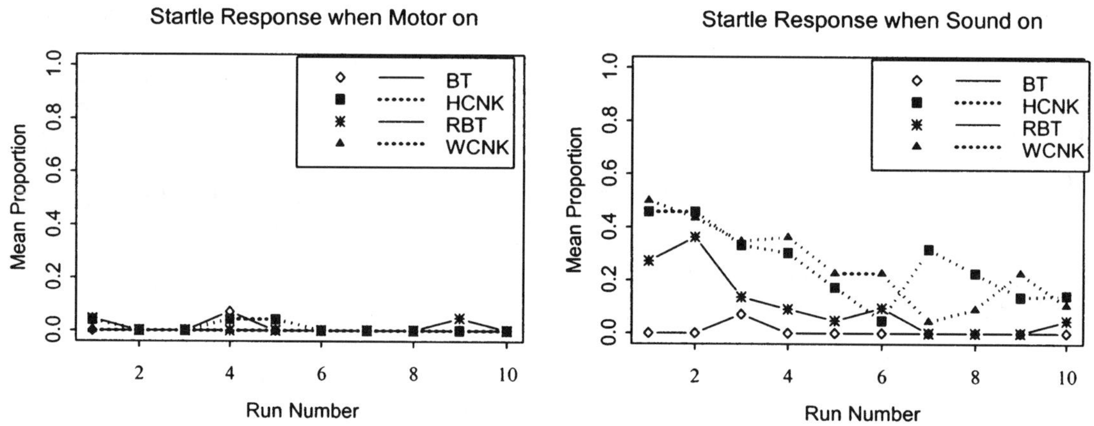

Figure 3. Proportion of tests that showed startle and flight behavior for each exposure (run) during the motor and sound testing. (BT = brook trout, HCNK = hatchery chinook salmon, RBT = rainbow trout, WCNK = wild chinook salmon).

period). Analysis was conducted for all species but, wild chinook salmon was the only species that maintained a consistent school and provided the largest sample size. Vector movements during the control measurements show general random movement patterns. During the sound on period, vector movement for wild chinook salmon was generally down and away for groups that were nearest the VDS and toward the VDS when fish were furthest from the VDS (Figure 4). The movement toward the VDS by wild chinook salmon was observed after repeated exposures.

Strobe Light

Avoidance Movement Responses

Responses to the strobe light stimulus also varied according to species. Chinook salmon and rainbow trout responded in predictable patterns to the strobe light, while small brook trout had no apparent response (Figure 5). Wild chinook salmon fry had greater avoidance responses when exposed to strobe lights than hatchery chinook salmon, however, this difference was not statistically significant (P = 0.22). When the strobe light was activated, their behavior was characterized by a startle response followed by a flight path away from the strobe usually to a lower corner furthest away from the light source. Rainbow trout showed a high proportion of movement when the first strobe was activated, while brook trout showed little movement from the strobe lights (one of five test groups showed movement; P = 0.001). Rainbow trout and chinook salmon fry moderate and great movements did not decrease as the number of exposures increased (up to 10, P = 0.81). The movement reaction (moderate and great) was more profound to the first strobe than the second strobe; however, both strobes caused significant proportions of movement for chinook salmon and rainbow trout. Neither the acclimation period nor times of day fish were tested resulted in any differences in fish response.

Startle Movement Responses

Startle responses elicited by the activation of the strobe lights were evident to some degree in all species tested, with the exception of brook trout. The proportion of each species that displayed any startle reactions during the first strobe light ranged from none for brook trout to near 30% for hatchery chinook (Figure 6). The startle reaction appeared small, but significant (P = 0.014), for the rainbow trout and hatchery and wild chinook salmon. Again, the startle responses did not decrease as the number of exposures increased (up to 10 exposures) (P = 0.98).

Vector Analysis

Only wild chinook salmon vectors are plotted due to a limited amount of data. Control plots show random movement patterns. During the strobe on tests, vector movements for rainbow trout and wild chinook salmon was characterized by a profound avoidance pattern generally down and away from the light source (Figure 7).

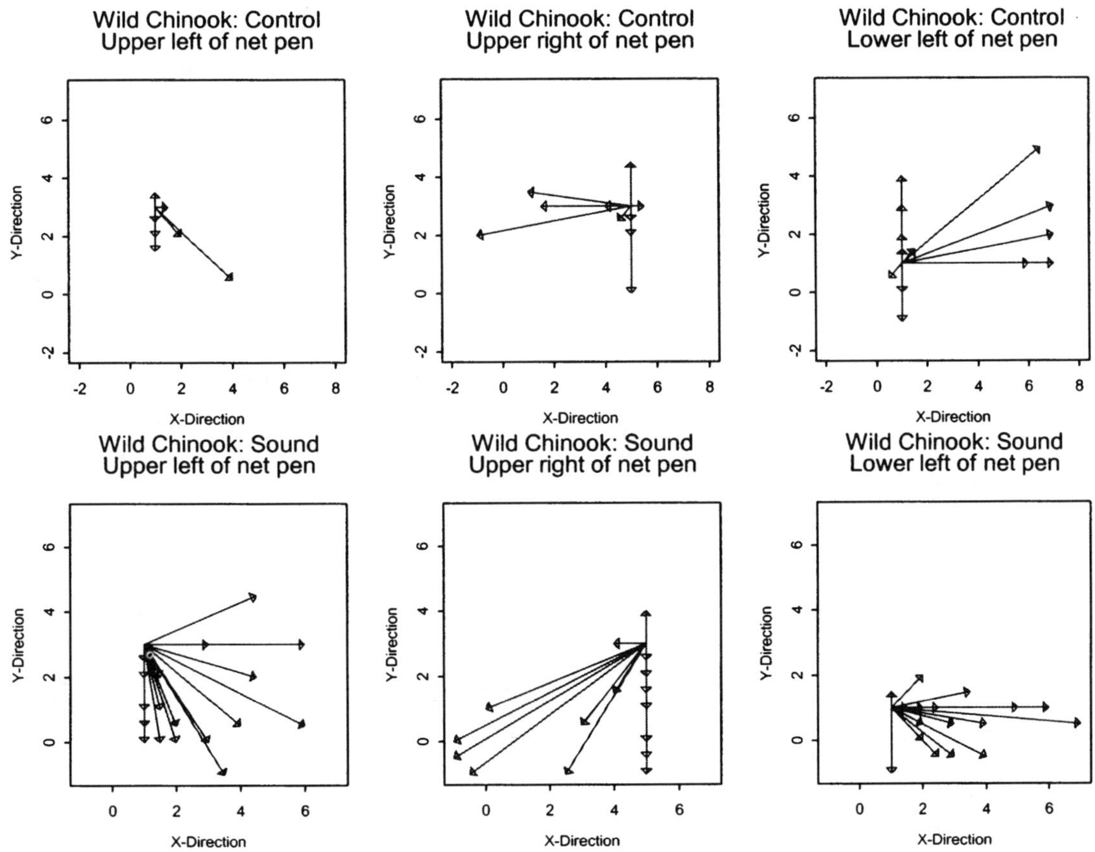

Figure 4. Mean vectors for wild chinook salmon centered locations within the net pen during the control and sound testing (arrows indicate number of schools).

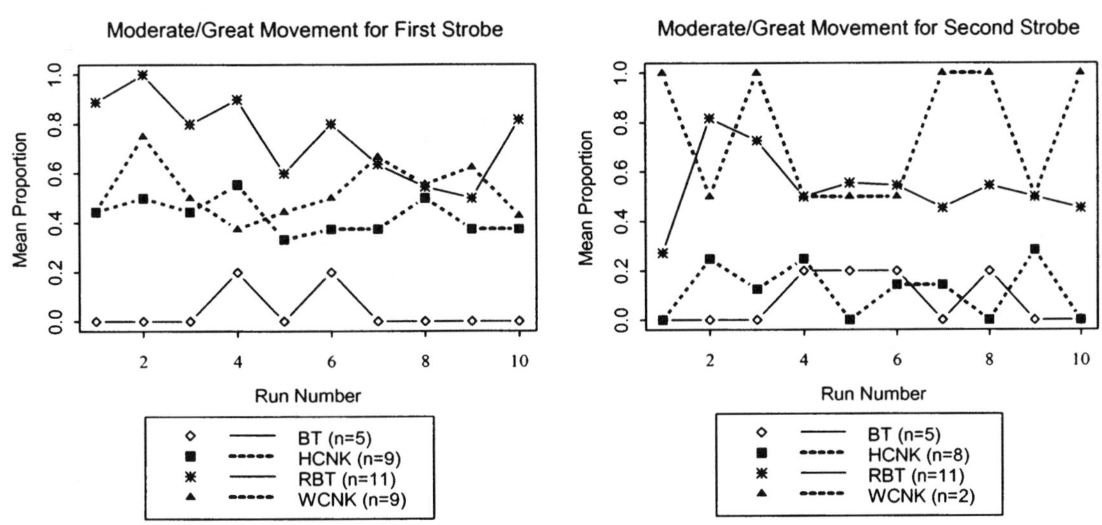

Figure 5. Proportion of tests that showed moderate or great movement for each exposure (run) during first and second strobe light tests. (BT = brook trout, HCNK = hatchery chinook salmon, RBT = rainbow trout, WCNK = wild chinook salmon).

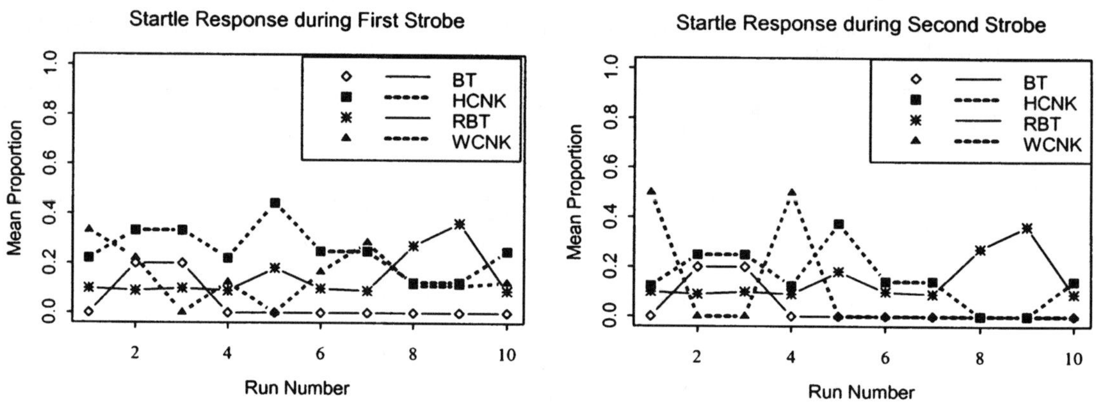

Figure 6. Proportion of tests that showed startle response for each exposure (run) during first and second strobe light tests. (BT = brook trout, HCNK = hatchery chinook salmon, RBT = rainbow trout, WCNK = wild chinook salmon).

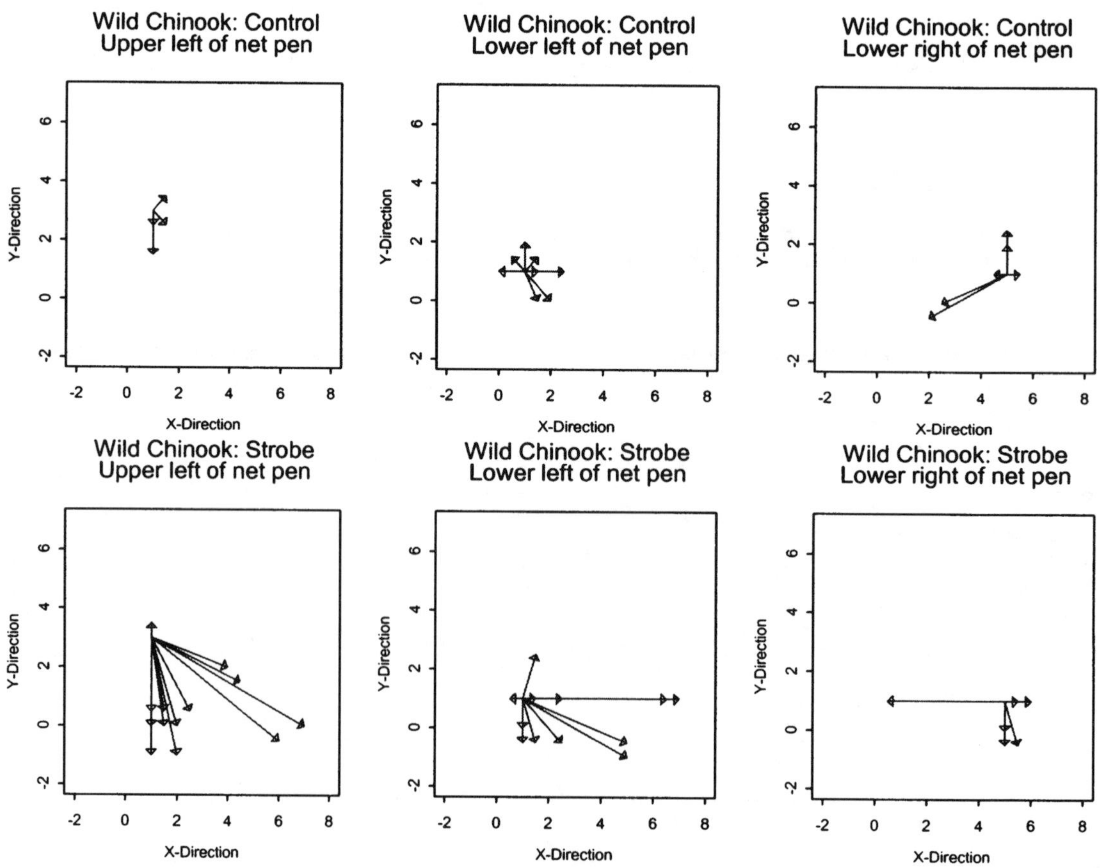

Figure 7. Control and movement vector plots during strobe light test, with strobe location at left. The sample sizes for the six centering locations are in the x-axis label for each plot.

Discussion

Our findings demonstrated that juvenile chinook salmon and brook trout fry exhibited different behavior patterns when subjected to infrasound and strobe lights in a laboratory setting. These behaviors were complex and often difficult to evaluate even under our controlled laboratory conditions. Because small fish are the most susceptible to passing through or around the screens, our tests were conducted using small (fry-sized) salmonids and char. The development of sensory perception to sound and light is, for the most part, incomplete for early life stages of teleost fish (Fernald 1988; Popper and Platt 1993). Thus, their behavior response to the subjected stimulus may change as the fish grows and develops.

Infrasound

The primary findings from the infrasound tests indicated wild and hatchery chinook salmon are most likely to respond in the near field with avoidance and startle behaviors. Their movement patterns were generally down and away from the VDS during initial exposures. In contrast, infrasound elicited little, if any, response in rainbow trout and brook trout. The responses of wild and hatchery chinook, as well as rainbow trout fry, were not as profound as observations made during similar tests conducted in 1996 using the identical VDS (Mueller et al. 1998). Chinook salmon showed a rapid decrease in avoidance responses to repeated exposures during short-term testing, suggesting that the interval of exposure was too short to acquire consistent responses. Although not a major objective of the experiment, our findings indicate that fish response was more profound for short acclimation periods at least for chinook salmon. Prolonging the acclimation period from 3 to 15 h reduced the initial avoidance responses for wild and hatchery chinook salmon. The rate at which the test groups habituated may not be a significant factor in the environment. The rationale is that emigrating salmonids would be expected to have a relatively long interval between encountering a series of behavior barriers if these devices were to be implemented.

Other studies have demonstrated that high particle acceleration infrasound was effective at producing startle and avoidance responses using hatchery reared chinook salmon, Atlantic salmon, and brown trout *salmo trutta* (Knudsen et al. 1992, 1994, 1997). Test fish used in these studies ranged in size from 100 to 150 mm TL. More recent field studies concluded that infrasound devices were not effective at deterring migrating and caged juvenile salmon smolts (Amaral et al. 1998; Ploskey et al. 1998).

Strobe Lights

Our experiments showed that, under controlled laboratory experiments, strobe lights are effective at producing avoidance responses in wild and hatchery chinook and rainbow trout fry 25–40 mm TL. In contrast, brook trout were unaffected by strobe lights. These results demonstrate that even during the early stage of development, certain test species were able to discern sudden changes in light intensity and were repelled. Why no discernible behavior was noted for brook trout may be related to incomplete eye development as the eye continues to develop new retinal cone cells as it enlarges (Northmore et al. 1978). In addition, the spectral sensitivity alike changes with age and within species (Fernald 1988). These facts suggest that species and age play an important role in how fish respond to artificial stimuli. Our studies indicate that species specific differences may play a more important role than life stage.

Several laboratory studies have been conducted on reaction of juvenile salmonids to strobe lights (e.g. Nemeth and Anderson 1992; Puckett and Anderson 1987). Most showed a moderate to strong avoidance response by juvenile chinook salmon smolts to strobe and mercury lights (Amaral et al. 1998; Ploskey et al. 1998). Migrating chinook juvenile salmonids were much more likely to respond to strobe lights when ambient light levels are low such as at dusk or during the nighttime or when fish migrate to deeper water (Amaral et al. 1998). Other studies have found that juvenile chinook salmon will avoid strobe lights during daytime testing near the water surface (Ploskey et al. 1998; Nemeth and Anderson 1992). Anderson et al. (1988) found the effective range of strobe lights to be dependant on environmental conditions (i.e., turbidity, time of day), age, and physical condition of the fish.

A primary goal of this research was to determine what specific behaviors could be elicited from juvenile salmonids using infrasound and strobe lights in a laboratory setting in order to establish how these systems may perform in the field. We conclude that under clear water, low ambient light conditions, strobe lights would be more effective at eliciting more consistent avoidance responses for juvenile chinook salmon and rainbow trout than infrasound. In the near term, the commercial availability of strobe lights and positive results from current research make their use a practical compliment to physical barriers (i.e., fish screening facilities).

The likely deployment location for behavior barriers would be near the point of diversion to prevent entrainment or near the screen face to direct fish to the bypass. Thus, the effective range of the devices is also an important aspect if these devices can be used in conjunction with the screening facilities. Based on the particle acceleration values for the VDS used in our tests, the effective range on axis is expected to be about 4 m in a controlled test tank (Carlson and Campana 1996). Studies using strobe lights have shown an effective range of 6 m for juvenile coho salmon *O. kisutch* under low ambient lighting (Ploskey and Johnson 1998; Ploskey et al. 1998). These ranges would be sufficient to prevent entrainment at most sites. Other factors such as water velocity, turbidity, and debris would need to be considered.

Acknowledgments

The success of this project depended on the involvement and cooperation of many people. The project was managed by Ken Barnhart of the Bonneville Power Administration under contract DE-AT79–94BP30373. We thank Jeff Condiotty and Frank Knudsen from Simrad Inc. for their assistance in leasing the infrasound device and Ron Brown from Flash Technology for loaning the strobe lights. We also thank Pacific Northwest National Laboratory staff Fenton Khan, Jeff Marco, and Anita Muzak for their assistance in analysis of the videotapes. Dennis Dauble and David Geist reviewed earlier drafts of this manuscript.

References

Amaral, S. V., F. C. Winchell, B. J. McMahon, and E. P. Taft. 1998. Evaluation of behavioral guidance technologies for diverting chinook salmon smolts at the Roza Dam screening facility. 218-98/E873F, Alden Research Laboratory Inc., Holden, Maine.

Anderson, J. J., K. J. Puckett, and R. S. Nemeth. 1988. Studies on the effect of behavior on fish guidance efficiency at Rocky Reach Dam: avoidance to strobe light and other stimuli. FRI-UW-8801, Fisheries Research Institute, University of Washington, Seattle.

Carlson T. J., and R. D. Campana. 1996. Evaluation of the characteristics of a volume displacement infrasound source as a means for implementing a behavioral barrier to reduce movement of juvenile salmonids through the causeway separating the inner and outer portions of the Burbank Slough. U.S. Department of the Interior, Pacific Northwest Division, Boise, Idaho.

Fernald, R. D. 1988. Aquatic adaptations in fish eyes. Pages 435–466 *in* J. Atema, R. R. Fay, A. N. Popper, and W. N. Tavolga, editors. Sensory Biology of Aquatic Animals. Springer-Verlag, New York.

Karlsen, H. E. 1992. The inner ear is responsible for detection of infrasound in the perch (*Perca fluviatilis*). Journal of Experimental Biology 171:163–172.

Knudsen, F. R., P. S. Enger, and O Sand. 1992. Awareness reactions and avoidance responses to sound in juvenile Atlantic Salmon (*Salmo salar L*). Journal of Fish Biology 40:523–534.

Knudsen, F. R., P. S. Enger, and O. Sand. 1994. Avoidance responses to low frequency sound in downstream migrating Atlantic salmon smolt (*Salmo salar L.*). Journal of Fish Biology 45:227–233.

Knudsen, F. R., C. B. Schreck, S. M. Knapp, P. S. Enger, and O. Sand. 1997. Infrasound produces flight and avoidance responses in Pacific juvenile salmonids. Journal of Fish Biology 51:824–829.

Mueller, R. P., C. S. Abernethy, and D. A. Neitzel. 1995. A fisheries evaluation of the Dryden Fish Screening Facility. DOE/BP-00029-2, Bonneville Power Administration, Portland, Oregon.

Mueller, R. P., D. A. Neitzel, W. V. Mavros, and T. J. Carlson. 1998. Evaluation of low and high frequency sound for enhancing fish screening facilities to protect outmigrating salmonids, DOE/BP-62611-13. Bonneville Power Administration, Portland, Oregon.

Neitzel, D. A., C. S. Abernethy, and E. W. Lusty. 1990a. A fisheries evaluation of the Wapato, Sunnyside, and Toppenish Creek Canal fish screening facilities spring 1988. DOE/BP-01830-6. Bonneville Power Administration, Portland, Oregon.

Nemeth, R. S., and J. J. Anderson. 1992. Response of juvenile coho and chinook salmon to strobe and mercury vapor lights. North American Journal of Fisheries Management 12:684–692.

Northwest Power Planning Council (NPPC). 1984. Columbia River Basin Fish and Wildlife Program: Measure 704(d). Portland, Oregon.

Northwest Power Planning Council (NPPC). 1987. Columbia River Basin Fish and Wildlife Program: Measure 1403.4. Portland, Oregon.

Northwest Power Planning Council (NPPC). 1994. Columbia River Basin Fish and Wildlife Program: Measure 7.10. Portland, Oregon.

Northmore, D. P. M., F. C. Volkmann, and D. Yager. 1978. Vision in fishes: color and pattern. Pages 79–136 *in* D. Mostofsky, editor. The Behavior of Fish and Other Aquatic Animals. Academic Press, New York.

Ploskey, G. R., and P. N. Johnson. 1998. Effectiveness of strobe lights for eliciting vertical avoidance, by juvenile salmon. Pages 421–430 *in* HydroVision 98. HCI Publications, Kansas City, Missouri.

Ploskey, G. R., P. N. Johnson, M. G. Burczynski, J. M. Nestler, and T. J. Carlson. 1998. Effectiveness of strobe lights, infrasound devices, and a sound transducer for eliciting avoidance by juvenile salmonids. Final Report by the U.S. Army Waterways Experiment Station. U.S. Army Engineer District, Portland, Oregon.

Popper, A. N., and C. Platt. 1993. Inner ear and lateral line of boney fishes. Pages 99–136 *in* D. H. Evans, editor. The Physiology of Fishes, CRC Press, Boca Raton, Florida.

Puckett, K. J., and J. J. Anderson. 1987. Behavioral responses of juvenile salmonids to strobe and mercury lights. FRI-UW-8717, Fisheries Research Institute, University of Washington, Seattle.

Sand, O., and H. E. Karlsen. 1986. Detection of infrasound by the Atlantic cod. Journal of Experimental Biology 125:197–204.

Sand, O., P. S. Enger, H. E. Karlsen, F. Knudsen, and T. Kvernstuen. 2000. Avoidance responses to infrasound in downstream migrating European silver eels (*Anguilla anguilla*). Environmental Biology of Fishes 57:327–336.

American Fisheries Society Symposium 26:91–104, 2001

Evaluation Of Low-Frequency Sound Transducers for Guiding Salmon Smolts Away from A Navigation Lock

FREDERICK A. GOETZ
U.S. Army Corps of Engineers, Seattle District Environmental Resources Section
Post Office Box 3755, 4733 East Marginal Way S., Seattle, Washington 98124, USA

JAMES J. DAWSON
Biosonics Inc., 4027 Leary Way NW, Seattle, Washington 98107, USA

TRAVIS SHAW AND JEFFREY DILLON
U.S. Army Corps of Engineers, Seattle District Environmental Resources Section
Post Office Box 3755, 4733 East Marginal Way S., Seattle, Washington 98124, USA

Abstract.—We evaluated the efficacy of a commercially manufactured, low-frequency (300–400Hz) sound transducer to guide yearling sockeye salmon *Oncorhynchus nerka*, coho salmon *O. kisutch*, and subyearling summer/fall chinook salmon *O. tshawytscha* smolts away from a lock and navigation channel. Tests were conducted during the spring of 1997 within the large lock chamber at the Hiram M. Chittenden Locks, a navigation project in Seattle, Washington. Two low-frequency acoustic transducers (300/400 Hz) were deployed at the entrance to the large lock chamber and were operated over a 29 d period with four daily treatments (4 h periods, from 0530 to 2130) of sound on and sound off. A randomized block design (with day as the blocking factor) was used to determine treatment periods. Treatment and control periods within a day were randomly selected with two treatments (sound-on) and two control (sound-off) periods per day. Smolt density (average number of fish/m^3) in the lock chamber (downstream of the transducers) during treatment and control was estimated from acoustic data collected with a split-beam echo sounder, with species composition verified by purse seining. The sound treatment did not impact the density of salmonids measured within the lock chamber. Analysis of variance (ANOVA) showed no significant difference in the density of salmonids in the lock chamber between sound treatments or between days of the study. Similarly, no difference was detected between sound treatments at different times of day. The results from our evaluation suggest that low-frequency sound is not an effective means of guiding salmon smolts.

The Lake Washington Ship Canal and Hiram M. Chittenden Locks (Locks) have been the focus for four years of pilot studies related to the behavior and survival of migrant salmon and steelhead *Oncorhynchus mykiss* smolts. These studies have been a cooperative effort between several resource agencies, including the Washington Department of Fish and Wildlife (WDFW), Muckleshoot Indian Tribe, King County, Waterways Experiment Station (WES), and Portland U.S. Army Engineer Corps of Engineers (USACE) and Seattle District USACE.

The Locks is a navigation project located at the outlet to the Lake Washington Ship Canal (Ship Canal) that has provided adult salmon and steelhead passage facilities since its inception in 1916. It is the egress point for all Lake Washington salmon and steelhead smolts. Historically, Lake Washington and its tributaries has been one of the most productive salmon watersheds in the Pacific Northwest. At full escapement the watershed can produce 375,000 adult salmon and steelhead and more than 7 million smolts. The out-migration season for the smolts to move through the Locks is from mid-April through the end of July. The greatest number of smolts migrating through the Locks occurs in May and June. While original design of the Locks included a fish ladder for upstream migrating adults, no features were included specifically to accommodate the passage of downstream migrating smolts and fry. In 1997, between 1 and 2 million smolts passed the Locks and were composed of sockeye salmon *O. nerka* and coho salmon *O. kisutch* yearlings, chinook *O. tshawytscha* subyearlings and steelhead *O. mykiss* yearlings and age-2 smolts.

Since the late 1970s to mid-1980s, Lake Washington salmon and steelhead populations have undergone a steep decline (Fresh 1994). By the mid 1990s, adult returns for all Lake Washington salmon and steelhead stocks declined to lows of less than 8%

relative to each stock's escapement goal. The rate of decline for each stock is the highest or nearly the highest of all stocks within their respective Puget Sound Evolutionary Significant Units (USACE 1999). The severe decline in all Lake Washington anadromous fish runs precipitated resource agency and tribal biologists to investigate the potential causes for the declines. In 1995, agency biologists began initial monitoring efforts at the Locks to assess the need for smolt passage facilities. Preliminary results suggest that more than 60% of all smolts use the large locks to leave Lake Washington even with a continuously operated small-scale prototype surface collector (1.7–2.9 m^3/s) on an adjacent spillway. Many smolts using the large locks become entrained in the filling culverts with injuries sustained by 5–60% of the fish and an unknown number of mortalities (WDFW 1996). The influence of the Locks as a potential source of smolt mortality can be demonstrated by relative comparison to nearby watersheds with known smolt-to-adult survival rates of coded-wire-tagged hatchery coho salmon. From 1978 to 1994, Lake Washington coho salmon have averaged less than 40% of the average smolt-to-adult (SAR) survival rate of two other nearby river system hatchery runs (Soos Creek and Skykomish). Over a more recent period, 1990–1994, Lake Washington coho salmon returns have averaged less than 10% of the other watersheds' average return rate. This period of low SAR for coho salmon corresponds to a period of increased use of the large lock chamber filling culverts. Since 1994, the large lock chamber has seen decreased use relative to previous years (USACE 1999).

In 1994, the Portland District, USACE, funded a Behavioral Technologies Program (BTP) to evaluate existing acoustic methods for guiding salmonids or redistributing fishes near dams on the Columbia River. The BTP was a joint research effort of the WES and Battelle Pacific Northwest National Laboratory. The Program has evaluated two very different acoustic technologies for guiding juvenile salmonids which included (1) small-particle-displacement low-frequency sound (Loeffelman et al. 1991a, 1991b) and (2) high-particle-displacement infrasound (Knudsen et al. 1992, 1994). In 1997, Seattle District was contracted by Portland District through the BTP to evaluate a commercially manufactured low-frequency sound guidance system at the Hiram M. Chittenden Locks.

A patented process has been developed that uses a tuning process to create signals to repel fish; this process (sound-system) was commercialized beginning in the early 1990s (Loeffelman et al. 1991a, b). The hypothesis underlying the tuning process is that fish are vocal, and that vocalizations of captive fish contain elements of sound (frequency, duration, amplitude) that are within the most sensitive part of the hearing range of a species and life stage. This sound-system has shown highly variable guidance efficiency for juvenile fall-run chinook salmon at Georgiana Slough off the Sacramento River within and among years of testing. Individual test and overall year results ranged from (1) −156 to +74% in 1993; (2) −49 to +87% with mean of 57% in 1994; and (3) −21 to +56% with mean of 15% in 1996 (Hanson et al. 1997). In that study, a blend of 300- and 400-Hz sounds (low-frequency) was synthesized from recordings of smolts and transmitted from commercial transducers (sound transducers) as a crescendo with maximum source levels of about 170–180 dB//1 μPa at 1 m (dB referenced to 1 μPa sound pressure 1 m from the sound source).

Our objective was to test and evaluate the effectiveness of commercially manufactured low-frequency sound transducers for excluding migrant juvenile salmon at the entrance to the large lock chamber at the Hiram M. Chittenden Locks. If successful, we planned on developing a permanent behavioral technology deployment.

Methods

Site Description and Study Plan

We conducted the low-frequency sound evaluation at the Hiram M. Chittenden Locks at the outlet of the Lake Washington Ship Canal in Seattle, Washington. The Project primarily functions as a navigation lock for vessels passing between the freshwater Lake Washington system and the salt water of Shilshole Bay and Puget Sound. From north to south, the Locks consists of a large lock, small lock, a spillway with one bay modified as a surface bypass, and an adult fish ladder (Figure 1). The large lock chamber is 24.4 m wide by 251.5 m long and accommodates vessels with drafts as deep as 9.1 m. It is enclosed by double-hinged miter gates. The chamber is divided into two half-chambers (upper and lower) by an intermediate (middle) gate. The small lock chamber is 9.1 m wide by 45.7 m long, with similar gates and accommodates vessels with drafts as deep as 4.9 m. The head differential from upstream to downstream of the lock chambers

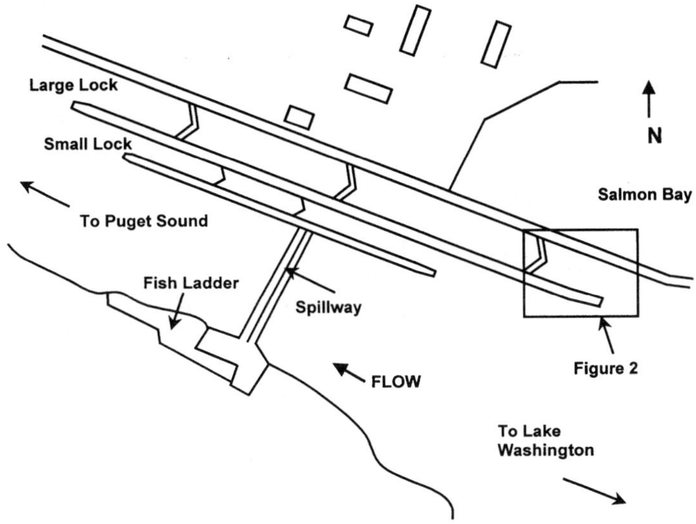

Figure 1. Plan view of the Hiram M. Chittenden Locks Project showing all major structures and location of sound guidance evaluation.

varies from 1.8 to 7.9 m depending on the tidal elevation and the level of Lake Washington. Tide levels measured just downstream from the Locks in Shilshole Bay fluctuate about 3.7 m over the course of each tidal cycle. The spillway is 71.6 m long with six, 9.8 m wide openings each capable of passing about 77 m^3/s at maximum discharge. Since 1995, a prototype low-flow (1.7–2.9 m^3/s) smolt bypass flume (smolt flume) has been installed in the second spillway north of the fish ladder. This smolt flume was built due to a decline in available water for spill during the smolt out-migration period.

The large lock chamber is filled via two culverts that originate as 4.9 m wide by 4.3 m tall openings located at the floor of the channel (Figure 2) in both walls just upstream of the upper miter gates. The culverts route water north and south, respectively, for about 4.5 m before turning westward (90° angle) and constricting to 2.6-m wide by 4.3-m tall. The culvert continues laterally along the chamber walls, emptying into the chamber through a series of 22 (1.2-m wide by 0.6-m high) filling ports. The chamber is filled by opening three fixed-wheel, vertical lift Stoney gate valves located just west of the upper miter gates and just east of the middle miter gates (Figure 2). For all sound tests, the "slow-continuous" valve opening procedure was used (USACE 1999). This procedure results in the Stoney gate valves opening completely in 4 min and 15 s. Maximum discharge at low tide into each culvert is approximately 62.3 m^3/s with discharges greater than 51.0 m^3/s lasting up to 3 min.

Two low-frequency acoustic transducers were deployed across the navigation channel at the upstream entrance to the large lock (Figures 1 and 2). They were operated over a 29 d period with four daily treatments (4 h periods) of sound-on and sound-off. A randomized block design (with day as the blocking factor) was used to determine treatment periods. Treatment and control periods within a day were randomly selected with two treatment (sound-on) and two control (sound-off) periods per day—only the principal investigator and the sound-system technical crew knew the final schedule of treatments. Differences in smolt density in the lock chamber downstream of the transducers under treatment (sound-on) and control (sound-off) were estimated using splitbeam mobile hydroacoustics with species verification using deep-draft purse seine. Mobile surveys were conducted four times daily, once per treatment with approximate times at 8:00 am, 12:00 p.m., 4:00 p.m., and 8:00 p.m.

Sound System

Sounds that were evaluated were synthesized from recordings of sounds made by juvenile chinook and sockeye salmon obtained from purse-seining sources in a portable acoustic recording studio. A blend of 300- and 400-Hz sounds was synthesized from recordings of juvenile salmon and transmitted from transducers as a crescendo with maximum source levels of about 170–180 dB//1 μPa at 1 m. The sound-system deployed at the Locks was composed of two 104.5-kg transducers (with a 3rd

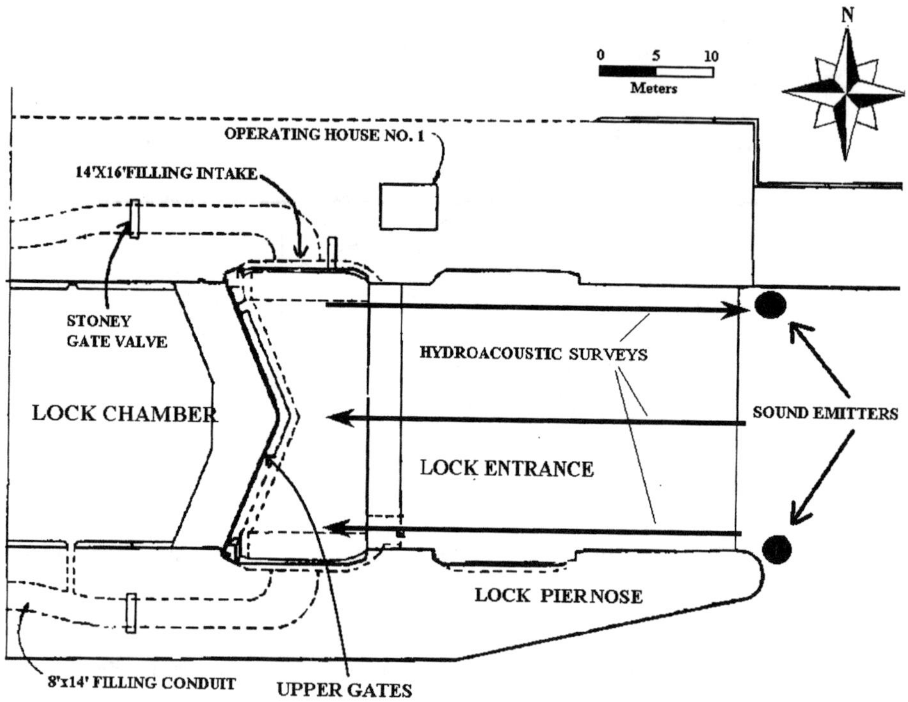

Figure 2. Plan view of the entrance to the large lock chamber showing major features and direction of hydroacoustic surveys.

available as back-up) anchored to the bottom of the navigation channel 24.4 m apart bracketing the entrance of the lock chamber (Figure 2). The equipment was deployed on 9 May 1997 using the Ship Canal's small barge-mounted crane. Transducers in the sound-system are bottom mounted (using water bladders that surround the transducer) to avoid possible collision with deep draft barges or from contact with "wayward" anchors dangling from large vessels. Power and control cables were channeled at the lock walls for protection against damage from passing vessels. Control equipment was installed in Operating House No. 1 (Figure 2).

According to the system manufacturer, the maximum anticipated range of effectiveness of the sound field is about 18.3 m from the transducer. The sound field, therefore, should have overlapped at the center of the lock entrance, 12.2 m from either transducer. During operation, the sound field was to repel smolts and guide them away from the entrance to the large lock chamber and towards the spillway dam and the surface collector (Figures 1 and 2). Mapping of the sound field to verify the acoustic barrier was performed by a technician from the manufacturer. During sound-on, recordings indicated pressures of (1) 170–180 dB at 1 m from the sources, (2) 154–156 dB near the center of the miter gate, and (3) 148–150 dB at 30 m upstream of the transducers. Ambient noise level at the gate was 110 dB, with a large variation (but not exceeding the above levels) as vessels approached and during operation of the saltwater barrier air compressors (P. Novakovic, letter dated 11 November 1997).

Standardized Lockages

The number and timing of large lockages within a treatment period was expected to have an effect on smolt density estimates in and near the Locks. To minimize this potentially confounding variable we designed the study to have a standard number of lockages per treatment period. We identified the maximum daily number of lockages in May and June for a typical year and then attempted to perform this number of lockages during every day and every treatment period over the course of the study.

Tests were conducted after an expected peak in lockages, during "boating weekend." Boating weekend is the peak in the pleasure boat season when the greatest number of lockages occurs—15–20 can occur in one day. The typical maximum daily lockage count in the large lock after boating weekend is 10–12. Using this as a standard daily number, we scheduled with the lockmasters to have 2 lockages per 4-hr treatment. Most of the lockages occurred at set times during each treatment period. Even during days when there would be only 4–5 lockages, the lockmasters performed "false" lockages at prescribed times where they opened the valves and filled the lock chamber just as if a boat were locking through. These "false lockages" were the primary method used to create a "flow stimulus" to guide smolts toward the lock entrance and the sound-system, although actual boat lockings occurred on an irregular basis throughout the study period and performed the same function. To minimize the influence of flow over the spillway gates on density estimates in the lock chamber, spill volume was controlled during the 29 d of the study. Spill other than the overflow weir was largely limited to weekends and nighttime periods (2130–0530), outside the scheduled treatment periods.

Hydroacoustic Surveys

Smolt densities in the lock chamber downstream of the sound-system under treatment (sound-on) and control (sound-off) conditions were estimated using splitbeam mobile hydroacoustics with species verification using deep-draft purse seine. The acoustic density estimates were collected blind, that is, with no prior knowledge of whether the sound was on or off. Mobile surveys were conducted four times daily, once per treatment approximately at 8:00 a.m., 12:00 noon, 4:00 p.m., and 8:00 p.m. These times were after a lock event near the mid-point of each 4-hr treatment (on or off).

We used a BioSonics split-beam survey system operating at 200 kHz to survey around and in the lock structures. The transducer had a 7° nominal beam angle to minimize reflections from side walls and the surface boundary. The transducer was mounted on a pipe extending down from the starboard side of the survey boat to a depth of about 0.7 m. During horizontal aiming, the transducer was aimed down about 5 degrees from horizontal. The acoustic survey system was operated at 5 pulses per second and transmitted a 0.4 ms pulse length signal into the water. Targets smaller than –60 dB target strength were excluded to minimize signals from false targets and turbulence. Both printed echograms and digital data were created. The system stored digitized data on a laptop computer for later analysis. Other system settings and parameters are detailed in Table 1 of Goetz et al. (2000).

The study area was surveyed in two regions. The first region (48.3 m-length, 24.4 m-width, and 11.6 m-depth) encompassed the area between the filling culverts and immediately east of the upper service gate (Figure 2). We referred to this sample area as "Study Area 1," or "the lock entrance." The second survey area included the entire length of the large lock and was referred to as "the lock chamber" (results from the lock chamber survey are discussed in Goetz et al. 2000). A standard sampling protocol was followed whenever possible for each of the four daily surveys in the lock entrance: the large lock was filled 30–60 min prior to the beginning of the survey, either as a real boat locking or as a false lockage. This flow stimulus occurred at least 1.5 h into the treatment period, with the hydroacoustic survey always occurring during the second half of the treatment. The survey boat moved at a constant speed of about 0.3 m per second. The first leg was run westerly along the south wall, with the transducer aimed vertically, straight down. The second leg was run easterly along the north wall. The third leg ran westerly up the center of the lock entrance and completed the survey. The three legs of the survey are shown in Figure 2.

After inspection of many sets of acoustic data from the locks, Echo Integration was chosen as the signal processing technique. This technique does not rely on resolving echoes from individual fish, but instead measures total reflected energy from the ensonified volume above the set threshold. The analysis software was programmed to amplify returning signals with a digital 20 Log R Time Varied Gain (TVG). The echo integration technique requires the operator to assign range and time bins for summing reflected energy. Seven 2-m range bins were chosen for all vertical surveys, while horizontal surveys were not echo integrated because of high volume reverberation (from bubbles and turbulence). Density estimates were tabulated once per leg in the lock entrance and five times per leg inside the lock chamber. Reflected acoustic energy was averaged in these range and distance bins and written to hard disk.

Echo Integration measures average intensity of sound reflected by fish (primarily), but these values must be converted to an estimate of fish density by calculating an echo integrator scaling constant. This constant is a function of system calibration parameters, operator-selected data collection parameters, and an estimate of the reflectivity (backscattering cross section) of an individual fish. This constant A is calculated as

$$A = 1/(c \times \tau \times \pi \times p^2 \times g^2 \times b_{av}{}^2 \times \sigma_{bs}),$$

where c = speed of sound in water (1500 m/s); τ = pulse length (seconds); π = pi (3.1416); p^2= pressure term calculated from system source level; g^2 = gain term calculated from system receive sensitivity; $b_{av}{}^2$ = average squared beam pattern (based on transducer directivity); σ_{bs} = sigma; back scattering cross section (acoustic size) of average fish.

A value of 1.0 E-5 was used for sigma, which corresponds to a target strength of –50 dB. When the average reflected intensity is multiplied by the scaling constant, an estimate of density (fish per cubic meter) is obtained.

The quantitative digital data were interpreted using information from the printed echogram records. During the design of analysis procedures for digital data, we looked at the printed echogram records from many of the surveys including when no fish targets were present. We observed several distinct acoustic patterns in these surveys, referred to as aggregates, dense schools, cloud-like formations, single targets, and hard-on-bottom versus midwater distributions. Based on this variety of patterns, we interpreted the acoustic returns as coming from a variety of different species. The aggregate formations were individual fish grouped together to form a school, but with the individuals still fairly distinct (BioSonics, Inc. 1997). These aggregate patterns were very similar to the distributions of fish observed the previous year with the fixed-location system mounted near the filling culverts, and led us to interpret these aggregates as salmonid smolts (Dillon and Goetz 1999). The aggregate patterns reformed after fish were drawn into the lock chamber within a period of a few minutes. Additionally, we interpreted single targets as salmonid smolts, except for obvious large individuals. Conversely, dense schools appeared during the survey and were either clustered on the bottom or observed in midwater later in the study. These schools, along with the "cloud-like" formations, were interpreted as nonsalmonids. The quantitative analysis includes density estimates of aggregates and single targets only to estimate salmonid abundance.

Purse Seining

The WDFW and the Muckleshoot Indian Tribe conducted a series of purse-seining sets in the upper half of the large lock chamber, to provide species verification of the mobile hydroacoustics density estimates. The WDFW operated, maintained and deployed a barge-mounted deep draft purse seine at the Locks in May and June of 1997. The seine was approximately 229 m in length, 11 m in depth and could sample almost the entire surface area of the upper lock chamber in one set. A depressor weight was used to increase the effective fishing depth. Two vessels deployed the net, a 10-m pontoon barge and 4-m skiff, each powered by outboard motors. Testing in 1996 showed a 33–50% capture efficiency of marked smolts (WDFW 1996). Capture efficiency was not calculated in 1997 as the middle miter gate doors were inoperable creating poor recapture conditions and negating any comparison of hydroacoustic density estimates with the purse-seine catch. Species composition in the large lock was identified by purse-seining 2–3 times per week during one daily treatment period in the first four weeks of the evaluation (May 12–June 4). Thirty-one successful net sets were performed over the four-week period: two sets were aborted (total N = 33). Sound guidance testing did not occur during any purse-seine period.

Experimental Design and Statistics

The experimental design utilized to evaluate the low frequency sound treatments was a randomized block design ANOVA with time of day acting as the blocking factor. Each day of the study contained two blocks consisting of morning and afternoon surveys. A blocked design was selected with the assumption that the number of smolts attempting to pass through the locks was likely to be highly variable. The locks are the final physical obstacle that migrating smolts encounter on their way to the sea. It was expected that smolt numbers would dramatically peak during the study at the height of the migration period before markedly decreasing. As a result, the density of smolts in different treatments on the same day are likely to be similar (correlated). A blocking factor allows for this type of variability over the duration of the study to be partitioned and accounted for in the analysis without the loss of statistical power needed to determine if a difference exists between sound treatments.

Transducers were turned on for 4 h and off for 4 h to create two test and two control treatments every 24 h, 5 d per week for 29 d. The sequence of on and off treatments was randomly selected between morning and afternoon blocks using Systat statistical experimental design software. Since smolt movement at the Locks appears to be diurnal, test periods were limited to daylight hours (dawn to dusk) with approximately equal hours of daylight per test period. Four-hour treatment periods were selected to assure that an adequate sample size of fish would be exposed to sound treatments (N = 58 periods for each sound treatment) and sampled with the mobile split-beam hydroacoustics.

The study design incorporated means to reduce the potential bias of counting smolts twice during previous treatments, including (1) performing at least two lockages between each treatment (and hydroacoustic survey) and (2) providing an alternate passage route (smolt flume) to the locks at all times.

Purse-seine data collected from the upper large lock in 1996 show a high degree of variability in smolt passage per lockage with an estimated range of 3 to greater than 3,960 smolts per lockage (WDFW 1996). The 1996 data were analyzed to determine required sample sizes and suggested that the 29 d, four 4-h treatments periods were necessary to obtain an 80% probability (β = 0.20) of detecting a biologically significant difference (60%) between treatments (G. Ploskey, WES, personal communication). The level of significance was selected for two reasons (1) this was the highest overall guidance efficiency previously demonstrated for this system—a 57% reduction in smolt density during testing in 1994 (San Luis and Delta-Mendota Water Authority, and Hanson 1996) and (2) in discussion with other Corps researchers a level of greater than 50% was considered meaningful for effective fish guidance (G. Ploskey and T. Carlson, WES, personal communication).

The initial mixed model for the resulting analysis will include two fixed factors, day and sound treatment, and a random time of day blocking factor (morning and afternoon). In the event that none of the interactions are significant, the Mean Square for these interaction terms were pooled to form a more accurate estimate of the population variance. The linear model for the planned analysis is presented below:

$$Y_{ijk} = \mu + D_i + T_j + DT_{ij} + S_k + DS_{ik} + TS_{jk} + \varepsilon_{ijk}$$

Factor	df	F
Day (D)	28	MSD/MSDS
Treatment (T)	1	MST/MSTS
Time of day (S)	1	MSS/MSe
D × T	28	MSDT/MSe
D × S	28	MSDS/MSe
T × S	1	MSTS/MSe
Residual	28	
Total	15	

Confidence intervals were constructed for the difference between mean smolt densities during sound on and sound off treatments using the log10 + 1 transformed data. Since the treatment was a fixed factor in a mixed model analysis, error mean square was used to estimate variance with v the within-cell degrees of freedom and N equal to the sample size of the level being analyzed. For example the 95% confidence interval was calculated using the formula:

$$95\% \text{ CI for } \mu_1 - \mu_2 = x_1 - x_2 \pm t_{0..05\,(2)}, v, k \sqrt{\frac{s^2}{bn}},$$

when n = 51; v = 72; b = 2; s^2 = 0.0009; $t_{0.05}$ = 2.009; $t_{0.10}$ = 1.676; $t_{0.20}$ = 1.299; x_1 = 0.018; and x_2 = 0.022.

ANOVA was used to evaluate other factors that may have confounded the analysis of sound treatment data. Other factors included spill, nighttime spill, and number of up-lockages.

The association between mean smolt densities and other physical factors was examined with a Pearson correlation analysis. Changes in smolt density with changes in precipitation, spill volume, nighttime spill volume and the salinity at the lock chamber pier nose at a depth of 13.7 m were all examined.

Results

Attainment of Experimental Design

A total of 340 fill events, or up-lockages (real and false), were performed during the 29 d of the study period. The original study plan called for at least two lockages per treatment period. We followed this schedule fairly well since at least two lockages were performed in 111 of the 116 scheduled treatment periods. There was little variation in the number of up-lockages per treatment with an average of 2.8–3.1 lockages by period: 2.8 lockages (SD = 0.7) for period 1; 3.0 lockages (SD = 0.5) for period 2; 3.1 lockages (SD = 1.1) for period 3; and 2.8 lockages (SD = 1.0) for period 4. By study day, there was an average of 11.7 lockages (SD = 1.8) with a range of 9–15. In addition to standardizing the number of up-lockages, the timing of lockages relative to hydroacoustic surveys of the lock entrance was considered an important part of the study protocol. Ideally, a fill event was to occur 30–60 min prior to a survey and at least 1.5 h after initiation of a treatment period. A fill event occurred prior to every hydroacoustic survey and all surveys were completed within 25–85 min of the start of a fill (101 of 106 occurred within 25–60 min).

Because of the high spring inflow and the restricted time available to spill, there were selected nontreatment periods of extremely high spill on the first and third weekends of the study (17 May, 31 May) when more than 128 m^3/s was released. During these spill periods, the pool elevation was drawn down to provide additional storage capacity so spill would not be required during treatments. During the course of the study, spill occurred in 7 of the 116 treatment periods. Water was spilled in 4 of these 7 periods during treatment period 1, 0530–0930 (study days 4, 19, 20, and 25), while the remaining 3 periods of spill were on one day, 5 June (study day 19). Spill volume during these 7 periods ranged from 5.7 to 85 m^3/s.

During the first four weeks of testing, anadromous juvenile salmonids were present during every purse-seine period with coho salmon present 100% of the time and chinook and sockeye salmon present all days except during the first day of testing. As a percent of the total juvenile salmonid catch, chinook salmon made up 48%; coho salmon 40%, sockeye salmon 11%, and steelhead 1%. The mean catch of juvenile salmonids by species were chinook salmon 72.1 fish/set (SD = 145.2); coho salmon 55.4 fish/set (SD = 74.5); sockeye salmon 16.6 fish/set (SD = 12.4); steelhead 2.9 fish/set (SD = 1.8), and chum salmon *O. keta* 1.5 fish/set (SD = 0.8). The total mean catch was 161.2 fish (SD = 213.4, total N = 4998) with juvenile salmonids making up 75.3% of the catch (121.5 fish/set, total N = 3768). Shiner perch *Cymatogaster aggregata* and longfin smelt *Spirinchus thaleichthys* were the two nonsalmonid species caught. Shiner perch were caught during the last two days of seining, 3 and 4 June.

Mobile surveys were performed at the lock entrance during 106 of the planned 116 survey periods (58 sound-on, 58 sound-off): missed surveys were a function of high boat traffic during mid-day and Fridays. Since the lock entrance was small, unconfined, and outside of the area of most boat traffic we were able to complete a high percentage of surveys. Survey performance (surveys completed /surveys scheduled) by treatment period was—93% period 1; 90% period 2; 90% period 3; and 93% for period 4. The acoustic average smolt density appeared to vary by survey period and day: by period, estimates were 0.0633 fish/m^3 in early morning; 0.033 fish/m^3 at mid-day; 0.0461 fish/m^3 in late afternoon; 0.0747 fish/m^3 in early evening; and 0.0184 fish/m^3 mean daily. Peak density estimates occurred on study days 1, 15 and 16–12 May, 2 and 3 June.

One of the assumptions of ANOVA is that the error terms of each treatment are normally distributed (Underwood 1981). As suggested by Keppel (1991), a rough test of this assumption is to plot the observed means within the two treatment groups against the expected frequency of a normal distribution. When plotted, the means for each day do not approach a normal distribution. However, a log10 + 1 transformation does appear to slightly reduce the departure from normality. As shown in Figure 3, with the log transformation, there was an apparent random pattern of fish density at the lock entrance with and without the sound treatment. The consequence of violating this assumption is the potential increase in the stated Type I error rate. Since ANOVA has been shown to be particularly robust to violations of this assumption (Keppel, 1991; Underwood 1981), further testing of this assumption will only be conducted for significant ANOVA results.

Statistical Results

The initial ANOVA model, complete with interaction terms, failed to detect any significant main factors or interactions. After pooling nonsignificant interactions Mean Squares with the residual Mean Square to form a better estimate of the population random error, the resulting model still showed no significant results for any of the three main factors (Table 1). Neither day, sound treatment, nor time of day had a significant effect on the mean number of smolts in the lock chamber. The total sample size was reduced from the completed 106 hydroacoustic surveys to 102 due to removal of 4 unpaired surveys (unmatched sound-on survey with sound-off or vice versa).

The confidence intervals for at all three confidence levels calculated (95%, 90%, and 80%) bracketed zero and were narrow: (1) 95% CI–

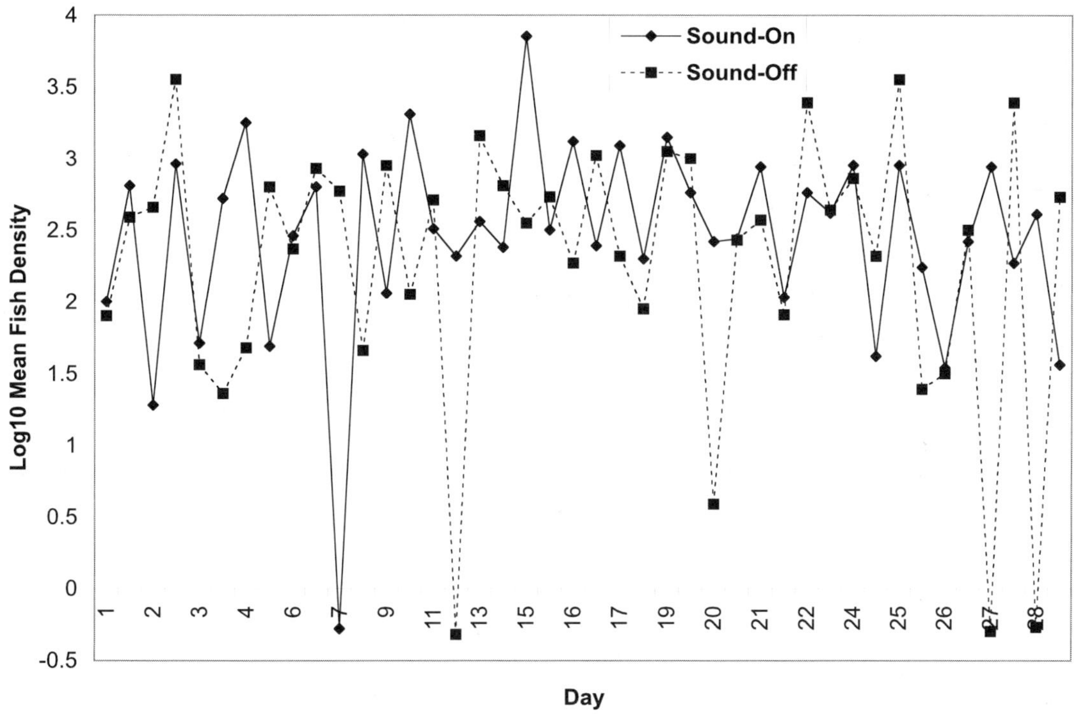

Figure 3. Comparison of log10 average smolt densities (fish/m^3) observed daily at the entrance to the large lock.

$0.0025 < \mu1-\mu2 < 0.0025$; (2) 90% CI–$0.0016 < \mu1-\mu2 < 0.0017$; and (3) 80% CI–$0.0003 < \mu1-\mu2 < 0.0038$. As stated previously, a 60% difference between sound treatments was considered meaningful for this study. In log10 + 1 transformed values, this criterion is represented by a difference between means of 0.0113. The calculated difference in means between sound on and sound off treatments in log10 + 1 transformed values was 0.0035. There is 95% confidence that the mean difference between sound treatments is contained in the interval 0.0035 ± 0.006. Similarly, the interval 0.0035 ± 0.0052 estimates the mean difference between sound treatments with 90% confidence and the interval 0.0035 ± 0.0038 estimates the difference between means at 80% confidence. Since the confidence intervals include zero and are narrow enough to not include a biologically meaningful difference, the data suggest that the sound treatment did not have a biologically meaningful effect on the mean density of smolts in the lock chamber. This interpretation is consistent with the results of the ANOVA.

Previous testing using low frequency transducers have shown a wide range of guidance efficiencies and have achieved at best a 57% reduction in smolt densities (Hanson et al. 1997). The estimated power of the ANOVA in this study to detect a 60% difference between means within the sound treatment main factor was calculated as greater than 0.98 using Pearson-Hartley tables. Thus, the analysis did meet the design criteria requiring detection of a 60% difference with a power of 80%.

The entire data set was used to evaluate whether the mean density of smolts was significantly different during periods with spill or between periods with varying numbers of up-lockages. In both cases, the one-way ANOVA failed to detect a significant difference in the mean density of smolts between levels of either of these factors (Table 1). The potential for nighttime spill volume to impact the density of smolts during the first survey of the day was also evaluated. Mean smolt densities measured during survey #1 were compared between surveys preceded by spill versus surveys preceded by no spill. Spill volume data preceding survey number one was available for 26 d of the study. No significant difference was detected in the mean smolt density between morning surveys preceded by spill or no spill (Table 1). Because 4-hr treatment periods usually included both routine lockages with vessels and false lockages, statistical

Table 1. Results of the ANOVA for the main factors of day, treatment (sound-on sound-off) and time of day (morning and afternoon) and secondary factors of spill volume, night spill, and number of up-lockages (N=102, α=0.05).

Source	Df	SS	MS	F	P
Day	27	0.0244	0.0009	0.925	0.575
Treatment	1	0.0001	0.0001	0.107	0.743
Time of Day	1	0.0006	0.0006	0.708	0.403
Error	72	0.0703	0.0009		
Total	101				
Spill Volume	2	0.006	0.003	0.398	0.673
Error	99	0.792	0.008		
Total	100				
Night Spill	1	0.006	0.006	1.698	0.204
Error	24	0.087	0.004		
Total	25				
Up-Lockages	4	0.043	0.011	1.366	0.251
Error	97	0.756	0.008		
Total	100				

analyses were not performed to evaluate changes in guidance efficiency due to differences in background noise within the lock.

Changes in smolt density with changes in precipitation, spill volume, nighttime spill volume and the salinity at the lock chamber pier nose at a depth of 13.7 m were all examined using a Pearson correlation analysis. The only significant correlation (r = 0.95) occurred between nighttime spill and spill volume. Spill at night increases during 24-h periods when spill occurred the following day. None of the other physical factors were associated with each other nor with changes in smolt density (Table 2).

Discussion

Despite not meeting the experimental design fully, the results clearly indicate that sound treatments were not effective. The sound treatment did not impact the density of salmonids measured in the lock chamber. ANOVA showed no significant difference in the density of salmonids in the lock chamber between sound treatments, between days of the study, or between treatments and different times of the day. Our ability to detect a minimum difference (10%) between log10 transformed means was good at an estimated power of 80% and excellent at the specified design guidance efficiency (60%). Our 4-hr treatment periods provided ample sample size and greater than 95% statistical power to detect a previously recorded level of guidance efficiency for the tested frequency.

There are several other potential factors explaining the failure of the sound-system. First, we had less control over fish densities than expected. Density estimates of salmonids and of all targets were highly variable. The expected bell-shaped run timing curve typical of salmonid smolt out-migrations was not observed. The random passage patterns may have resulted from the low numbers of sockeye smolts during 1997 (200,000 versus normal

Table 2. Association between selected physical factors and mean smolt density in the large lock chamber measured during the 28 days of this study using the Pearson correlation coefficient (N = 28; a = 0.05).

Factors	Mean smolt density	Precipitation	Spill volume	Night spill volume
Mean smolt density	–	–	–	–
Precipitation	0.0648	–	–	–
Spill volume	0.0394	–0.1885	–	–
Night spill volume	0.0974	–0.1968	0.9561	–
Salinity	–0.1343	0.0561	0.0343	0.0864

1–4 million), the overlapping arrival of three distinct species with differences in abundance, or the semi- or uncontrollable variables at the site, such as spill, boat traffic and weather.

Our characterization of acoustic survey patterns as salmon smolts may have been inaccurate. We observed that fish aggregates that we interpreted as salmonids accumulated in specific locations in the lock entrance, and that these aggregates reformed inside the lock minutes after filling of the chamber. We also observed nonsalmonid species in the study areas. We believe that these potential sources of error should not have influenced our density estimates. Smolts predominated in the purse-seine catch in the lock chamber (75% of total catch) and the acoustic patterns of bottom-oriented and midwater fish schools were very different from the aggregate patterns and should not have influenced density estimates. We also have visual observations of fish in the water column and of bird catches to confirm periods when salmonids and nonsalmonids are present. Lastly, the overflow weir counts provide a secondary, but more accurate picture of the presence and relative abundance of smolts passing the Locks. The peak in mean smolt counts at the lock entrance and lock chamber and peak daily smolt flume counts show a clear overlap (Figure 4).

Flow may have reduced the ability of smolts to respond to the sound. Ploskey et al. (2000) considered the possibility that a sound-system might fail if smolts are entrained by flows and cannot avoid the sound-system. The Locks provides a highly controlled flow environment that should negate this possibility. The velocity at the face of the sound-system (<0.25 m/second) is well within the swimming ability of the smallest smolts presented to the sound-system (90 mm chinook salmon) and should not have entrained smolts through the sound-system (Bell 1991). Our ability to largely standardize the timing and number of flow events (up-lockages) provided a strong control on this potential influence on smolt behavior. For all survey periods there was an average of 15 min of flow across the sound-system (3 up-lockages) and for any one period—never any less than

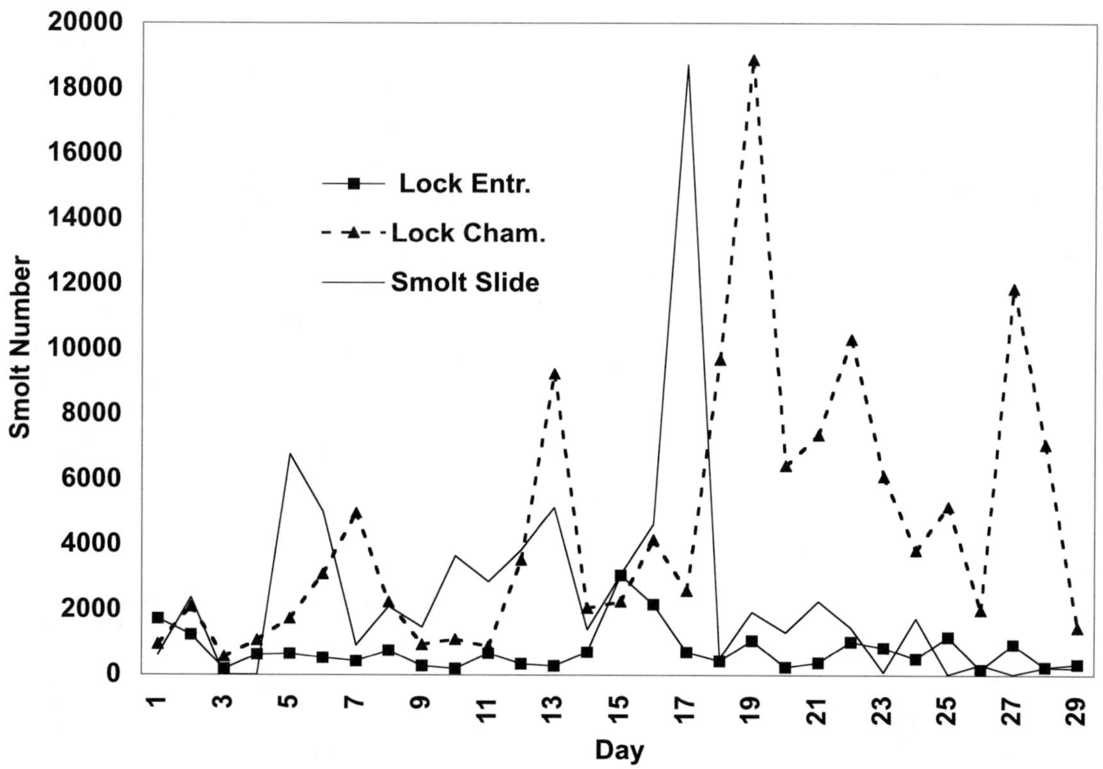

Figure 4. Comparison of smolt number observed daily at the lock entrance (average), lock chamber (average), and smolt flume (total).

5 min (1 up-lockage) or more than 25 min (5 up-lockages) of flow. The timing of a flow event prior to a survey was always within 25–85 min of the survey. The ANOVA of up-lockages showed that this factor had no significant influence on the mean density of smolts per survey period.

At the Locks we had the unusual concern of double counting smolts (downstream of the sound-system) in consecutive survey periods. The large lock creates a discontinuous obstructed flow path for migratory salmonids to follow. If a fill event did not occur between survey periods there is the possibility that smolts counted in the first survey period would be present in the second period, especially if they were avoiding the sound field. Our protocol of having 2 fill events per period was intended to provide a multiple number of flow events between surveys to keep smolts moving through the lock entrance. The lock chamber is 5 times longer and 3.7-m deeper than the lock entrance and each fill event uses several times the volume of water found in the area of the lock entrance. We believe that our protocol performance, 105 of 106 survey periods with two or more lockages, addresses this concern of any bias from double counting smolts in our density estimates.

In the evaluation of the performance of low-frequency sound at Georgiana Slough, investigators found differences in guidance efficiency by time of day and during changes in tidal flow (ebb or flood tides) (Hanson et al. 1997). We controlled for the influence of diel behavior differences by conducting all surveys during daylight hours. Previous monitoring has shown that smolts pass the smolt flume only during daylight hours and that smolt presence in front of the large lock filling culverts is diurnal (Goetz et al. 2000; Dillon and Goetz 1999). The ANOVA showed no differences in mean densities by time of day or by study day. There is no tidal influence at the Locks, however we do have a brief period with reverse flows (flood) when boats lockup and the upper miter gates are opened. As there was no clear pattern to when boat lockings occur, the randomized block design of our study should have eliminated this as an influence on smolt densities—with no trend of greater boat lockings during sound-on or sound-off. Even if there was a correlation of boat lockages with treatment or control periods, we have no evidence to date that many smolts move back upstream into the Ship Canal; in fact, the reverse should be true. When a boat up-lockage occurs, there is a deep current of denser salt water that intrudes (floods) into the lock entrance while a localized countercurrent of less dense freshwater flows into the lock chamber at velocities of 1–1.5 m/s. If smolts do migrate when the miter gates are open and since smolts in the lock entrance were predominately found at shallower depths, in low salinity water, it seems likely that these fish would follow the freshwater downstream flow.

Another study concurrent with our own study has shown that juvenile salmonids do not respond to the type of low-frequency sounds used in our study. In April and June of 1997, WES conducted a series of net-pen tests in Salmon Bay, on the far side of the large lock north wall (Figure 1); (Ploskey and Johnson 2001, this volume). The responses of juvenile coho, chinook and sockeye salmon to infrasound, low frequency sound, and strobe lights were observed by underwater video camera during a series a series of horizontal and vertical test and control periods. This study used the same type of transducer to produce the same blended sound (300/400 Hz) used in the large lock evaluation. The 300/400 Hz sounds failed to elicit a startle or directed avoidance by both hatchery sub-yearling coho and chinook or wild yearling sockeye salmon (Ploskey and Johnson 2001, this volume). Other researchers have also tested the response of juvenile salmonids to low-frequency sound in net pens. At Georgiana Slough, net-pen experiments were conducted using the acoustic transducers at Georgiana Slough. The researchers failed to observe a consistent startle response among fall-run Chinook salmon exposed to the sound frequencies tested (Hanson et al. 1997).

In discussing biological engineering in development of sound-deterrence systems, Carlson (1995) said "the essence of the biological engineering approach is 1) exposure of test fish to sound at various discrete frequencies or bands of frequencies, intensities, and presentation patterns, and 2) observation of fish for spontaneous avoidance responses, typically called 'startle' responses by investigators. The search is for something that 'works,' and the criterion is a response that can be used to deter, exclude or in some other way modify fish behavior." The work performed by previous investigators on the response of juvenile salmonids to low-frequency sound greater than 280 Hz has shown that even if salmonids may hear a sound they are not reacting to the sound either with a startle response or directed movement.

In a California study, VanDerwalker (1967) conducted laboratory work on the response of juvenile chinook salmon to frequencies of 10–500 Hz. He found that there was an observable response at frequencies from 10 to 280 Hz, with no response at higher frequencies. In particular, there appeared to be a consistent response from 35 to 170 Hz with a large drop on either end of this range. All observed responses were within 0.7 m of the sound source. Knudsen et al. (1992) conducted laboratory and field tests to measure the avoidance response of Atlantic salmon parr exposed to frequencies of 10 and 150 Hz. There was no measured response at 150 Hz, even when the fish was within 10 cm of the transducer at intensities 114 dB above the hearing threshold of Atlantic salmon (Knudsen et al. 1997; Hawkins and Johnstone 1978). In continuing his initial work in the laboratory, Knudsen et al. (1994) did demonstrate that Atlantic salmon *Salmo salar* could be diverted in field locales at short range using 10 Hz infrasound. EPRI conducted laboratory tests on the response of Atlantic salmon, chinook salmon, and rainbow trout juveniles to low-frequency sound (20–800 Hz) generated by a hammer device. There was no response noted at any of the tested frequencies (Taft and Brown 1997).

Evaluations of the commercially manufactured, low-frequency sound transducers we used have shown mixed results. At Georgiana Slough, Hanson et al. (1997) reported a wide range of guidance efficiencies within and among years of testing. During the last two years of testing (1994 and 1996), individual tests ranged from a 49–74% change in juvenile salmon densities while mean guidance efficiency ranged from 57% in 1994 to 15% in 1996. The senior researcher of this study concluded that testing of the acoustic barrier has shown that it "does not provide a reliable method for achieving juvenile salmonid guidance over a range of site-specific and environmental conditions" (C. Hanson, Hanson Environmental Inc., personal communication).

The 1997 work at the Hiram M. Chittenden Locks provides the most powerful evaluation of low-frequency sound sources to date. The WES net-pen tests and the large lock avoidance test is a nicely paired, small scale and larger scale test of this sound source under similar conditions with the same species of Pacific salmonids. The parallel nature of these two studies meets the classic definition of bioengineering described by Carlson (1995). Based on the results of our study and that of WES (Ploskey et al. 2000; Ploskey and Johnson 2001, this volume), we would recommend against future testing of low-frequency sound for guiding juvenile Pacific salmon *Oncorhynchus* spp.

Acknowledgments

The authors thank and acknowledge the contributions of Bob Dach and John Ferguson, Portland District USACE (now with National Marine Fisheries Service); Lyle Calvin; Gene Ploskey and Peter Johnson, Waterways Experiment Station; and Dave Seiler, Washington Department of Fish and Wildlife. This study was funded by the U.S. Army Corps of Engineers Portland District under the Behavioral Technologies Program. Additional funding was provided by the USACE Seattle District and King County Department of Land and Water Resources under Section 22, Planning Assistance to the States authority.

References

Bell, M. 1991. Fisheries handbook of engineering requirements and biological criteria, 3rd edition. U.S. Army Corps of Engineers, North Pacific Division, Fish Passage Development and Evaluation Program.

Biosonics, Inc. 1997. Evaluation of salmonid smolt passage into the Hiram M. Chittenden Locks, Spring, 1997. Report to U.S. Army Corps of Engineers, Seattle District.

Carlson, T. J. 1995. Use of sound for fish protection at power facilities: a historical perspective of the state of the art, Phase I. Phase I Final Report on the Evaluation of the Use of Sound to Modify the Behavior of Fish. Report DOE/BP-62611–4, U.S. Department of Energy, Bonneville Power Administration, Portland, Oregon.

Dillon, J., and F. Goetz. 1999. Manual analysis of fish presence and abundance at a large lock using fixed hydroacoustics techniques. Seattle District, U.S. Army Corps of Engineers, Report to the Lake Washington Ecological Studies Group, Seattle.

Goetz, F. A., J. J. Dawson, T. Shaw, and J. Dillon. 2000. Evaluation of low-frequency underwater sound transducers for guiding salmon, and steelhead smolts away from a navigation lock at the Hiram M. Chittenden Locks, Seattle, Washington. Seattle District, U.S. Army Corps of Engineers, Report to the Portland District, U.S. Army Corps of Engineers, and the Lake Washington Ecological Studies Group.

Hanson, C. H., D. Hayes, and K. A. F. Urquhart. 1997. Biological evaluations of the Georgiana Slough experimental acoustical fish barrier. Phases I–IV during 1993–1996. Fish Passage Workshop in Milwaukee, Wisconsin, May 6–8, 1997. Co-sponsored by Alden Research Laboratory Inc., Conte

Anadromous Fish Research Center, Electric Power Research Institute, and Wisconsin Electric.

Hawkins, A. D., and A. D. F. Johnstone. 1978. The hearing of the Atlantic salmon, *Salmo salar*. Journal of Fish Biology 13:655–674.

Keppel, G., 1991. Design and analysis, 3rd edition. Prentice Hall, Englewood Cliffs, New Jersey.

Knudsen, F. R., P. S. Enger, and O. Sand. 1992. Awareness reactions and avoidance responses to sound in juvenile Atlantic salmon, *Salmo salar*. Journal of Fish Biology 40:523–534.

Knudsen, F. R., P. S. Enger, and O. Sand. 1994. Avoidance responses of low frequency sound in downstream migrating Atlantic salmon smolt, *Salmo salar*. Journal of Fish Biology 45:227–233.

Knudsen, F. R., C. B. Schreck, S. M. Knapp, P. S. Enger, and O. Sand. 1997. Infrasound produces flight and avoidance responses in Pacific juvenile salmonids. Journal of Fisheries Biology 51:824-829.

Loeffelman, P. H., D. A. Klinect, and J. H. VanHassel. 1991a. Fish protection at water intakes using a new signal development process and sound system. Pages 355–365 *in* WaterPower '91. Proceedings of the International Conference on Hydropower, Volume I. American Society of Civil Engineers, Denver.

Loeffelman, P. H., J. H. Van Hassel, and D. A. Klinect. 1991b. Using sound to divert fish from turbine intakes. Hydro Review 10(6):30–43.

Ploskey, G. R., P. N. Johnson, and T. J. Carlson. 2000. Evaluation of a low-frequency, sound-pressure-system for guiding juvenile salmon away from turbines at Bonneville Dam, Columbia River. North American Journal of Fisheries Management 20:951–967.

Ploskey, G. R., and P. N. Johnson. 2001. Effectiveness of strobe lights, and an infrasound device for eliciting avoidance by juvenile salmon. Pages 37–55 *in* C. C. Coutant, editor. Behavioral Technologies for Fish Guidance. American Fisheries Society, Symposium 26, Bethesda, Maryland.

San Luis and Delta-Mendota Water Authority, and C. H. Hanson. 1996. Georgiana Slough acoustic barrier applied research project: results of 1994 phase II field tests. San Luis and Delta-Mendota Water Authority, Department of Water Resources Technical Report 44, Sacramento, California.

Taft, N., and N. Brown. 1997. Sonic fish deterrence: EPRI/Alden laboratory's experience. Pages 247–264, *in* T. J. Carlson and A. N. Popper, editors. Using sound to modify fish behavior at power-production and water-control facilities: A workshop assessment of technologies to improve measurement capabilities in the Columbia River Fish and Wildlife Program. Task: Evaluation of the use of sound to modify the behavior of fish Phase II, Report DOE/BP-62611–111, U.S. Department of Energy, Bonneville Power Administration, Portland, Oregon.

Underwood, A. J. 1981. Techniques of analysis of variance in experimental marine biology and ecology. Oceanography and Marine Biology: An Annual Review 19:513–605.

U.S. Army Corps of Engineers. 1999. Lake Washington Ship Canal Section 1135 Environmental Restoration Feasibility Study. Seattle District, Seattle.

VanDerwalker, J. G. 1967. Response of salmonids to low frequency sound. Pages 45–58 *in* W. N. Tavolga, editor. Marine bioacoustics, Volume 2. Pergamon, New York.

Washington Department of Fish and Wildlife. 1996. Ballard Locks 1996 smolt passage assessment: Summary Statement. Report prepared by Fish Management Section, Olympia, Washington.

American Fisheries Society Symposium 26:105–113, 2001

Integrated, Multi-Sensory, Behavioral Guidance Systems for Fish Diversions

Charles C. Coutant[1]
Environmental Sciences Division, Oak Ridge National Laboratory
Post Office Box 2008, Oak Ridge, Tennessee 37831-6036, USA

Abstract.—Behavioral guidance includes, at least in principle, both repulsion and attraction. Although they are usually used separately, this paper encourages use of both to develop integrated guidance systems using a multi-sensory approach to fish diversion. Research in the 1940s and 1950s with adult salmon and prototype fish ladders emphasized attraction of fish to turbulent currents. As a result, attraction flows to entice fish to fish ladder entrances are now a usual feature of adult dam bypasses. In contrast, techniques for fish repulsion have dominated research on guidance of fish away from risks of water intakes in the hydropower and steam electric utility industries and agriculture (irrigation). The objective has been to find devices that would guide fish away from the hazards of entrainment and impingement. Thus, strobe lights, sound, bubble curtains, electric fields, and the like have been evaluated as ways to repel fish from intakes. None of these technologies has become an industry standard and all are usually considered experimental. I encourage consideration of both repulsion and attraction together to both guide fish away from a hazard and attract them toward an alternative safe location. In addition, more than one technology for either (or both) may be usefully combined. Fish bypasses at dams, for example, could use strobe lights or ultrasound to divert (repel) fish from a trajectory aimed at turbine entrances and concurrently use attraction flows to entice them toward surface bypass entrances. Behavioral and structural technologies can be integrated as hybrid systems. Innovative combinations tailored to a site and specific behaviors of the fish to be guided are likely to show better overall results than one technique alone.

There has been a tendency in fish guidance research and development to seek the proverbial "silver bullet." Whether it has been research teams striving to perfect a particular technology or commercial firms intent on selling their particular product, the focus often has been on one technology at a time. There has been outright competition (e.g., strobe lights versus sound as deterrents at water intakes). The results of research and demonstration of single technologies have been spotty (EPRI 1998a, 1998b; Taft 2000). Although hybrid systems have been tried, in which two or more devices are used to achieve a greater level of success than with one device alone, these, too, have had spotty success (EPRI 1998b; Michaud and Taft 2000). The sad effect has been that few behavioral guidance technologies have gained sufficient acceptance to be universally recommended by government regulatory agencies.

It is time, perhaps, to take off the single-technology blinders, to examine the multi-sensory real world we are trying to influence, and to redouble efforts at examining hybrid or multi-sensory guidance systems. The purpose of this paper is to take a brief conceptual look at the diverse problems we are attempting to solve and to suggest that integrated use of a diversity of sensory technologies, applied together to match the environmental and species diversity of a particular locale, should offer promising opportunities for behavioral fish guidance

Attraction and Repulsion

There are two main categories of behavioral guidance techniques when viewed in terms of their effect—those that attract and those that repel. Both guide in the sense of influencing animal behavior, but attraction seeks to turn the animal toward the stimulus whereas repulsion seeks to turn it away. Fish have a variety of sensory capabilities and a rich repertoire of behavioral responses to sensory stimuli. Some, like the capabilities of the lateral line organ for sensing localized pressure differentials and of the air bladder for sensing hydrostatic pressure, are foreign to our usual mammalian thinking. The anatomical, physiological, and

[1]The author may be reached at 865-576-6830 or coutantcc@ornl.gov.

behavioral basis for development of fish guidance systems is described in academic texts (e.g., Pitcher 1986; Evans 1993; Vogel 1994) and reviews (e.g., Popper and Carlson 1998; papers in this volume). Both attraction and repulsion techniques can mimic natural stimuli and responses. Fish are adapted to finding certain temperatures, light levels, current velocities, and the like that are suitable for survival, feeding, growth, or migration and tend to be attracted to them. Likewise, they recognize that certain stimuli signal predators or other risks to be avoided, and thus they are repelled. Some stimuli are unnatural (such as loud sounds or strobe lights); many such "foreign" stimuli will trigger an abrupt avoidance response. The diversity of sensation and behavior in fishes belies the notion that in guidance systems "one fits all." If we recognize the diversity of adaptations in fish, we should feel compelled to make use of those adaptations in multiple ways to design multi-sensory approaches to integrated passage systems (Table 1).

Perhaps the most-used technology for attraction is the flow regime at the entrance of fish ladders on mainstem dams such as in the Columbia River basin (Figure 1 of Coutant 2001, this volume). There, a complex set of water diversions and outlets extending from the tailwater of the dam to the lower reaches of the fish ladder itself creates flows that simulate the turbulent flows of a natural river and entice upstream-migrating adult salmonids first toward the fish ladder and away from the turbines and spillways, and then into the ladder entrance. This system for guidance by attraction was developed through hydraulic and biological research at the bioengineering laboratory at Bonneville Dam primarily in the 1940s and 1950s (Mighetto and Ebel 1995). Clearly, that particular behavioral guidance technology was well demonstrated, was accepted, and is standard practice for adult salmon passage. It is used for other upstream migrants, as well, including American shad.

In contrast, technologies that rely on fish repulsion have proliferated in the past couple of decades. Their development has been stimulated by fish kills at several types of water intakes, including those for hydropower, cooling water for steam electricity generation, and agricultural diversions for irrigation. The goal is to keep animals, usually fish, away from the risks of being drawn into the intakes (entrained), damaged by capture on intake screens (impingement), or stresses of passing through turbines, steam condensers, or irrigation systems. Example technologies include strobe lights, sound, electric currents and bubble curtains (EPRI 1998a, 1998b; Popper and Carlson 1998; papers in this volume). Alternative passage routes for the repulsed fish have not necessarily been part of the test installations. Guidance results in field tests have been mixed (EPRI 1998a, 1998b; Taft 2000). The result has been that government regulatory agencies look upon these technologies with skepticism.

There also are technologies that rely on both structural and behavioral features (most structural technologies probably could be considered partial-

Table 1. Behavioral guidance technologies potentially useful in multi-sensory systems, grouped by sensory mechanism, with an example literature citation suitable for further literature searching. Also, see papers in this volume, EPRI 1998a,b, and Taft 2000. Species tested differed.

Mechanism	Technology	Reference
Acceleration	Infrasound (repulsion)	Sand 2001, this volume
Chemoreception	Homing (attraction)	Quinn and Dittman 1990
	Fright reaction (repulsion)	Chivers and Smith 1998
Electricity	Electrified grids (repulsion)	Savino et al. 2001, this volume
Hearing	Sound (repulsion)	Popper and Carlson 1998
Hydrostatic pressure	Curtain walls (attraction)	Skorobogatov et al. 1993
Localized pressure	Induced turbulent flow	Coutant 2001, this volume
Temperature	Thermal preference (att./rep.)	Coutant 1977
Vision	Incandescent lights (attraction/rep.)	Jahn and Herbinson 2000
	Mercury lights (attraction/repulsion)	Hadderingh and Smythe 1997
	Strobe lights (repulsion)	Brown 2000
	Fluorescent lights (repulsion)	Hadderingh and Smythe 1997
	Sodium lights (attraction)	Ploskey et al. 1995
	Bubble curtains (repulsion)	McCauley et al. 1996

ly behavioral). An example is the submerged turbine-intake screen commonly used on the mainstem Columbia and Snake river dams (Figure 1). A physical structure (screen) is placed in the trajectory of juvenile salmon movement into the turbines. The surface-oriented, downstream-migrating fish are entrained with the mass water flow into the deep intake, with resulting abnormal increase in hydrostatic pressure. The fish attempt to resist sinking (pressure change) by swimming upwards, as shown by detailed records of fish movement in dam forebays (Johnson et al. 1998, 2000). Once drawn by flow into the intake, fish tend to stay close to the ceiling in attempts to return to the surface. Their first opportunity to swim upward is the dam's gatewell (a vertical slot designed for inserting water-blocking gates into the turbine intake). The screen directs those fish passing deeper in the intake toward the gatewell. Repulsion from the high hydrostatic pressure causes fish to rise in the gatewell and into fish bypass channels close to the surface at near atmospheric pressure. Skorobogatov et al. 1993 (summarized in Coutant and Whitney 2000) have used this response of fish to hydrostatic pressure changes induced by structural components of a hydropower intake to design a new passive fish separator.

Another example of structural and behavioral integration is the "behavioral guidance structure" at Lower Granite Dam on the Snake River, that consists of a physical wall inserted into the migration corridor, designed to guide fish along it to a bypass (Adams et al. 2001, this volume). The sequence of behavioral events in response to the structures of hydropower intakes is discussed further by Coutant and Whitney (2000) and in a series of papers and a book by Russian investigators (e.g., Skorobogatov et al. 1993; Pavlov et al. 1999).

Diversity of the Environment and Target Resources

Why would we expect there to be a universal "silver bullet" behavioral technology to solve our fish-guidance problems, considering the complexity and diversity of both the physical and biological environment? The physical settings of water intakes are diverse and include main channel structures, shoreline structures, and structures at the ends of headraces and canals of varying lengths and dimensions. Intakes are shallow (top 10 m), deep (35–50 m), and everything in between. Hydraulic patterns in forebays range from smooth, straight streamlines in otherwise quiescent waters to high

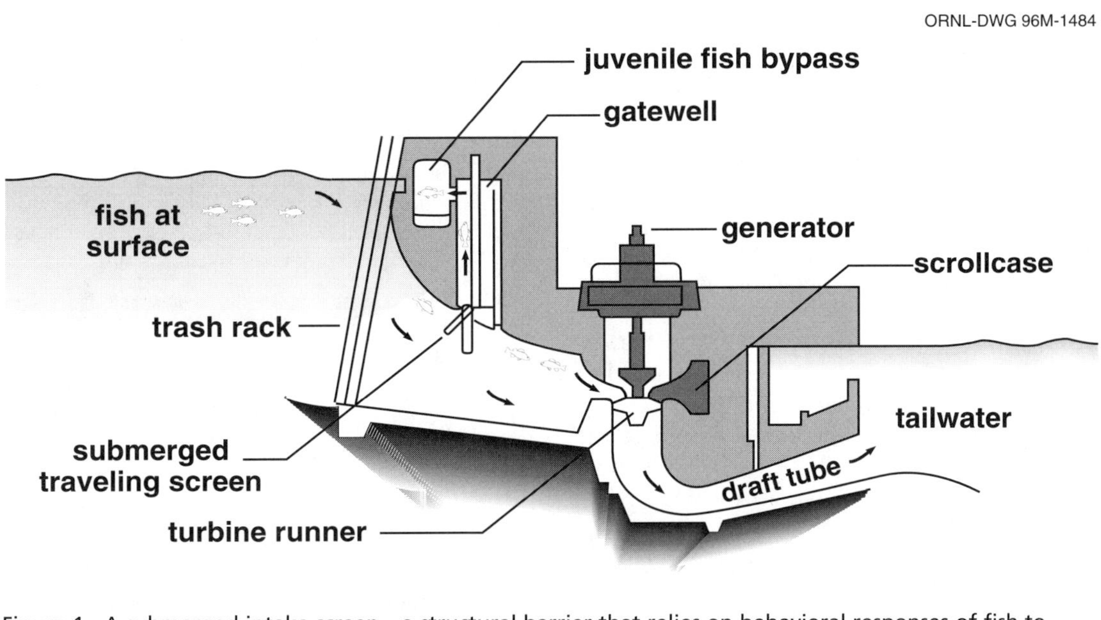

Figure 1. A submerged intake screen—a structural barrier that relies on behavioral responses of fish to hydrostatic pressure for its effectiveness—shown in a cross section of a typical Columbia River basin mainstem dam. Fish at the surface in the forebay resist being drawn deeply into the turbine intake, and follow the intake ceiling where many are diverted to gatewells. In the gatewells, fish move toward the surface, and into bypass channels. (From Coutant and Whitney 2000).

turbulence with strong vortices and cross flows. Flow volumes and water velocities vary greatly. Waters are turbid, clear, oxygenated, polluted, hot and cold. On the biological side, we are trying to protect a wide range of species and stocks with different ecological niches and behavioral characteristics—downstream-migrating salmonids, alosids (shads), and eels; resident warm-water fish; schooling reservoir species; bottom and surface dwellers; and all sizes of each. Common sense says that what works well for one species and environmental situation is unlikely to work well for another.

The divergent research results represented in the fish-guidance literature (e.g., EPRI 1998a, 1998b) probably reflect the suitability of the applications actually tested more than the inherent abilities of the behavioral technologies to produce favorable results. The test species or environmental setting may not have been appropriate to best demonstrate the technology. Yet the technology may be condemned for all uses based on negative results obtained in an inappropriate or otherwise poor test situation.

The importance for overall success is not only what works technologically, but what is needed ecologically. Technological fixes that target a narrow part of the spectrum of biological diversity actually function to reduce that diversity to the detriment of the continuation of the target species (Williams et al. 1996, 2000). For example, the success of salmon species is more and more believed to be due to the diversity of genetic traits among various stocks. Environmental conditions (such as wet and dry years) that favor one stock at one time will, at a later time, change and favor another stock. As the environment fluctuates, so too will the fortunes of the various stocks. A diverse scene of ups and downs among stocks keeps the species going as a whole. If diversity is reduced to one stock alone (such as by using only technologies that work best for that stock), environmental fluctuation has a greater likelihood of wiping out the whole species. It follows logically that we should seek not just one "silver bullet" technology but a diversity of technologies that simultaneously accommodate the diverse needs of the species and stocks to be protected.

Integration of Behavioral Devices

The old saying that "two heads are better than one" has its parallel in the fish guidance business. The most straightforward example is the combination of repulsion and attraction. If we are to guide animals efficiently, we need to consider both where we don't want them and where we would prefer that they go. We may repel them from the vicinity of an intake, but it makes sense to also attract them to where we do want them, such as to a bypass. Thus, technologies for avoidance and attraction can be used together. Strobe lights, for example, can guide (repel) fish away from turbine intakes but may work better if attraction flows to bypasses are arrayed in the unlighted zones where fish may congregate.

There are published papers on integrating behavioral systems based on more than one sensory capability, but they tend to be specific to multiple repulsion systems. Patrick et al. (1985) and McCauley et al. (1996) investigated a combination of strobe lights and air bubble curtains as a repellant, for example. Multiple repellants may be necessary at a site, for example when fish have very divergent repulsion behavior in darkness and light (as affected by day/night cycles or differences in turbidity). Some hybrid systems are summarized in EPRI (1998b).

Great Northern Paper Company's Weldon Dam on the Penobscot River, Maine, is an example of where a combination of repulsion and attraction would be useful (one of many sites that could be used as examples). Studies have been conducted there since 1993 to divert juvenile Atlantic salmon from turbine intakes using strobe lights (Brown 2000). A strobe-light system (Flash Technology AGL 4100) was positioned to produce a "wall of light" barrier completely covering turbine bays 1 and 2 (Figure 2). Additional lights in turbine bays 3 and 4 duplicated the wall of light in just the lower half of those intakes. The top halves of bays 3 and 4 were left dark and were provided with surface water outlets for downstream fish passage. Fish behavior was observed with radiotelemetry. Control data on fish behavior without strobe lights were collected in 1993 and 1994. Strobe lights were used experimentally in 1995 and 1997–1999 (heavy flooding in spring 1996 prevented testing). Fish were successfully guided away from the fully lighted zones of intake bays 1 and 2 and the lower portions of bays 3 and 4 (Bernier 1999). However, smolts did not readily enter the surface bypass inlets. Underwater video in 1997 confirmed that fish milled in the surface waters of intake bays 3 and 4. Attempts to improve entry into bypasses included reduction of turbulence near the inlets by rounding the trash-rack bars. This actually reduced

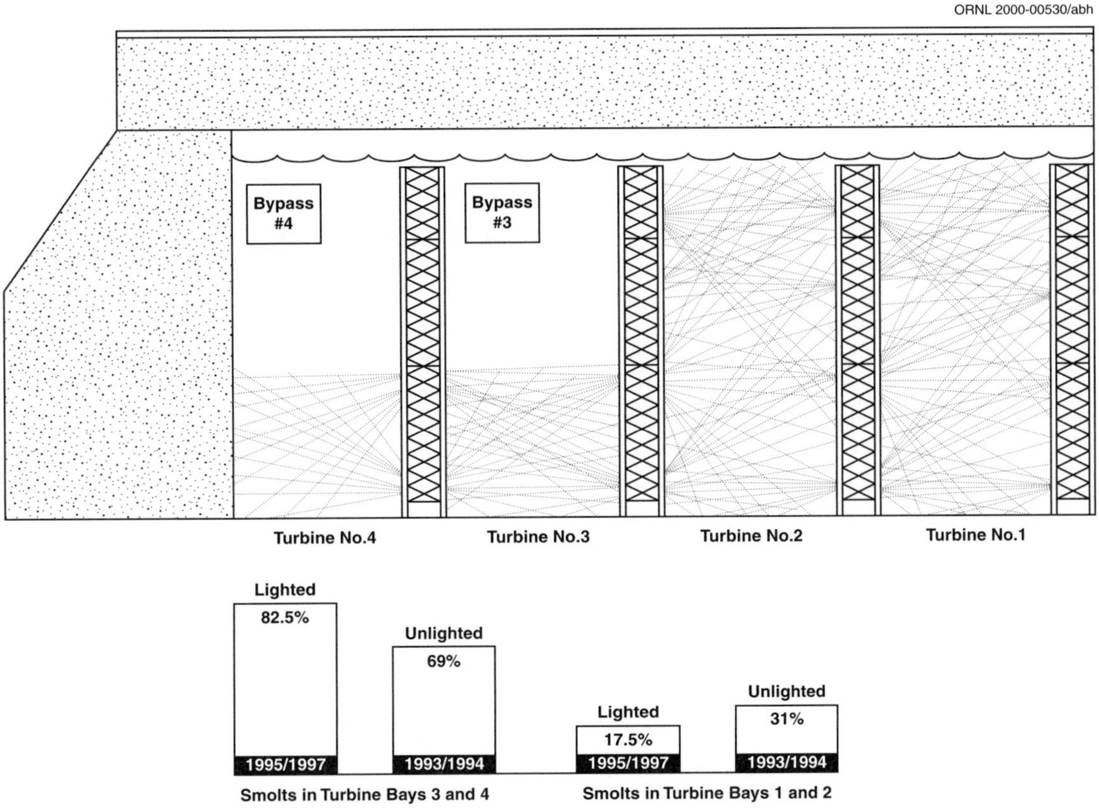

Figure 2. Weldon Dam, Penobscot River, Maine, as seen looking downstream at the turbine entrances from the forebay (upper panel). Hatched lines indicate turbine bay zones illuminated experimentally with strobe lights. Surface bypass entrances are located in turbine bays 3 and 4. Smolt passage in the fully illuminated bays 1 and 2 decreased between control (1993, 1994) and illuminated years (1995, 1997), with more fish appearing in bays 3 and 4 (lower panel). However, congregation in the unlighted top parts of turbine bays 3 and 4 did not ensure passage through bypasses. (Modified from Brown 2000).

use of the bypasses (Bernier 1999; R. Brown, Flash Technology, personal communication). Fish that found and entered the bypasses, however, passed through quickly.

The 6-year record of control and test observations at Weldon Dam suggests that an attractant is needed at the bypass entrances to go along with the demonstrated effectiveness of strobe lights in repelling fish from the turbine entrances. Furthermore, attempts to reduce turbulence at the bypass entrances has yielded progressively poorer bypass use, suggesting that the turbulent waters may have initially aided fish's discovery of the inlets. It might be fruitful to test increased turbulent flow as an attractant (see Coutant 1998; 2001, this volume). While turbulence may provide a mechanism for discovery of the bypass (in the sense of Rainey 1997), the hydraulics of the bypass opening and the stimuli provided fish there need to be considered, for Haro et al. (1998) found that a uniform acceleration was needed for acceptance and passage over a weir.

Behavioral diversity among fish species and individual fish can be accommodated by using several guidance devices. On the Columbia and Snake rivers, for example, intake screens (discussed above) alone have difficulty meeting the criterion of passing 80% of the fish by routes other than through turbines (Whitney et al. 1997). Yet experiments operating several passage routes simultaneously (screens, spill of water over spillways, and surface bypasses) show that the criterion can be more than met by the combination (Johnson et al. 1999, 2000). Johnson et al. (2000) describe the "hybrid" system at Lower Granite Dam and indicate that 94% of the fish passed through the dam using non-turbine routes. About half of those were spilled, screens diverted 34%, and 12% passed through the experimental surface bypass installed over turbines 4–6. Considering just the fish near turbines 4–6 that were equipped with the surface bypass, 37% went through the surface bypass, 55% were diverted by the intake screens, and 8% passed through the turbines. The full passage budgets are given in the paper (Johnson et al. 2000).

Another example illustrates the use of multiple technologies where diversion of fish away from a hydropower intake canal is the objective (Figure 3). Juvenile salmonids are initially migrating through the full cross section of a river that is dammed to divert water to the power canal. Induced flow could be used, as demonstrated by Darland et al. (2001), to shift migrants from the upper shoreline in the figure toward the lower one. This shifts the downstream trajectory of most fish to the side of the river containing the dam's spillway (where they can pass in managed spill) and away from the power canal. Strobe lights (or perhaps ultrasound) at the entrance of the power canal could repel remaining fish from that shoreline and discourage a trajectory that would take them into the canal. A second zone of induced flow near the canal entrance would create a gentle trail of turbulent flow that could guide migrants to a trash sluiceway in the spillway near the canal.

There are other published examples that lead to speculation about the potential value of a multisensory approach. Hanson (1999) provided several examples using louvers, diversion panel, fish tube, sluice, and a fish diversion boom. Louvers

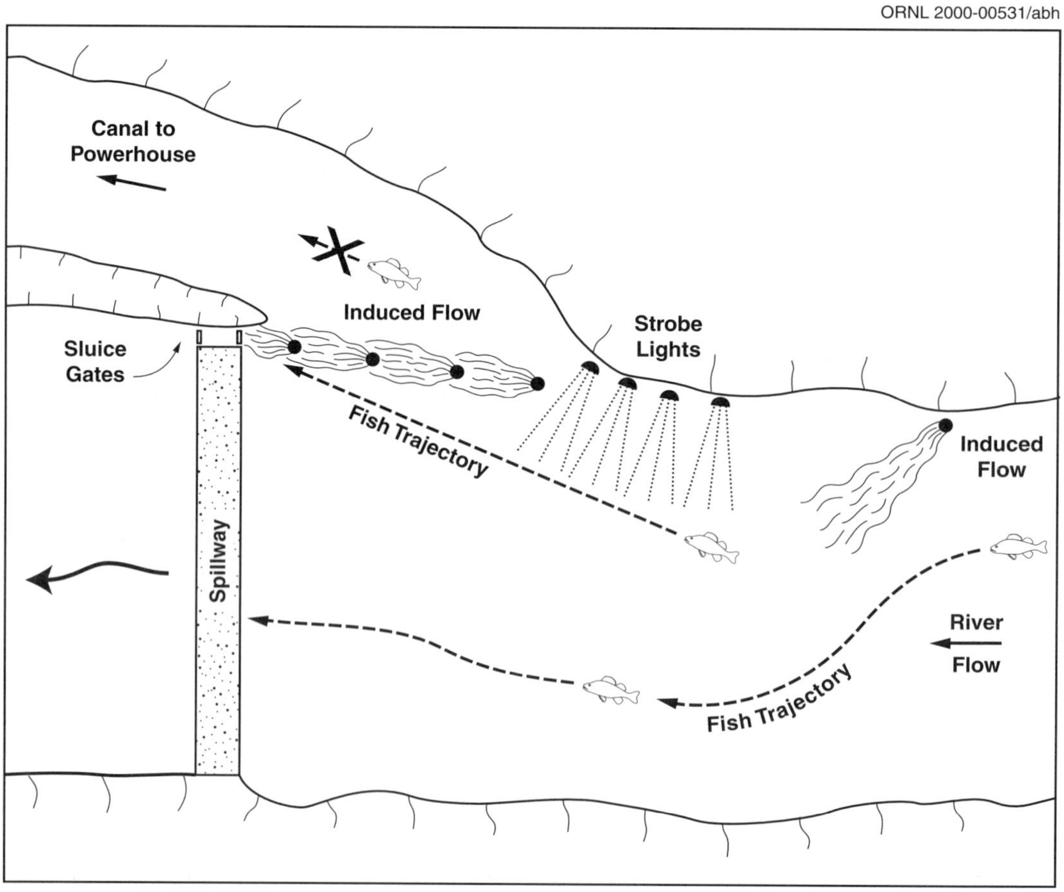

Figure 3. Diagrammatic view of potential integrated use of induced flow, and strobe lights to reduce numbers of downstream-migrating salmon smolts entering a power canal of a hydroelectric facility. Fish would be diverted toward the spillway side of the river by induced flow, repelled from the canal entrance by strobe lights, and guided to sluice gates by a second zone of induced flow.

have been successful for diverting salmon (average 73 mm) and striped bass (20–34 mm; Bates and Vinsonhaler 1956) but generally have been rejected for most new installations because of entrainment of juvenile fish (Rainey 1985), which may require concurrent application of another diversion technique.

The "integrated" of integrated behavioral guidance systems also should include integration with the rest of the dam operations. Inclusion of spill in the passage budget at Lower Granite Dam (Johnson et al. 2000) makes the point well. Also, the performance of other volitional passage routes can be dependent on operation of turbines and the resultant forebay flow patterns. Returning to Weldon Dam, for example, in 1998 the number of radio-tagged smolts passing turbines 1 and 2, those equipped with a behavioral barrier in the form of a "wall of light" from strobe lights, doubled compared with the 1995/1997 baseline period. The increase was attributed to a 50% reduction in flow through turbine 3 and additional reductions at turbine 4 (Great Northern Paper, Inc. 1998). The forebay flow patterns were changed significantly by these changes in turbine operations, with more of the total bulk flow passing through turbines 1 and 2. The need to manage passage of debris and trash as well as fish can lead to both obstacles and possible opportunities for fish passage routes (Clay 1994). Managers, using knowledge of the performance of each passage route in an integrated guidance system (structural and behavioral diversion) can make dam operating decisions to balance the need for fish protection with other needs (e.g., electricity generation). Every time a fish does not pass through a turbine, but rather an alternative route with a higher survival, the overall successful fish passage at the dam project is improved.

It is useful to map an area where movement of fish or other aquatic life is to be managed along with water flows. In doing so, we need to "think like a fish" (Coutant 1999). With the repertoire of many fish behaviors and guidance devices in mind, one can visualize where behavioral technologies (and structural ones, as well) can be used in combination, under the constraints of the operating dam system. A large-scale example is given in Figure 10 of Coutant (2001, this volume) for guidance of juvenile salmonids through a reservoir and toward a bypass at a dam. In the near field of the intake, induced flow and strobe lights set in banks horizontally were suggested as a combination that could help fish stay in surface layers as they are guided to surface bypasses. Sound might be used to further deter their entrance into the turbine. Submerged in the powerhouse is the screen and its gatewell bypass system (perhaps already in place) that would capture those not already guided away. These options can be considered in view of the distribution and trajectories of fish in the forebay, which may show a preponderance along one shoreline or the other, or in the middle, and at the surface or near the bottom.

Many uses of multiple behavioral technologies involving a multi-sensory approach have not yet been tested, but appear reasonable from what we know about the performance of each technology alone. Each application situation would be different, but the principle is the same: use multiple technologies to match the environment and diversity of biota to be protected.

Room for Innovation

There is room for innovation in both the development of new and improved behavioral technologies and in the creative deployment of them in varying combinations to match environmental and biological diversity. No one technology is necessarily right or wrong (and we divert scarce resources by carrying out competitive exclusion). We may have done about all that we can do now to develop certain technologies alone. The leaps of innovation likely will come in devising the most appropriate *combinations* to meet overall protection goals. A multi-sensory approach seems ripe for further development. Test facilities and experimental sites should be selected and outfitted to use various combinations of technologies. These combinations should be tested rigorously with a wide array of species and life stages (together, if possible, reflecting their natural occurrence). Above all, "failed" tests in one situation should not squelch our search for other, more appropriate situations where the technology or its particular deployment could possibly excel by better matching biological needs. In many cases, clues to the appropriate combination of technologies will come from close scrutiny of the animals' natural behavior patterns. Developers, vendors, and regulators would do well to replace any fixation on one technology with a broader vision of many technologies applied together that more closely match the physical and biological diversity of aquatic species and their habitats.

Conclusion

My bottom line is this: I encourage integration of our several behavioral guidance technologies in a multi sensory approach for better combined performance and to better match environmental and biological diversity. With a multi-sensory and multiple-technology perspective, not as alternatives but as components to be used together to manage animal behavior, I believe we can make an innovative leap and more likely meet overall protection goals through behavioral technologies.

Acknowledgments

I thank D. Rondorf and P. Patrick for critical reviews of the draft and for many useful suggestions. Oak Ridge National Laboratory is managed and operated by UT-Battelle LLC under Contract No. DE-AC05–00OR22725 with the U.S. Department of Energy. This is Environmental Sciences Division publication 5078.

References

Adams, N. S., G. E. Johnson, D. W. Rondorf, S. M. Anglea, and T. Wik. 2001. Biological evaluation of the behavioral guidance structure at Lower Granite Dam on the Snake River, Washington. Pages 145–160 *in* C. C. Coutant, editor. Behavorial technologies for fish guidance. American Fisheries Society, Symposium 26, Bethesda, Maryland.

Bates, D. W., and R. Vinsonhaler. 1956. Use of louvers for guiding fish. Transactions of the American Fisheries Society 86:38–57.

Brown, R. 2000. The potential of strobe lighting as a cost-effective means for reducing impingement, and entrainment. Environmental Science and Policy 3 (Supplement 1):S405–S416.

Bernier, K. 1999. 1999 Report on the effectiveness of the permanent downstream passage system for Atlantic salmon at Weldon Dam. Inexon/Great Northern Paper, Inc., Millinocket, Maine.

Chivers, D. P., and R. J. F. Smith. 1998. Chemical alarm signalling in aquatic predator-prey systems: a review and prospectus. Ecoscience 5:338–352.

Clay, C. H. 1994. Design of Fishways and Other Fish Facilities, 2nd edition. Lewis Publishers, Boca Raton, Florida.

Coutant, C. C. 1977. Compilation of temperature preference data. Journal of the Fisheries Research Board of Canada 34:739–745.

Coutant, C. C. 1998. Turbulent attraction flows for juvenile salmonid passage at dams. Report ORNL/TM-13608, Oak Ridge National Laboratory, Oak Ridge, Tennessee.

Coutant, C. C. 1999. Think like a fish! Emphasizing the "behavior" in behavioral guidance systems. Hydro Review 18(3):18–24.

Coutant, C. C. 2001. Turbulent attraction flows for guiding juvenile salmonids at dams. Pages 57–77 *in* C. C. Coutant, editor. Behavorial Technologies for fish guidance. American Fisheries Society, Symposium 26, Bethesda, Maryland.

Coutant, C. C., and R. R. Whitney. 2000. Fish behavior in relation to passage through hydroelectric turbines: a review. Transactions of the American Fisheries Society 129:351–380.

Darland, T. J., S. D. Evans, D. W. Rondorf, G. L. Rutz, D. H. Feil, N. S. Adams, C. F. Morrill, and J. D. Serl. 2001. Test of concept to improve fish guidance using induced water currents at Cowlitz Falls Dam, Washington. Annual report for 1999. U.S. Geological Survey, Columbia River Research Laboratory, Cook, Washington.

EPRI. 1998a. Evaluation of fish behavioral barriers. Report TR-109483. EPRI, Palo Alto, California.

EPRI. 1998b. Review of downstream fish passage and protection technology. Evaluations and effectiveness. Report TR-111517. EPRI, Palo Alto, California.

Evans, D. H. 1993. The Physiology of Fishes. CRC Press, Boca Raton, Florida.

Great Northern Paper, Inc. 1998. 1998 report on the effectiveness of the permanent downstream passage system for Atlantic salmon at Weldon Dam. Mattaceunk Project, FERC No. 2520. Great Northern Paper, Inc., Millinocket, Maine.

Hadderingh, R. H., and A. G. Smythe. 1997. Deflecting eels from power stations with light. In Proceedings of Fish Passage Workshop. Alden Research Laboratory, Inc., Holden, Massachusetts.

Hanson, B. N. 1999. Effectiveness of two surface bypass facilities on the Connecticut River to pass emigrating Atlantic salmon smolts. Pages 43–60 *in* M. Odeh, editor. Innovations in fish passage technology. American Fisheries Society, Bethesda, Maryland.

Haro, A., M. Odeh, J. Noreika, and T. Castro-Santos. 1998. Effect of water acceleration on downstream migratory behavior and passage of Atlantic salmon smolts and juvenile American shad at surface bypasses. Transactions of the American Fisheries Society 127:118–127.

Jahn, A. E., and K. T. Herbinson. 2000. Designing a light-mediated behavioral barrier to fish impingement, and a monitoring program to test its effectiveness at a coastal power plant. Environmental Science and Policy 3 (Supplement):S383-S391.

Johnson, G. E., N. S. Adams, R. L. Johnson, D. W. Rondorf, D. D. Dauble, and T. Y. Barila. 2000. Evaluation of the prototype surface bypass for salmonid smolts in spring 1996 and 1997 at Lower Granite Dam on the Snake River, Washington. Transactions of the American Fisheries Society 129:381–397.

Johnson, R. L., and eight coauthors. 1999. Hydroacoustic evaluation of fish passage and behavior at Lower Granite Dam in spring 1998. Summary of final report. Report PNWD-2448, U.S. Army Corps of Engineers, Walla Walla District, Walla Walla, Washington.

Johnson, R. L., and eight coauthors. 1998. Fixed-location hydroacoustic evaluation of the prototype surface bypass structures at Lower Granite Dam in spring 1998. Preliminary draft report. U.S. Army Corps of Engineers, Walla Walla District, Walla Walla, Washington.

McCauley, D. J., L. Montuori, J. E. Navarro, and A. R. Blystra. 1996. Using strobe lights, air bubble curtains for cost-effective fish diversion. Hydro Review 15(2):42–51.

Michaud, D. T., and E. P. Taft. 2000. Recent evaluations of physical, and behavioral barriers for reducing fish entrainment at hydroelectric plants in the upper Midwest. Environmental Science and Policy 3 (Supplement):S499–S512.

Mighetto, L., and W. J. Ebel. 1995. Saving the salmon: A history of the U.S. Army Corps of Engineers' efforts to protect anadromous fish on the Columbia and Snake rivers. Historical Research Associates, Inc. for the U.S. Army Corps of Engineers, North Pacific Division, Portland, Oregon.

Patrick, P. H., A. E. Christie, D. Sager, C. Hocutt, and J. Stauffer, Jr. 1985. Responses of fish to a strobe light/air bubble barrier. Fisheries Research 3:157–172.

Pavlov, D. S., A. I. Lupandin, and V. V. Kostin. 1999. Downstream migration of fish through hydropower dams. Nauka Publishers, Russian Academy of Sciences, Moscow (in Russian).

Pitcher, T. J. 1986. The behavior of teleost fishes. The Johns Hopkins University Press, Baltimore, Maryland.

Ploskey, G. R., J. M. Nestler, G. N. Weeks, and C. Schilt. 1995. Evaluation of an integrated fish-protection system. Pages 162–171 *in* J. L. Cassidy, editor. Waterpower '95. American Society of Civil Engineers, New York.

Popper, A. N., and T. J. Carlson. 1998. Application of sound and other stimuli to control fish behavior. Transactions of the American Fisheries Society 127:673–707.

Quinn, T. P., and A. H. Dittman. 1990. Pacific salmon migrations and homing-mechanisms and adaptive significance. Trends in Ecology and Evolution 5:174–177.

Rainey, S. 1985. Considerations in the design of juvenile bypass systems. Proceedings of the Symposium on small hydropower and fisheries. American Fisheries Society, Bethesda, Maryland.

Rainey, S. 1997. "Opportunity for discovery:" An important tool for assessment of surface collection technology potential for improving juvenile salmon passage at Columbia River dams. Proceedings, Fish Passage Workshop (Milwaukee, Wisconsin, May 6–8, 1997). Alden Research Laboratory, Inc., Holden, Massachusetts.

Sand, O., P. S. Enger, H. E. Karlsen, and F. R. Knudsen. 2001. Detection of infrasound in fish and behavioral responses to intense infrasound in juvenile salmonids and European silver eels: a minireview. Pages 183–193 *in* C. C. Coutant, editor. Behavorial Technologies for fish guidance. American Fisheries Society, Symposium 26, Bethesda, Maryland.

Savino, J. F., D. J. Jude, and M. J. Kostich. 2001. Use of electrical barriers to deter movement of round gobies. Pages 171–182 *in* C. C. Coutant, editor. Behavorial Technologies for fish guidance. American Fisheries Society, Symposium 26, Bethesda, Maryland.

Skorogobogatov, M. A., A. A. Mosiyevskiy, V. G. Korolyov, and S. N. Tyuryukov. 1993. Hydraulic and biological research of fish bypasses for high-head hydropower systems. Energeticheskoye stroitelstovo December 1993:67–71 (in Russian; translation available from C. C. Coutant).

Taft, E. P. 2000. Fish protection technologies: as status report. Environmental Science and Policy 3 (Supplement):S349–S359.

Vogel, S. 1994. Life in moving fluids. The physical biology of flow. 2nd edition. Princeton University Press, Princeton, New Jersey.

Whitney, R. R., L. D. Calvin, M. W. Erho, Jr., and C. C. Coutant. 1997. Downstream passage for salmon at hydroelectric projects in the Columbia River basin: development, installation, and evaluation. Report 97–15, Northwest Power Planning Council, Portland, Oregon.

Williams, R. N., L. D. Calvin, C. C. Coutant, M. W. Erho, Jr., J. A. Lichatowich, W. J. Liss, W. E. McConnaha, P. R. Mundy, J. A. Stanford, R. R. Whitney, D. L. Bottom, and C. A. Frissell. 1996. Return to the river: Restoration of salmonid fishes in the Columbia River ecosystem. Report No. 96–6. Independent Scientific Group, Northwest Power Planning Council, Portland, Oregon.

Williams, R. N., L. D. Calvin, C. C. Coutant, M. W. Erho, Jr., J. A. Lichatowich, W. J. Liss, W. E. McConnaha, P. R. Mundy, J. A. Stanford, R. R. Whitney, D. L. Bottom, and C. A. Frissell. 2000. Return to the river 2000: Restoration of salmonid fishes in the Columbia River ecosystem. Report No. 2000–12. Northwest Power Planning Council, Portland, Oregon.

American Fisheries Society Symposium 26:115–124, 2001

Electric Power Research Institute's (EPRI) Research on Behavioral Technologies

Edward P. Taft
Alden Research Laboratory, Inc., 30 Shrewsbury Street, Holden, Massachusetts 01520, USA

Douglas A. Dixon
Electric Power Research Institute, 7905 Berkeley Drive, Gloucester Point, Virginia 23062, USA

Charles W. Sullivan
Electric Power Research Institute, retired

Abstract.—Laboratory and field evaluations of behavioral fish protection devices sponsored by the Electric Power Research Institute (EPRI) since 1987 are reviewed. Devices evaluated include strobe lights, mercury lights, incandescent lights, transducer-based sound systems, and infrasound generators. Responses of nearly 40 fish species (anadromous salmonids, Alosa species, and a variety of riverine species) have been evaluated in cage and field studies conducted at sites throughout North America, as well as in laboratory studies conducted at university and private research facilities. Results have varied by species, device, and location. In general, strobe lights have been effective with the largest number of species. Transducer-based sound systems have been particularly effective with Alosa species but less effective with most other species. Results of tests with other light types have been equivocal and appear to be influenced strongly by site-specific conditions. Infrasound devices tested by EPRI during cage tests failed to elicit responses in riverine fishes.

The Electric Power Research Institute (EPRI) has sponsored a variety of behavioral barrier evaluations throughout North America over the past 14 years. EPRI has also prepared four comprehensive reviews of fish passage technologies that include behavioral technologies (EPRI 1986, 1994a, 1994b, 1998a, 1999). The EPRI laboratory and field studies have encompassed a wide variety of anadromous and riverine species and have included evaluations of strobe lights, mercury lights, high-pressure sodium lights, acoustic sound systems, and infrasound generators. The following discussion presents a site-by-site summary of the studies conducted and the results obtained. A concluding section integrates the EPRI-sponsored work with other studies and addresses current guidelines for testing and deployment of behavioral devices.

Laboratory Studies

University of Washington

Laboratory experiments were conducted at the University of Washington fish hatchery to determine behavioral avoidance and attraction responses of hatchery-reared, sub-yearling chinook *Oncorhynchus tshawytscha*, coho *O. kisutch*, and Atlantic salmon *Salmo salar* and steelhead *O. mykiss* to strobe and mercury vapor lights (EPRI 1990). Experiments were performed in an 8.8 m long by 1.6 m wide by 1.2 m deep outdoor cement raceway. The tests were conducted under no-flow conditions at a water depth of 0.65 m and ranging in temperature from 12 to 17.5°C. An EG&G model SS-122 underwater strobe light was installed at both ends of the raceway. The strobe was operated at 300 flashes per minute. Video cameras were suspended above the raceway to monitor and record behavioral responses.

Fifty fish of one species were used for each test. The fish were introduced into the raceway and allowed to acclimate for 30 min. After the adaptation period ended, a light stimulus was turned on and behavior was monitored. Testing was conducted using a paired sampling design. At any point in time, only one strobe or one mercury light was operating. The following sequences were evaluated: "upstream" strobe on followed by "downstream" strobe on; downstream strobe on followed by upstream strobe on; upstream mercury light on followed by downstream mercury light on; and downstream mercury light on followed by upstream mercury light on.

A test involved illuminating one light for 30 min, then illuminating a light of the same type (i.e., strobe or mercury) on the opposite end of the raceway for an equal period of time. To ensure that prior exposure to one type of light did not influence the behavioral response to the second light, separate groups of test fish were used in each pair of experiments. Four "trials" (the response of each fish group to a pair of lights) were completed per day. Five replicates of each trial set were completed for each species under daytime (9:00 a.m. to 7:00 p.m.) and nighttime (6:00 p.m. to 3:00 a.m.) conditions. Direct observations yielded three types of information: qualitative observations, attraction and repulsion responses verses time, and the equilibrium response to light.

All species, except for steelhead trout tested during the day, hid from or avoided the strobe light. During strobe light tests, most fish sought shelter at the far end of the test tank for the duration of each test. These results are consistent with the avoidance responses shown by other species to strobe light both in the laboratory and in the field.

The strobe light avoidance demonstrated by chinook and coho salmon during the daytime tests was unexpected; it was anticipated that the reduced contrast between ambient daylight and test light would decrease avoidance. The very high contrast at night (due to the small test space and the reflection of light off the white walls of the tank) appeared to stun fish, resulting in a cessation of movement. Consequently, the response observed might not have been indicative of expected behavior under more natural background conditions. In contrast, fish were not stunned during the day and swam immediately away from the light. As a result, fish were better able to avoid strobe light during the day than at night. This situation may not occur in the field since natural turbidity and open space could allow fish to respond before they become stunned. Therefore, the authors concluded that strobe lights could be effective in repelling these species under nighttime field conditions. During the day, lack of contrast in the natural environment might render the lights less effective than they were in laboratory tests.

Steelhead fry tested under nighttime conditions demonstrated the greatest attraction to mercury lights. Approximately 80% of steelhead fry congregated at the end of the 9.1 m long test tank that had an active mercury light. Responses of chinook and coho salmon to mercury lights were variable. These species tended to swim the length of the raceway demonstrating no strong attraction or repulsion response.

University of Iowa

In laboratory studies conducted at the University of Iowa, behavioral responses to strobe lights and mercury lights were examined with bluegill *Lepomis macrochirus*, adult and juvenile channel catfish *Ictalurus punctatus*, walleye *Stizostedion vitreum*, hybrid striped/white bass *Morone saxatilis* x *M. chrysops*, and largemouth bass *Micropterus salmoides* (EPRI 1990). The equipment and experimental design were similar to those described above for the University of Washington studies. Bluegill, channel catfish, juvenile walleye, and hybrid bass avoided the strobe light. Juvenile channel catfish avoided only the brightest sections of the test flume. Largemouth bass exhibited no distinct response. With mercury light, largemouth bass, channel catfish, and walleye were repelled. In contrast, bluegill tended to slowly swim in the middle of the tank showing no startle or avoidance behavior, indicating little attraction or repulsion to mercury light. Hybrid striped bass tended to swim end to end in large groups displaying no defined reaction to mercury light.

Ontario Hydro

EPRI sponsored studies evaluating the effectiveness of hammers in repelling adult blueback herring *Alosa aestivalis* and juvenile Atlantic salmon, chinook salmon, and rainbow trout *O. mykiss* in Ontario Hydro's laboratory in 1987 (EPRI 1990). The tests were conducted in a semianechoic tank designed to minimize echoes and reverberations. Each experiment included three, 30-min trials consisting of one control trial and two experimental trials with the hammers operating at different frequencies ranging from 20 to 1,000 Hz. For all frequencies, sound pressure levels were 190 db one meter from the source. The test tank was partitioned into a gridwork of 3-by-3 m zones. The distribution of fish among the zones was recorded every 10 s during the first 5 min of each trial and once every 5 min thereafter. Test results were highly variable and were not subjected to statistical analysis. Observations indicated that the hammers elicited no obvious or consistent avoidance or attraction responses among the species evaluated.

Field Studies

Hadley Falls

In 1987, EPRI sponsored studies examining the use of strobe lights to repel adult American shad *Alosa sapidissima* at the Hadley Falls Hydroelectric Project located on the Connecticut River in Holyoke, Massachusetts (EPRI 1990). The project's powerhouse is located on the west side of the dam. The dam also diverts water into a canal system that supplies flow to another hydroelectric project and various other water users. Adult and juvenile American shad encounter the Hadley Falls project during out-migrations in the spring and fall, respectively. Once in the canal system, fish can follow one of two passage routes: downstream to the Boatlock Hydro Station (utilizing a fish bypass and return pipe) or down a branch canal (where they become trapped).

The strobe lights were modified Fish Avoidance Xenon System (FAXS) designed by EG&G. All the strobe heads were synchronized and operated at a flash rate of 300 flashes per minute (fpm). Strobe lights were suspended into the water from a canal bridge in attempts to repel adult shad away from a first level canal branch. The effectiveness of the strobe lights in excluding shad from the branch canal was evaluated by video monitoring. Strobes were sequenced on and off hourly for eight hours daily, and monitoring was conducted during both on and off conditions. Video recordings were made during the first 15 min of each test period and for several minutes at half-hour intervals throughout the test period. The tests indicated that the strobe lights were ineffective in repelling adult shad during daylight hours, which was when most fish movement occurred. Although the point sources (flash tubes) were highly visible to the human eye, the level of illumination of a strobe light is minor compared with daylight illumination. Fish were observed moving in schools that reacted to the bridge's shadow by maintaining position upstream of the shadow. The addition of strobe lights had no apparent effect on fish behavior as they approached the bridge. Fish were observed swimming within a few inches of the light even though the light was flashing directly towards their eyes. This lack of reaction indicated that strobe light, as an indirect or direct (point source) stimulus, had no effect on adult American shad during daylight hours.

York Haven

Juvenile American shad demonstrated a strong avoidance response to strobe lights during field tests conducted over a 5-year period at the York Haven Hydroelectric Project (EPRI 1990; 1992a; Martin et al. 1991; Martin and Sullivan 1992). The project is located on the Susquehanna River 24 km south of Harrisburg, Pennsylvania. It consists of a 2.4 km long by 7 m high dam and powerhouse containing 6 Kaplan turbines and 14 Francis turbines each with a capacity of approximately 22.7 m^3/s. Strobe lights were evaluated to determine their ability to divert juvenile shad away from the plant turbines and into an existing trash sluiceway.

From 1988 to 1991, various arrangements of strobe lights were evaluated (flash rate = 300 flashes per minute). In all years, the lights were attached to various floats located in front of the intake trash racks upstream of an ice/trash sluiceway (considered a safe bypass route). The configuration tested in 1991 is shown in Figure 1.

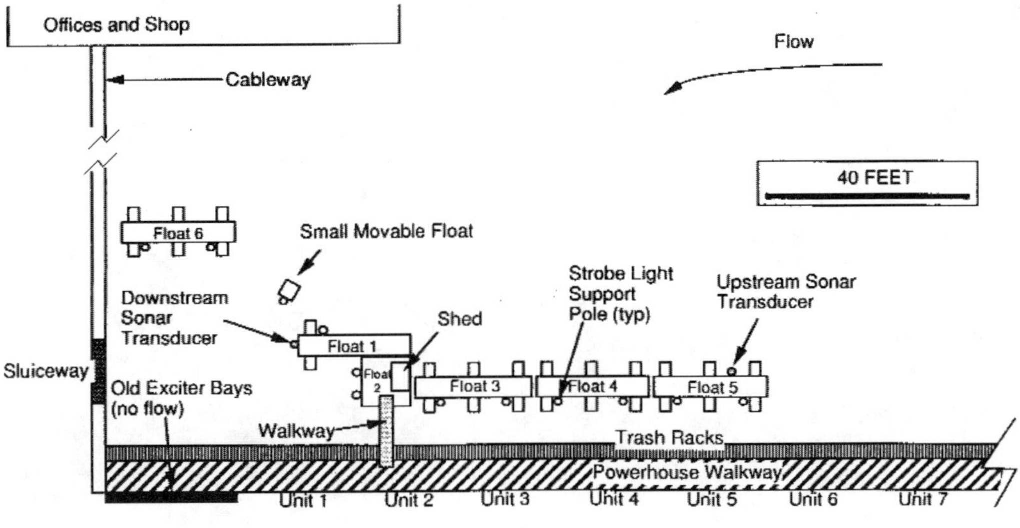

Figure 1. 1991 Forebay and float positions, York Haven hydroelectric project.

This configuration was highly effective in preventing fish passage into Units 1 through 6 and repelling them to the sluiceway bypass. However, the float system created maintenance problems. Therefore, the system was redesigned for tests conducted in 1992. Four newly designed floats were arranged in an arc around the sluiceway opening with none positioned at the trashracks (Figure 2). Two Flash Technology strobe lights were fastened to steel poles under each of the floats at depths of 0.9 m and 2.7 m.

The strobe light system was evaluated by using scanning sonar and a netting program. The scanning sonar system (WES-MAR model SS390) was used to monitor fish behavior and responses to the strobe lights. Tailrace nets were used to quantify the passage of fish through the turbines and sluiceway. All tests were performed at night when the juvenile shad were actively migrating (approximately 6:00 p.m. to 2:00 a.m.). The strobe lights strongly and consistently repelled fish in 1988 tests. When the strobes were turned on, large aggregations of fish were observed to pass immediately through the sluiceway. The effect of the lights did not diminish over time (up to 1 h); the illuminated area remained void of fish as long as the strobes were in operation. When the strobes were turned off, fish moved quickly into the previously lit area. Data collected during 1989 and 1990 were limited by the occurrence of major floods, which resulted in most out-migrants passing over the dam. During these years, however, hydroacoustic observations and limited net sampling indicated that the light array effectively deterred fish that were approaching the turbine intakes. In 1991, favorable flow conditions allowed extensive sampling to be conducted, including netting in the tailrace and at the sluice gate. In 156 paired tests (strobe light and control), the estimated number of juvenile American shad passing through the sluiceway averaged 1,712 fish per test with the strobe lights activated and 38 fish per test with the lights off. The estimated total number of fish passed through the sluiceway over the entire out-migration period with the lights on and off was 243,144 and 5,354, respectively. Passage through the turbine also increased during strobe light tests; however, the increase only represented about 6% of the fish passed via the sluiceway (EPRI 1992). The increase in turbine passage during strobe light illumination demonstrates the need to determine relationships between behavioral fish bypass systems and site-specific hydraulics in order to maximize bypass efficiency. Similar results were obtained with the new strobe light configuration deployed in 1992 (Figure 2).

Ludington

The effects of strobe lights, mercury lights and hammers on fish densities were examined at the Ludington Pumped Storage Project (EPRI 1990). The devices were tested as potential means to reduce entrainment of Lake Michigan fish into the reservoir during pumpback operation. The project

Figure 2. 1992 Forebay and float positions, York Haven hydroelectric project.

is located near Ludington, Michigan on the east shore of Lake Michigan. The powerhouse consists of six turbine units that also serve as pumpback units. Flows were 1,869 m^3/s during pumpback operation. Key species entrained during pumpback operation include steelhead, brown trout *Salmo trutta*, lake trout *Salvelinus namaycush*, coho salmon, chinook salmon, alewife *Alosa pseudoharengus*, yellow perch *Perca flavescens*, smelt (Osmeridae), and chubs (Cyprinidae).

A boat, secured by piles on both sides, served as a work platform in the test zone. Four strobe lights were used, two set at a depth of 1.5 m and two set at a depth of 4.6 m, with 9 m of horizontal distance between the pairs. The strobe lights used were modified Fish Avoidance Xenon System (FAXS) designed by EG&G. The 1000-W mercury lights were set in a similar configuration. Ontario Hydro supplied the two hammers that were tested.

Over a 4-month period in 1987, the behavioral devices were evaluated during nighttime hours near one of the project's intake jetties (where fish were known to congregate). Each test lasted for 2 h. During this time, three test conditions or two test conditions and one control conditions were evaluated. When the plant was operating in pumpback mode, fish abundance was monitored using hydroacoustic techniques during periods with and without test devices operating. Gill-net sampling was performed during testing to identify fish species in the test area.

Statistical analysis was used to determine the influence of the test devices and test periods on fish abundance. Fish abundance was lower during periods of strobe light illumination than periods with the lights off. When the monthly data are combined, the average fish abundance during strobe light illumination periods (abundance index [A.I.] = 2.99) was significantly less than the average abundance for the control condition (A.I. = 8.03). The mercury lights achieved a consistent attraction of resident fish. The average abundance of fish present with the mercury lights on (A.I. = 15.38 was nearly twice that of the control condition A.I. = 8.03). The difference in fish abundance between test and control conditions was significantly different in all test months. The results of hammer tests indicated that the hammers were ineffective in consistently repelling fish. In all tests combined, the mean abundance of fish in the test area with the hammers operating was significantly ($p \leq 0.05$) less than the control condition. The results for individual months were variable; in one month, the mean abundance of fish was greater when the hammers were operating (EPRI 1990).

Wanapum

Studies were conducted at Wanapum Dam to determine the potential effectiveness of mercury lights in improving the passage of out-migrating salmon via the spillway gates (EPRI 1990). The project is located on the Columbia River in Grant County, Washington and consists of 10 turbines, 12 bottom-spill tainter gates, and 1 top-spill sluice gate.

The mercury lights were installed on two adjacent pier noses of one spillway gate. Lights were set at two depths: 3 m for the first half of the test period and 6 m for the second. Hydroacoustics were used to monitor fish passage. Monitoring the spill 7 d prior to initiation of the mercury light treatment generated baseline data. The lights were tested by turning them on and off for equal time intervals each night during the period of active migration. The on (test) and off (control) sequence was reversed from one night to the next.

Results showed great variability in passage rates from day to day over the 29 d of sampling, typical of normal migratory variability. Statistical analysis indicated the lights had no effect on fish passage rates between test and control conditions. The authors suggest the results should be considered inconclusive due to a short out-migration period and that the light illuminated only a small portion of the spillway relative to the total spillway area.

Wapatox

Mercury lights were evaluated to assess their effectiveness at maximizing the number of juvenile salmon and steelhead using the diversion system at Wapatox Canal Fish Screening Facility (EPRI 1990). The canal is located on the Naches River 0.8 km from the confluence of the Naches and Tieton River in Washington. Flow rate in the canal ranges from 8.5 to 14.2 m^3/s. Water from the diversion is used for irrigation and hydroelectric power. The diversion system is made up of six drum screens, 2 m in diameter and 3.4 m wide, oriented perpendicular to the river flow (Figure 3). Five of the six screens are covered with 0.64 cm mesh; the sixth is covered in 0.32 cm mesh. A vertical slot fish bypass entrance was located on the

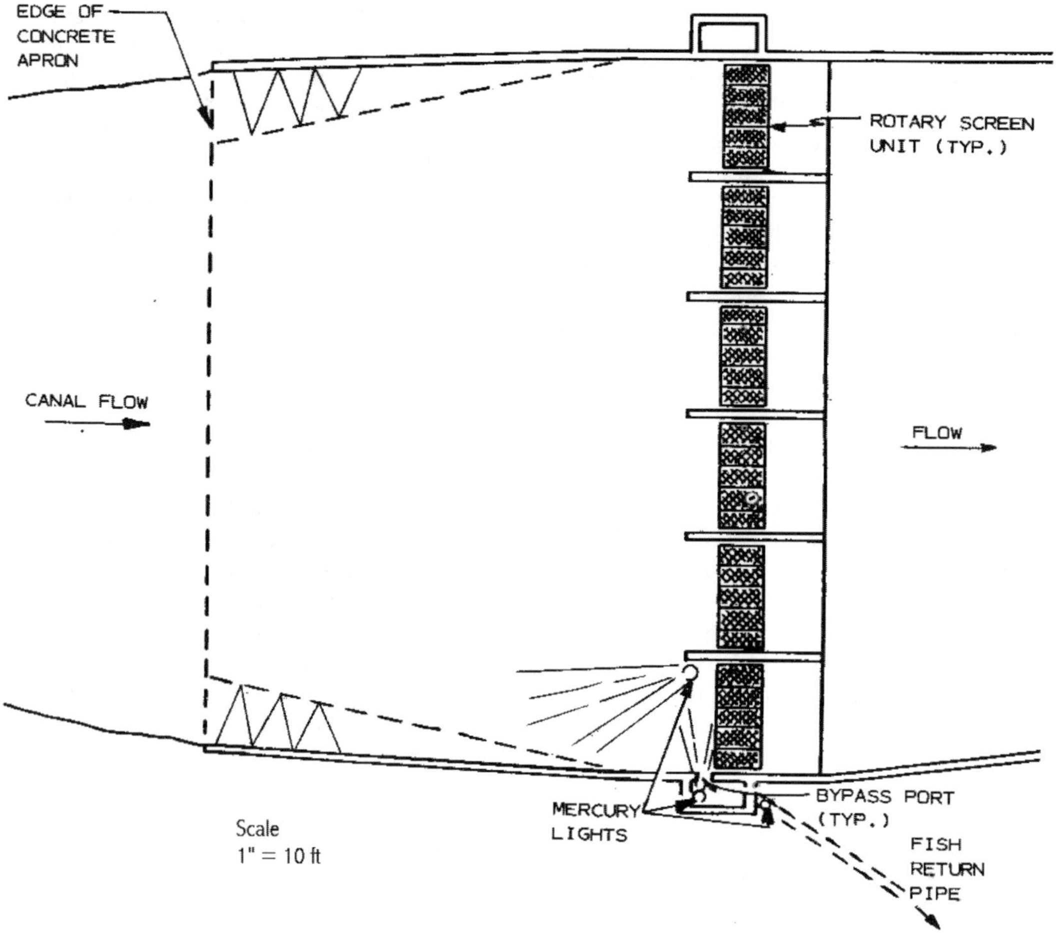

Figure 3. Modified light orientation, Wapatox Canal, plan view.

left bank of the canal. A fish return pipe transported the diverted fish to the river 366 m downstream of the canal diversion.

Mercury light effectiveness was determined by locating a 1,000-W underwater mercury light on the left side of the canal at the bypass entrance and an additional mercury light inside the bypass entrance in order to illuminate the bypass slot (Figure 3). Tests compared fish passage rates for the bypass during periods with the lights on and off. A paired testing strategy was employed. Fish movement was monitored in several ways. Rate of passage was monitored by removing and counting fish collected in the smolt trap every 10–15 min. Impingement on the drum screens was also monitored occasionally and visual observations of fish behavior were recorded.

During spring tests, overall fish bypass rates did not differ between light and dark conditions. Based on increases in passage rates immediately after a light condition change, it was believed that fish were attracted to the mercury light but were avoiding the area of brightest illumination at the bypass openings. Typically, when the mercury lights were turned on, fish were attracted to, but would not enter, the bypass. When the lights were turned off, fish would move into the bypass. Consequently, the mercury light orientation and operation scheme was modified for fall tests (Figure 3). Attempts were made to attract fish to the bypass area by maintaining illumination of the bypass well, but moving the canal light suspected of repelling fish from the bypass opening during the spring testing. Test and control periods were alternated between the first half (7:30 p.m. to 11:30 p.m.) and last half (11:30 p.m. to 3:30 a.m.) of each night. During the test periods, the light was cycled off for 5 min each hour in order to "pulse" fish into the bypass. A substantial increase in out-migrant (chinook salmon and kokanee *Oncorhynchus nerka*) passage rates was observed during fall tests with the new array and operation scheme; nearly twice as many fish were bypassed with the lights on (863 fish) than with the lights off (451 fish).

Kingsford

The response of riverine fishes to behavioral devices was evaluated during cage tests conducted at the Kingsford Hydroelectric Project on the Menominee River in Wisconsin (EPRI 1998b; Winchell et al. 1997; Michaud and Taft 1999). The cage tests were designed to determine whether stimuli produced by strobe lights, mercury lights, high-pressure sodium lights, sound transducers, and infrasound generators could elicit avoidance or attraction responses that might be useful for designing or enhancing the effectiveness of fish passage and protection systems.

Fish response to strobe light was assessed by placing a group of fish in the test cage (1 m by 3.6 m long by 1 m deep) and alternating the operation of single lights located at either end of the cage. Each light was activated for periods up to 10 min. An underwater video system was used to observe fish during each test. Qualitative observations describing fish response were based on spatial and temporal movements (i.e., distance, direction, and speed of movement). The species that were evaluated with strobe lights included walleye, largemouth bass, smallmouth bass *Micropterus dolomieu*, yellow perch, sunfish species (Centrarchidae), and rainbow trout. The target size range for test specimens was 50–250 mm in length. Test fish were collected from the Menominee River or obtained from private hatchery sources. During the cage tests, strobe lights elicited consistent avoidance reactions from walleye, and weak reactions from largemouth bass and yellow perch. No responses to strobe light were observed for smallmouth bass, sunfish species, and rainbow trout. Based on observations for the three species that exhibited avoidance, strobe lights were selected to be evaluated by EPRI during the next phase of field study at the White Rapids Project.

The responses of black crappie *Pomoxis nigromaculatus*, golden shiner *Notemigonus crysoleucas*, bullhead (Ictaluridae), smallmouth bass, largemouth bass, rainbow trout, walleye, and sunfish species to acoustic signals were evaluated at Kingsford. With the exception of golden shiner and black crappie, the test species demonstrated some level of avoidance to various acoustic signals during the Kingsford cage tests. Avoidance reactions of bullhead, sunfish, and smallmouth bass were classified as weak. Avoidance behaviors exhibited by rainbow trout, walleye, yellow perch, and largemouth bass were classified as moderate. The center frequency of signals that produced avoidance reactions from rainbow trout was 6,000 Hz. Center frequencies of signals that elicited avoidance from walleye included 566, 673, 1,350, and 2,990 Hz. Effective signals for yellow perch were centered at 673, 953, 1,000, and 2,000 Hz. Largemouth bass demonstrated avoidance to signals with center frequencies of 283, 600, 673, 2,000, 2,500, 2,990, and 5,500 Hz. Based on these results, sound was carried forward to the field tests at White Rapids.

Mercury lights, high-pressure sodium lights, and infrasound generators were not found to elicit any meaningful avoidance or attraction responses among fish tested during the Kingsford cage tests. Therefore, these devices were not carried forward to the field trials at White Rapids.

White Rapids

Based on the results of the 1996 Kingsford cage tests, field tests were conducted at the White Rapids Hydroelectric Project to evaluate the ability of strobe light and acoustic signals to repel fish away from a turbine intake (EPRI 1998b; Michaud and Taft 1999). The White Rapids Project is located on the Menominee River, which borders Wisconsin and the upper peninsula of Michigan. The project has three Francis turbines, Units 1 and 3 are rated at 3 MW each with discharge rates of 43.7 m^3/s, and Unit 2 is rated at 2 MW with a discharge of 25.8 m^3/s. The strobe lights and sound transducers were mounted on two steel frames that were deployed over the trash racks of the two intake bays of Unit 1 (Figure 4).

A total of 24 strobe lights operating at 300 flashes per minute were deployed at the Unit 1 intake. Based on the cage test results, sound signals with center frequencies of 673, 2,000, 2,990, and 5,000 Hz were selected for evaluation at White Rapids.

Effectiveness was measured by comparing entrainment numbers during treatment and control periods. Entrained fish were collected in full-flow tailrace nets that sampled the two discharge bays of Unit 1 (Figure 4). The field evaluation was conducted during three test periods in July, September, and October of 1997. The following four test conditions were evaluated during the field evaluation:

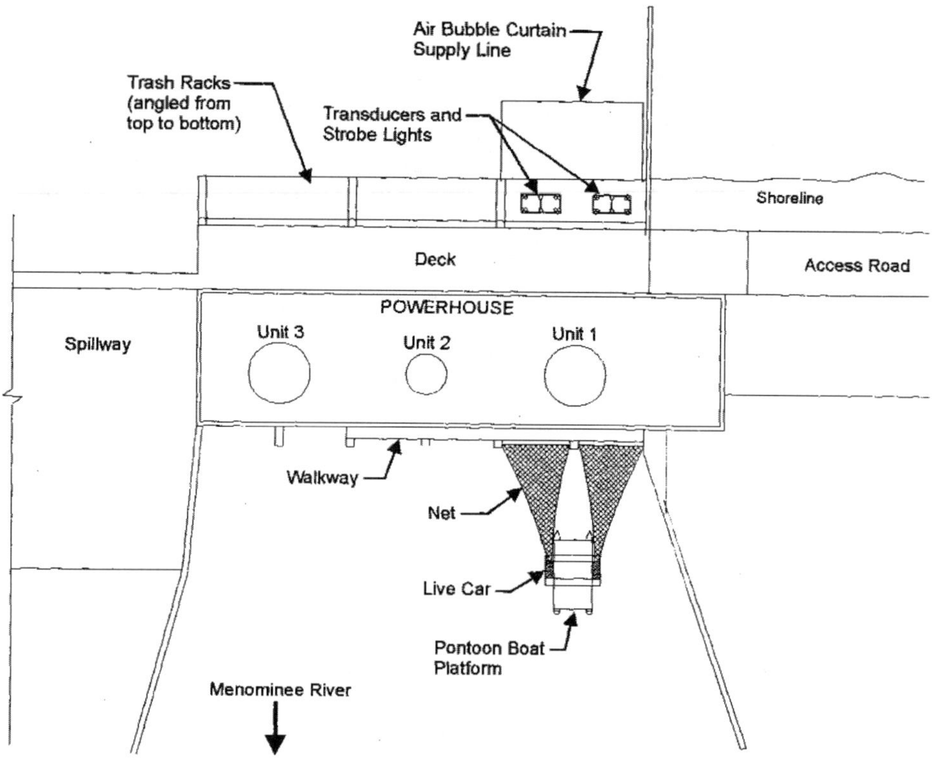

Figure 4. White Rapids hydroelectric project behavioral barrier evaluation.

(1) strobe lights on, (2) acoustic system on, (3) both lights and sound on, and (4) control (no devices operating). An air bubble curtain also was evaluated during the October test period. At least 33 species were collected in the Unit 1 tailrace nets during the three sample periods. There were no significant differences in entrainment numbers between strobe light and control periods for any species, family, or size-group analyzed, or for all fish combined, during any of the three sample periods. The statistical analysis of the collection data accounted for diel differences in entrainment rates, which typically were greater during evening hours. Based on these results, it was concluded that strobe lights, sound, and air bubble curtains, operating with the parameters tested, may not be applicable to projects that are similar to White Rapids with respect to project design and fish species occurrence (EPRI 1998b).

Discussion

Laboratory and field evaluations of behavioral fish protection devices sponsored by EPRI have been presented. Devices evaluated included strobe lights, mercury lights, incandescent lights, transducer-based sound systems, and infrasound generators. Responses of nearly 40 fish species (anadromous salmonids, Alosa species, and a variety of riverine species) have been evaluated in cage and field studies conducted at sites throughout North America, as well as in laboratory studies conducted at university and private research facilities. Consistent with current National Marine Fisheries Service (NMFS 1994) and American Fisheries Society (2000) recommendations for studying behavioral guidance devices, most of these studies were conducted with oversight by an expert review panel. The panels were composed of personnel representing the FERC, federal and state resource agencies, the participating utilities, and the project team who are knowledgeable in the scientific and engineering issues associated with fish protection and associated research.

Consistent with studies of behavioral devices by other researchers, the results of the EPRI studies have varied by species, device, and location. Variability in observed responses may be simply the result of species-specific differences in behavior. Alternatively, responses may vary as a result of differences in other variables between test locations. For example, it is reasonable to expect that light systems may be less effective in repelling or attracting a given species at sites with high turbidity levels. Similarly, the propagation of sound in water is strongly affected by the physical environment in which it is transmitted. Two identical sound generators can, therefore, create a different pattern of sound intensity at two different sites with different physical boundaries. Given that many of the EPRI studies were conducted at operating hydro projects with specific design and operational characteristics, it is possible that the results obtained were influenced by site-specific conditions. Therefore, the results should be viewed in the broader context of the available data on behavioral barriers.

In general, strobe lights have shown promise for protecting a variety of species. These lights have been shown to effectively and repeatedly repel Alosa species under prototype conditions. As presented elsewhere in these proceedings, strobe lights continue to be the subject of considerable research. Results of tests with other light types have been equivocal and appear to be influenced strongly by site-specific conditions.

Transducer-based sound systems also have been particularly effective with Alosa species but less effective with most other species. Other studies have not shown sound to be consistently effective in repelling species such as largemouth bass, smallmouth bass, yellow perch, walleye, rainbow trout, gizzard shad, and Atlantic herring. Given the species-specific responses to different frequencies that have been evaluated and the variable results that often have been produced, additional research is warranted at any site where there is little or no data to indicate that the species of concern may respond to sound.

Infrasound devices tested by EPRI during cage tests failed to elicit responses in riverine fishes. Results of tests with several salmon species have shown variable responses, as presented elsewhere in these proceedings. Given these results, it appears that infrasound sources need to be further developed and evaluated before they can be considered for larger scale application.

Continuing EPRI Research on Behavioral Technologies

Since the fall of 1999, EPRI has sponsored laboratory evaluations of angled bar racks and louvers which are behavioral guidance devices. These studies are investigating the effects of bar rack spacing, angle, and orientation on the guiding capabilities of juvenile fish. Species tested include American eel *Anguilla rostrata*, lake sturgeon *Acipenser fulvescens*, shortnosed sturgeon *A. brevirostrum*, channel catfish, walleye, smallmouth and largemouth bass and golden shiner. The project is being conducted to provide data on angled bar rack effectiveness with these previously-untested species. An EPRI project report is due to be published in the first quarter of 2001. EPRI also is intending to pursue research on turbulent attraction flows to enhance downstream passage, as proposed by Coutant (1998 and elsewhere in this proceedings).

References

American Fisheries Society (AFS) 2000. Guidelines for Evaluating Fish Passage Technologies (Initiative 2: Fish Passage Technologies Research Development Process). Prepared by the American Fisheries Society Bioengineering Section, Bethesda, Maryland.

Coutant, C. C. 1998. Turbulent attraction flows for juvenile salmonid passage at dams. Report No. ORNL/TM-13608 U.S. Department of Energy, Oak Ridge National Laboratory, Oak Ridge, Tennessee.

EPRI (Electric Power Research Institute). 1986. Assessment of downstream migrant fish protection technologies for hydroelectric application. EPRI AP-4711. Palo Alto, California.

EPRI (Electric Power Research Institute). 1990. Fish protection systems for hydro plants test results. EPRI GS-6712. Palo Alto, California.

EPRI (Electric Power Research Institute). 1992. Evaluation of strobe lights for fish diversion at the York Haven hydroelectric project. EPRI TR-101703. Palo Alto, California.

EPRI (Electric Power Research Institute). 1994a. Research update on fish protection technologies for water intakes. EPRI TR-104122. Palo Alto, California.

EPRI (Electric Power Research Institute). 1994b. Fish protection/passage technologies evaluated by EPRI and guidelines for their application. EPRI TR-104120. Palo Alto, California.

EPRI (Electric Power Research Institute). 1998a. Review of downstream fish passage and protection technology evaluations and effectiveness. EPRI TR-111517. Palo Alto, California.

EPRI (Electric Power Research Institute). 1998b. Evaluation of fish behavioral barriers. EPRI TR-109483. Palo Alto, California.

EPRI (Electric Power Research Institute). 1999. Fish protection at cooling water intakes: status report. EPRI TR-114014. Palo Alto, California.

Martin P., and C. Sullivan. 1992. Guiding American shad with strobe lights. Hydro Review (July Issue). Vol. 11(7):52–59.

Martin, P., J. Downing, E. P. Taft, and C. Sullivan. 1991. A demonstration of strobe lights to repel fish. Pages 103–112 *in* Waterpower '91 Proceedings of the International Conference on Hydropower, Denver, Colorado, July 24–26, 1991. Sponsored by American Society of Civil Engineers. TP 91–5.

Michaud, D. T., and E. P. Taft. 1999. Recent evaluations of physical and behavioral barriers for reducing fish entrainment at hydroelectric plants in the upper Midwest. *In* Power Impacts on Aquatic Resources Conference, Atlanta, Georgia, April 12–15, 1999. Sponsored by Electric Power Research Institute (EPRI).

National Marine Fisheries Service (NMFS). 1994. NMFS Southwest Region position paper on experimental technology for managing downstream salmonid passage. January 1994.

Winchell, F. C., E. P. Taft, S. V. Amaral, L. Everhart, D. Michaud, and C. W. Sullivan. 1997. Evaluation of behavioral devices for attracting/repelling fishes commonly-entrained at Midwest hydro projects. *In* Fish Passage Workshop, Milwaukee, WI, May 6–8, 1997. Sponsored by Alden Research Laboratory, Conte Anadromous Fish Research Laboratory, Electric Power Research Institute, and Wisconsin Electric Power Company. Published by Alden Research Laboratory, Inc., Holden, Massachusetts.

American Fisheries Society Symposium 26:125–144, 2001

Reaction of Chinook Salmon, Northern Pikeminnow, and Smallmouth Bass to Behavioral Guidance Stimuli

Stephen V. Amaral
Alden Research Laboratory, Inc., 30 Shrewsbury Street, Holden, Massachusetts 01520, USA

Frederick C. Winchell
Alden Research Laboratory, Inc., 30 Shrewsbury Street, Holden, Massachusetts 01520, USA

Todd N. Pearsons
Washington Department of Fish & Wildlife, 201 North Pearl Street
Ellensburg, Washington 98926, USA

Abstract.—We examined the response of three fish species to various behavioral stimuli during cage tests conducted in the forebay of the Roza Dam irrigation diversion on the Yakima River, Washington. Wild yearling chinook salmon *Oncorhynchus tshawytscha* were evaluated for responses to strobe light, drop light, and infrasound stimuli that might be used to guide smolts to screening facilities at Roza Dam. We tested northern pikeminnow *Ptychocheilus oregonensis* (strobe light, infrasound) and smallmouth bass *Micropterus dolomieu* (strobe light) to determine if these predators could be repelled. Behavioral responses of chinook salmon to all stimuli were evaluated under daylight, dusk, and nighttime light conditions. All tests with northern pikeminnow and smallmouth bass were conducted at night. We observed strong avoidance by chinook salmon smolts to strobe light during 60-min exposures at night. Little or no reaction was observed in similar tests conducted during daylight hours and at dusk. Avoidance responses during nighttime tests were characterized by rapid movement to the end of the cage opposite the active strobe light. Chinook salmon responses were greatest when the downstream light was operated and when the strobe lights were positioned closer to the test cage. Smallmouth bass response to strobe light was similar to chinook salmon. We observed less avoidance of strobe light by northern pikeminnow. Chinook salmon exhibited minimal response to the drop light and did not respond to infrasound. Northern pikeminnow strongly avoided three of the four infrasound frequencies that were evaluated. These results indicate that strobe lights have potential to repel chinook salmon smolts and smallmouth bass at night. Infrasound generators and drop lights are not considered viable alternatives for guiding chinook salmon smolts at Roza Dam. Infrasound may have potential for repelling northern pikeminnow from areas where smolt predation may be high (e.g., bypass outfalls).

The Yakima/Klickitat Fisheries Project (YKFP) is exploring techniques for increasing the passage rate of wild chinook salmon *Oncorhynchus tshawytscha* smolts into the fish screening and collection facilities at Roza Dam on the Yakima River. These facilities are designed to bypass outmigrating smolts around an irrigation canal intake. The Washington Department of Fish & Wildlife is using the screening facilities to collect out-migrants for a PIT-tagging program that is being conducted to assess spring chinook supplementation in the Yakima River basin. The feasibility of techniques to repel predators, including northern pikeminnow *Ptychocheilus oregonensis* and smallmouth bass *Micropterus dolomieu*, from bypass outfalls in the river also is of interest. Based on recent favorable research, the YKFP selected three behavioral devices, strobe lights, a drop light and an infrasound generator, for evaluation of their abilities to guide fish into the screening facilities at Roza.

Strobe light stimuli have produced avoidance responses from several species of anadromous salmonids during controlled experiments and field studies (Puckett and Anderson 1988a; Nemeth and Anderson 1992; Bernier 1995; Brown and Bernier 2000; Ploskey et al. 1998; Mueller et al. 1999). The drop light source was chosen for evaluation at Roza based on avoidance responses that were observed during cage tests conducted with juvenile chinook salmon (Gene Ploskey, U.S. Army Corps of Engineers, personal communication). Similar to strobe light, infrasound stimuli have been shown to elicit avoidance responses from salmonids during laboratory and field tests (Knudsen et al. 1992, 1994, 1997; Taft et al. 1996; Mueller et al. 1998, 1999).

The primary goal of our research was to determine if strobe light, drop light, and infrasound stimuli have potential to effectively divert out-migrants to the Roza collection facilities. An additional goal was to develop design and operating criteria for a field application with any of the tested devices that elicited strong avoidance reactions. In conjunction with the evaluation of chinook salmon responses to each behavioral device, a limited number of tests was conducted with northern pikeminnow and smallmouth bass to assess the feasibility of using behavioral deterrents to repel predators from critical areas where smolts may be especially vulnerable to predation.

Methods

Study Site and Test Equipment

Experiments were conducted at the Roza Diversion Dam located on the Yakima River approximately 208 km upstream from its confluence with the Columbia River. The dam is located within the lower end of a 37 km long basalt canyon. The Roza facility consists of the headworks to an irrigation canal, a concrete dam, and a screening facility that diverts entrained salmon out-migrants through a bypass system. The bypass facilities return fish to the main-stem river approximately 300 m below the dam.

Fish responses to behavioral stimuli were evaluated using a floating test platform (Figure 1). The test platform was moored downstream of the screening facilities and upstream of the entrance to the irrigation canal on the western shoreline. On-site testing is advantageous because it allows for the evaluation of fish that occur naturally at a site and under environmental conditions (e.g., water temperature, pH, DO level, and turbidity) that are experienced by the populations targeted for diversion. The test platform was a modified pontoon boat with a third pontoon in the form of an outrigger that was attached to the main deck. An enclosed shed on the main deck housed all electronic equipment associated with the behavioral devices and video observation system.

A cage was positioned between the boat and the pontoon outrigger (Figure 1). The behavioral devices were suspended from steel beams attached to the outrigger and main decks. Position and orientation of each device in relation to the cage could be manipulated both horizontally and vertically in most directions and within certain distances, which were determined by support beam lengths, suspension lines, and water depths. The cage was an aluminum frame fitted with a clear vinyl exterior shell and contained a separate test channel in which fish were placed for evaluation of behavioral responses (Figure 2). The separation of the main cage compartment from the test channel

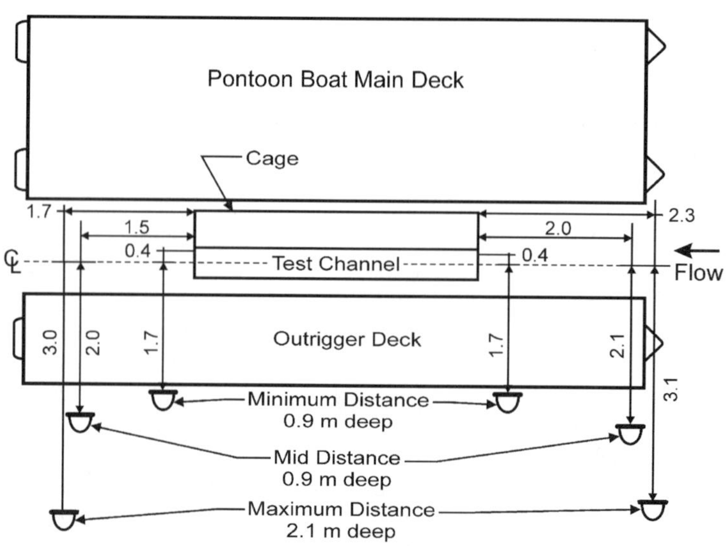

Figure 1. Schematic plan view of test platform and strobe light positions relative to the cage and test channel centerline (distances are in meters).

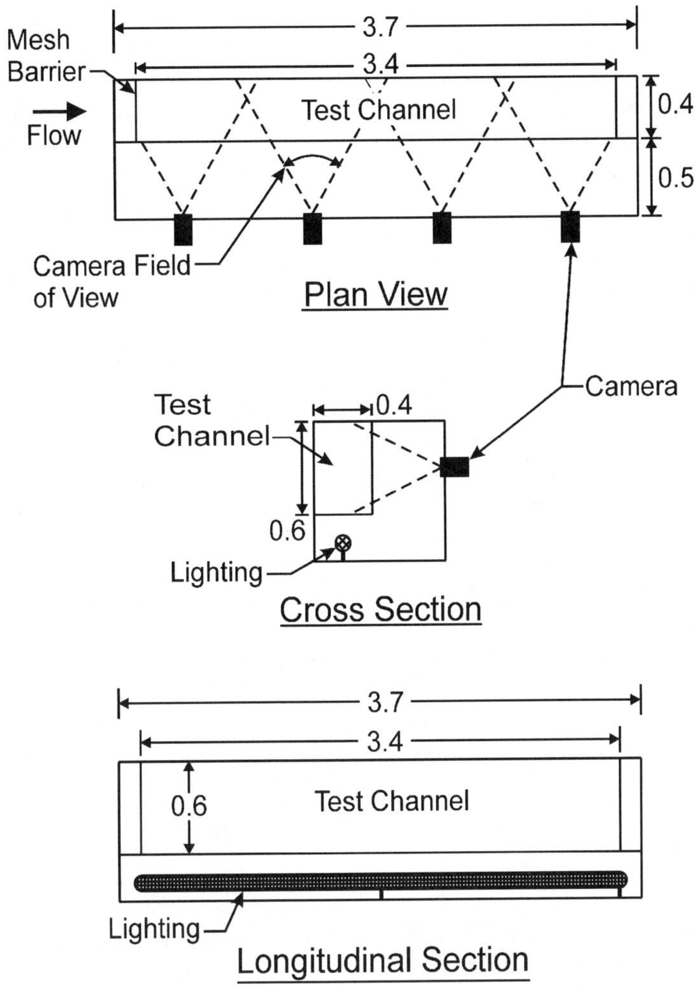

Figure 2. Cage design and specifications (dimensions are in meters).

allowed for filtered water to be used in the main compartment. The filtered water was needed for unobstructed observation of test fish with underwater cameras. Underwater cameras and supplemental lighting were positioned in the cage outside of the test channel. The dimensions of the cage were approximately 0.9 m wide by 0.9 m deep by 3.7 m long. The cage was suspended with cables from a lifting apparatus that was used to position it at desired depths and to lift it clear of the water when modifications or repairs were necessary.

The test channel (0.4 m wide by 0.6 m deep by 3.4 m long) was constructed within the cage (Figure 2) and was enclosed on the top and sides using the same clear vinyl material that enclosed the entire cage. The ends of the test channel were covered with a black plastic mesh material to allow ambient river water to pass through.

Four underwater cameras were mounted on the inside of the cage across from the test channel (Figure 2). The cameras were evenly spaced along the horizontal centerline of the test channel. A four-channel Pulsar multi-image processor (i.e., multiplexer) was used to allow viewing of all four video-camera images on a single monitor. Video images from all tests were recorded with a VCR. A single camera view or a quad-image display from all four cameras could be replayed from the VHS tapes on which video images were recorded. Four underwater lights with diffusers were mounted under the test channel to create uniform light conditions during periods when ambient light was not adequate for video observation of fish (i.e., nighttime hours). The lights were powered through a rheostat so that illumination could be set at the minimum level required to observe fish response.

Two Flash Technology model AGL 901 underwater strobe lights, each mounted on an aluminum pole, were deployed from either end of the outrigger deck (i.e., bow and stern) in three alternative configurations (Figure 1). For reference purposes, the strobe lights were designated as upstream (bow) and downstream (stern) light sources to correspond to their position in relation to the flow through the test cage. Power supplies for the two strobe lights were located on the main deck and were equipped with controllers capable of varying flash rates between 300 and 450 flashes per minute. During a test, the strobe lights were positioned at one of three locations designated as "minimum," "mid," and "maximum" based on the distance from the test channel (Figure 1). The minimum and maximum distances were the actual limits at which the strobe lights could be placed in relation to the ends of the cage. When both lights were used during a test (activation of each light was alternated over a specified time period), they were placed at the same position in reference to their respective ends of the cage (i.e., both lights were located at either the minimum, mid, or maximum distance).

A drop light (500-Watt Subsea light) with a variable output transformer was also deployed on the platform for fish testing. This light was positioned on the center line of the test channel about 0.6 m from either the upstream or downstream end of the cage, depending on the location of fish at the beginning of a test. Two on/off intervals (1 and 15 s) and continuous operation were evaluated during drop light tests. The on/off intervals were tested by manually ramping the light on and off at the specified intervals (i.e., every 1 or 15 s).

A single particle motion generator (PMG) was used to evaluate fish response to infrasound. The main features of the PMG include a rotary valve, an air-filled expansion chamber, and a submersible pump. Water supplied by the pump is forced through the rotary valve to create the infrasound signal. The volume of water pumped through the valve controls the amplitude of the output signal and its rotation speed determines the frequency. The PMG was deployed from the bow of the pontoon boat between the two decks and upstream of the test channel at a depth of about 0.9 m. The PMG is capable of producing frequencies between 10 and 60 Hz. Although the amplitude (energy output level) of PMG emissions can be varied, the signals that were tested at Roza were produced at full power to maximize the area of coverage. The frequencies that were evaluated ranged from 10 to 50 Hz.

Test Fish

Yearling spring chinook salmon smolts were collected from the juvenile bypass collection facility at the Roza Dam. Generally, fish were collected during late morning or early afternoon hours and used in tests scheduled for that day or night. Most fish were held in the collection box that receives fish diverted from the bypass channel. Fish that were retained for more than a day prior to testing were kept in holding bins supplied with a continuous flow of water from the Roza forebay. The mean fork length of chinook salmon test groups ranged from 111 mm to 129 mm, with an overall mean of 123 mm (SD = 11). Northern pikeminnow were collected by rod and reel from the Roza forebay and a small pond near the Yakima River; smallmouth bass were collected by Washington Department of Fish and Wildlife personnel from nearby ponds. The average lengths of pikeminnow and smallmouth bass were 327 mm (SD = 53 mm) and 351 mm (SD = 6 mm), respectively.

Experimental Design

Initial tests with chinook salmon smolts consisted of exposing a group of fish to the five different light stimuli (three drop light on/off intervals and two strobe light flash rates) during 1 min treatment periods (i.e., one treatment period per stimulus; Table 1; Figure 3). Control periods were designated as the 1 min prior to a treatment period (Figure 3). A 13 min ambient period separated the end of a stimulus treatment period from the beginning of the next stimulus control period (Figure 3). Therefore, a complete trial consisted of five paired 1-min control and treatment periods separated by 13-min ambient periods (Figure 3). Fish position data were recorded from the videotapes at 15-s intervals for each control and treatment period. A total of nine trials was conducted using this design, three each during daylight, dusk, and nighttime hours (Table 1). Each trial was conducted with a different group of fish (i.e., nine separate groups of fish were used) and the number of fish per trial ranged from 12 to 17. The strobe lights were located at the mid-distance position during these trials.

Table 1. Summary of tests conducted with strobe and drop lights. Sixty-min strobe light tests consisted of alternating the operation of the upstream and downstream lights every 10 min, resulting in three, 10-min periods for the light at each location.

Trial no.	Strobe flash rates (flashes/min)	Drop light on/off intervals (s)	Duration of each stimulus (min)	Duration of ambient period between stimuli (min)	Strobe light location	Ambient light condition
			Chinook Salmon Smolts			
1, 4, 7	300, 450	1, 15, constant	1	13	mid	day
2, 5, 8	300, 450	1, 15, constant	1	13	mid	dusk
3, 6, 9 a	300, 450	1, 15, constant	1	13	mid	night
10, 11, 15	300, 450	1, 15, constant	2	13	mid	night
27, 36, 38	300	–	60	–	min	day
14, 19, 26	300	–	60	–	mid	night
29, 31, 34	300	–	60	–	max	night
40, 41, 44	450	–	60	–	min	day
28, 37, 39	450	–	60	–	min	dusk
45	450	–	60	–	mid	dusk
16, 21, 24	450	–	60	–	mid	night
30, 33, 35	450	–	60	–	max	night
			Northern pikeminnow			
20	300	–	60	–	mid	night
25	450	–	60	–	mid	night
			Smallmouth bass			
46	450	–	60	–	max	night

[a]Supplemental lighting was used for the first three nighttime tests with chinook salmon smolts (trial numbers 3, 6, and 9); no additional lighting was used during all subsequent nighttime tests.

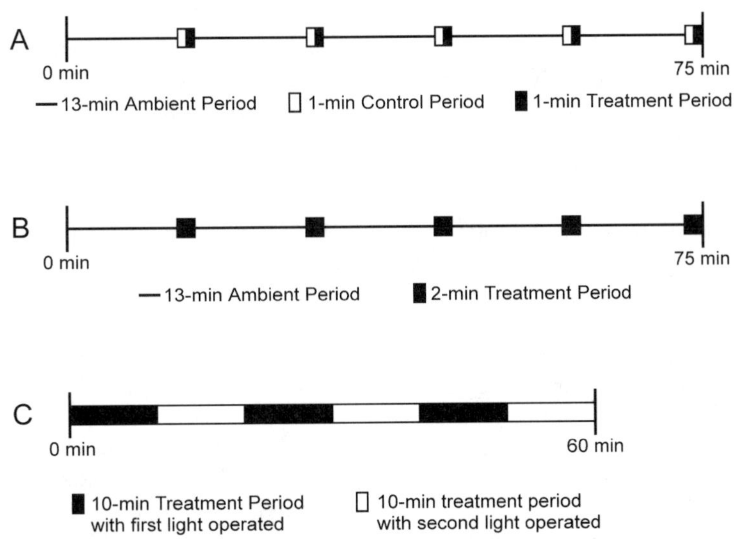

Figure 3. Testing sequence of 1-min exposure trials (A), 2-min exposure trials (B), and 60-min exposure trials (C). Fish position data were recorded at 15-s intervals during control and treatment periods of the 1-min exposure trials, at 30-s intervals during treatment periods of the 2-min exposure trials, and at 30-s intervals during the first minute of each 10 min period of the 60-min exposure periods and at 1 min intervals thereafter. Light stimuli evaluated during 1-min and 2-min exposure periods included two strobe light flash rates (300 and 450 flashes/min) and three drop light on/off intervals (1-s, 15-s, and continuous operation). Trials with the infrasound device were conducted using the 2-min exposure protocol. The two strobe light flash rates were evaluated during 60-min exposure trials.

To eliminate the possibility that the use of artificial light was affecting fish behavior during the initial nighttime trials (i.e., lack of response was an artifact of experimental design), a series of tests was conducted without supplemental lighting. Direct observation from above the test cage was used to record fish positions during these trials. For this approach, fish positions were observed and recorded at 30-s intervals during 2-min strobe and drop light treatment periods (the same five light stimuli were used). The 2-min treatments were separated in time by 13-min ambient periods (Figure 3). Treatment periods and data recording intervals were extended because it was difficult to manually record fish positions in real time at 15-s intervals. Three trials were conducted in this manner during nighttime hours with three different groups of fish and with the strobe lights at the mid-distance position (Table 1). Control periods were not included in these trials because fish were only visible when the strobe or drop lights were on.

Observations during the 2-min trials without supplemental lighting indicated that fish were responding to the strobe light. These responses appeared to be limited by the time duration of stimulus exposure and by the location of fish in relation to the operating light (i.e., the starting position of fish was typically near the end of the cage opposite the light source; this restricted the ability of fish to move greater distances away from the light source). Consequently, additional modifications to the testing protocols were made to assess chinook salmon response during longer exposures to strobe light, and to identity more definitive behavioral patterns that may not be evident in trials with shorter exposure periods and with the light coming from only one direction. For these trials, fish positions were recorded over a 60-min period during which the operation of the upstream and downstream strobe lights was alternated every 10 min (i.e., each light was operated for three, 10-min periods; Figure 3). In addition to having a strobe light at either end of the cage, the distance of the lights also was manipulated to assess fish responses under varying strobe light intensities. Three light locations, as described previously, were incorporated into the test program for this purpose. Control periods were not included in these trials mainly because fish could not be seen under nighttime conditions without additional light, but also because the longer treatment periods were considered adequate to identify any avoidance response patterns that may occur with the alternating light operation. Additionally, drop light stimuli were not evaluated during 60-min trials due to a lack of response observed during the 1- and 2-min exposure trials. Strobe light tests conducted during daytime and early evening hours also were performed using the 60-min format.

Three 60-min nighttime trials with chinook salmon were conducted at each flash rate and with the strobe lights at the mid and maximum distances (i.e., 12 separate tests; Table 1). Three trials also were conducted during the daytime with both flash rates (six separate tests) and during dusk with 450 flashes per min at the minimum distance (Table 1). One trial was conducted at dusk with 450 flashes per min and the lights at the mid distance (Table 1). Fish position data for nighttime trials were recorded 30 and 60 s after a light source was activated and at 1 min intervals thereafter. The number of chinook salmon used per trial during the 60-min exposures ranged from 13 to 24 fish.

Two 60-min strobe light trials also were conducted with adult northern pikeminnow, and one trial was conducted with adult smallmouth bass. Northern pikeminnow were evaluated with 300 and 450 flashes per minute, and smallmouth bass with 450 flashes per minute. All trials with these species were conducted during nighttime hours with four fish per test. The lights were located at the mid-distance position for northern pikeminnow trials and at the max-distance position for the smallmouth bass trial.

Five frequencies, 10, 20, 30, 40, and 50 Hz, were evaluated during infrasound trials with chinook salmon smolts. Each trial consisted of one 2-min exposure per frequency. The five frequencies were emitted in random order with a 13-min ambient period between each 2-min exposure period. Six trials were conducted during daytime hours and two trials were done at night. The number of smolts used in these tests ranged from 18 to 21. One trial was conducted during the day with three northern pikeminnow. The underwater supplemental lighting was used for PMG tests conducted at night, as there was no other means for determining fish position during these tests.

Environmental Conditions and Light Field Measurements

Light levels, water temperature, turbidity, water velocity in the test channel, and weather conditions were recorded throughout the two-week test program. Ambient light, water temperature, and test channel water velocity were measured at the beginning and end of each test. Turbidity and Secchi disk measurements were recorded daily in the late morning or early afternoon. Strobe light intensities were measured inside the test channel under each of the three ambient light conditions for which tests were conducted, and at night along three different transects extending from a flash head. Test channel measurements were made in each quadrant for the three strobe light positions and two flash rates that were tested to determine the light intensity gradient along the length of the cage. To determine the effect of nylon sheeting on light penetration into the test channel, light measurements also were recorded with the sensor directly outside and inside the test cage. The two measurements were made virtually at the same location, separated only by the nylon sheeting.

Ambient light levels and the strobe light intensities were measured using a Li-Cor model 185B Quantum Radiometer/Photometer fitted with a LI-193SB Spherical Quantum omnidirectional sensor and a directional sensor. All light measurements were recorded in quantum energy units of $\mu E/m^2/s$. Turbidity was recorded in nepholometric turbidity units (NTU's) using a turbidity meter. Water velocities were measured with an Ott velocity meter that was deployed behind the fish screen at the downstream end of the test channel.

Data Analysis

We analxyzed fish position data in two ways, one as an aggregate center-of-school for an entire test group and the other as individual fish positions. Procedures used to analyze aggregate fish position and movement relative to the stimulus source were similar to methods employed by Nestler et al. (1992) for cage tests conducted with alewife and various sound stimuli. Fish position data for each 15-s interval in the 1-min strobe and drop light control and treatment periods were determined during a review of video recordings from each trial. To standardize the location of fish relative to the active source, the number of fish in each quadrant of the test channel was multiplied by the distance from the end of the channel nearest the stimulus source to the midpoint of the corresponding quadrant. For example, if 10 fish were in quadrant 1, then this count was multiplied by 0.419 m (i.e., midpoint of quadrant 1 from end of the test channel closest to the source) producing a weighted distance value of 4.2; fish counts for quadrants 2, 3, and 4 were multiplied by 1.258, 2.097, and 2.936 m, respectively. The weighted distances that were calculated for each quadrant were summed and divided by the total number of fish in the pen to estimate a center-of-school location for each 15-s interval. Similar to the weighted distances calculated for each quadrant, the center-of-school estimates were relative to the location of the active source. The same procedures were used to calculate center-of-school positions for the 2-min exposure trials with strobe and drop lights and the 60-min trials with strobe light.

Center-of-school positions for the 60-min strobe light trials conducted at night were plotted against time to evaluate the response of fish to light from both source locations (i.e., to determine if fish demonstrated a discernible pattern of behavior over the entire 60 min). Fish movements during 60-min daytime and early evening trials with strobe lights and during all infrasound trials were evaluated based on real-time observations which consisted of descriptions of fish behavior and visual estimates of changes in the center-of-school position. Qualitative descriptions of response (no response, weak, moderate, strong) were assigned based on the distance and quickness of fish movements.

For the 1- and 2-min exposure trials with strobe and drop lights, the distance and direction of fish movement were determined for control and treatment periods by subtracting the center-of-school position at the end of a period from the position at the beginning. Positive displacement values represented movement of the center-of-school position towards a light source and negative values represented movement away from the source. The center-of-school displacement data provide a relative measure of fish movement that we used to evaluate the strength of responses and to examine differences between treatment and control fish movements. However, the center-of-school position can mask the variability in individual fish positions, making it an inappropriate parameter for statistical comparisons. For example, the center-of-school position for a time interval where the test fish are evenly split between quadrants 1 and 4 would be the same as it would be for an interval where all fish were evenly divided between quadrants 2 and 3.

To account for the variability in individual fish movements, we used Mann–Whitney Rank Sum tests to determine statistical differences between individual fish positions (the location of a fish was represented by the midpoint of the quadrant it was in) during treatment and control periods of the 1-min exposure trials. A nonparametric statistical test was used because the standardized fish position data did not meet assumptions of normality or equality of variances (Zar 1984). For these tests, the mean fish position for a stimulus exposure period was calculated by summing the individual fish locations (i.e., quadrant midpoint distance from end of cage nearest a light source) from all observation intervals and then dividing by the sample size. Sample size was the number of individual fish position observations for a control or treatment period, which was equivalent to the number of fish in the cage multiplied by the number of time intervals for which fish position data were recorded (i.e., five intervals).

Results

1- and 2-min Exposure Trials

Chinook salmon response to strobe and drop lights during trials with 1-min exposure periods varied among the different stimuli and among the time periods (i.e., daytime, early evening, and nighttime). In general, no startle responses and few dramatic movements away from the light sources were observed for any of the drop light on/off intervals or for either of the two strobe flash rates (Table 2).

Chinook salmon appeared to respond slightly to the drop light illuminated at 1-s intervals during one daytime and two early evening tests. Despite significant differences in fish position when compared with the control periods ($P < 0.05$), the change in center-of-school position during the treatment periods of these three tests was only between 0.4 and 1.0 m away from the light source (Table 2). With the drop light alternated on and off at 15-s intervals and with the light operated continuously, the mean fish distance from the active light during treatment periods was significantly different from the control in one evening trial and one nighttime trial (Table 2). During these tests, shifts in the center-of-school position away from the light source were minimal (less than 0.5 m), except for the nighttime test with the light operated continuously, when the shift in center-of-school position was 1.7 m away from source.

For strobe lights, mean fish distance from the active source did not differ significantly between control and treatment periods for any of the tests conducted with a strobe flash rate of 300 flashes per minute and for only one test conducted with a rate of 450 flashes per minute (Table 2). Shifts in the center-school-position away from the active source occurred during several treatment periods at both flash rates, but these shifts never exceeded a distance of 0.7 m (Table 2). The low number of significant differences between treatment and control periods, variability in responses among and within time periods, and minimal distance of movement away from an active source, all indicated that the drop light and strobe light stimuli were not affecting chinook salmon behavior under the conditions evaluated.

During the 2-min exposure trials, no discernable avoidance responses were exhibited by chinook salmon to any of the three on/off intervals evaluated with the drop light (Table 3). The mean distance of fish from the end of the cage nearest the drop light ranged from 1.7 to 2.5 m (Table 3). This distance range demonstrates that most fish remained in the third quadrant of the test channel for all drop light exposure periods during these trials. Center-of-school movement was either toward the drop light or stagnant during all trials with the 1- and 15-s on/off intervals. Center-of-school position shifted away from the active source during two of the three tests when the drop light was operated continuously, but the distance of displacement was only 0.1 and 0.2 m (Table 3).

The direction of center-of-school displacement was away from the active source during all 2-min exposure trials with the two strobe light flash rates (Table 3). With the exception of two of these trials, displacement of the center-of-school position was between 0.5 and 1.0 m away from the light source. At the start of the strobe light exposure periods, most fish were positioned in the half of the test channel furthest from the light source, often in quadrant 3. The mean individual fish distance from the end of the channel nearest the light source ranged from 2.4 to 2.8 m during the 2-min strobe light exposures. This indicates that most fish were located in the quadrant farthest from the strobe light after the light was activated.

60-min Exposure Trials

Chinook salmon demonstrated a strong pattern of avoidance to both strobe light flash rates during the 60-min exposure tests conducted at night (Figure 3). During tests with the strobe lights at the mid distance, avoidance responses often were characterized by shifts in center-of-school position from the quadrant closest to the source to the quadrant furthest away. This end-to-end cage movement typically occurred within three minutes after a

Table 2. Results from 1-min exposure trials with chinook salmon and five drop and strobe light stimuli. *P*-values with an asterisk indicate a statistically significant difference in the mean distance of fish from the end of the cage nearest the source during control and treatment periods. Center-of-school displacement was calculated by subtracting the center-of-school position at the beginning of a control or treatment period from the position at the end. Negative displacement values indicate movement away from the stimulus source and positive values indicate movement towards it. Stimulus abbreviations represent the following: D1, drop light activated at 1-s intervals; D15, drop light activated at 15-s intervals; D60, drop light operated continuously; S300, strobe light operated at 300 flashes per minute; and S450, strobe light operated at 450 flashes per minute.

				Mean and SD of individual fish distance (m) from end of cage nearest active light source and Mann-Whitney Rank Sum Test *P*-value			Center-of-school displacement (m)	
Stimulus	Time period	Trial	*N*	Control	Treatment	*P*	Control	Treatment
D1	day	1	70	2.2 (0.8)	1.2 (0.7)	0.001*	1.8	−1.0
		4	60	1.5 (1.3)	1.5 (1.3)	0.998	0.0	0.0
		7	75	1.2 (1.0)	1.0 (0.8)	0.736	0.3	0.2
	eve	2	80	0.5 (0.3)	0.6 (0.3)	0.413	−0.1	0.0
		5	75	0.7 (0.5)	1.0 (0.7)	0.001*	0.0	−0.4
		8	75	0.6 (0.6)	0.9 (0.7)	0.001*	0.0	−0.6
	night	3	70	1.5 (1.2)	1.5 (1.1)	0.741	0.4	−0.2
		6	85	0.9 (0.6)	0.9 (0.5)	0.730	0.1	−0.1
		9	80	1.0 (0.5)	1.2 (0.4)	0.125	−0.1	−0.2
D15	day	1	70	2.7 (0.8)	2.6 (0.8)	0.434	−0.1	0.8
		4	60	1.7 (1.2)	1.5 (1.3)	0.379	0.2	0.2
		7	75	2.6 (0.8)	2.7 (0.7)	0.862	−0.3	0.1
	eve	2	80	0.5 (0.3)	0.5 (0.3)	0.786	0.0	0.0
		5	75	0.9 (0.5)	1.6 (0.9)	0.001*	0.1	−0.3
		8	75	0.6 (0.3)	0.5 (0.3)	0.889	0.0	0.0
	night	3	70	1.5 (1.1)	1.5 (1.2)	0.863	0.0	0.2
		6	85	0.9 (0.8)	0.9 (0.8)	0.828	0.0	0.0
		9	80	0.7 (0.6)	0.9 (0.5)	0.005*	0.1	−0.4
D60	day	1	70	1.3 (1.2)	1.3 (1.2)	0.927	−0.1	0.0
		4	60	1.7 (1.3)	1.6 (1.2)	0.629	0.0	0.2
		7	75	1.4 (1.2)	1.2 (1.1)	0.326	−0.1	0.5
	eve	2	80	1.3 (1.2)	1.2 (1.1)	0.572	0.2	0.2
		5	75	0.9 (0.4)	1.1 (0.5)	0.015*	0.1	−0.4
		8	75	2.0 (0.3)	2.1 (0.3)	0.232	−0.1	0.0
	night	3	70	1.5 (1.2)	1.4 (1.1)	0.996	−0.3	0.5
		6	85	0.9 (0.8)	1.6 (1.2)	0.001*	−0.1	−1.7
		9	80	0.9 (0.5)	1.0 (0.5)	0.371	0.2	−0.3
S300	day	1	70	2.9 (0.1)	2.9 (0.0)	0.885	0.0	0.0
		4	60	0.6 (0.6)	0.7 (0.7)	0.630	−0.2	0.0
		7	75	2.6 (0.8)	2.6 (0.8)	0.721	0.2	0.2
		7	75	0.8 (0.7)	0.9 (0.7)	0.765	−0.2	0.0
	eve	2	80	0.8 (0.8)	0.9 (0.9)	0.568	−0.1	−0.2
		5	75	0.5 (0.3)	0.6 (0.3)	0.398	−0.1	−0.1
		8	75	0.4 (0.0)	0.5 (0.2)	0.481	0.0	−0.1
	night	3	70	1.2 (1.1)	1.4 (1.2)	0.208	−0.1	−0.4
		6	85	1.3 (0.8)	1.4 (0.9)	0.481	0.1	−0.7
		9	80	1.0 (0.5)	1.1 (0.5)	0.799	0.0	0.0

Table 2. Continued.

Stimulus	Time period	Trial	N	Mean and SD of individual fish distance (m) from end of cage nearest active light source and Mann-Whitney Rank Sum Test P-value			Center-of-school displacement (m)	
				Control	Treatment	P	Control	Treatment
S450	day	1	60	1.8 (1.1)	1.4 (1.2)	0.030*	1.6	−0.6
		4	75	1.5 (1.3)	1.5 (1.3)	0.998	0.0	0.0
		7	80	0.7 (0.5)	1.1 (1.0)	0.144	0.2	−0.7
	eve	2	75	0.6 (0.6)	0.7 (0.7)	0.364	−0.1	−0.1
		5	75	1.3 (0.7)	1.5 (0.7)	0.083	−0.4	−0.5
		8	70	2.1 (0.1)	2.1 (0.1)	0.998	0.0	0.0
	night	3	85	1.6 (1.2)	1.6 (1.2)	0.977	0.1	0.1
		6	80	0.9 (0.6)	1.0 (0.6)	0.161	0.0	−0.4
		9		1.1 (0.4)	1.3 (0.4)	0.060	0.1	−0.3

Table 3. Results from 2-min exposure trials with chinook salmon and five drop and strobe light stimuli. Center-of-school displacement was calculated by subtracting the center-of-school position at the beginning of a treatment period from the position at the end. Negative values indicate movement away from the stimulus source and positive values indicate movement towards it. Stimulus abbreviations represent the following: D1, drop light activated at 1-s intervals; D15, drop light activated at 15-s intervals; D120, drop light operated continuously; S300, strobe light operated at 300 flashes per minute; and S450, strobe light operated at 450 flashes per minute.

Stimulus	Trial	N	Mean (SD) individual fish distance (m) from end of cage nearest active light source	Center-of-school displacement (m)
D1	1	70	2.1 (0.9)	0.2
	2	60	1.7 (0.9)	0.1
	3	75	2.2 (0.7)	0.1
D15	1	70	2.4 (0.8)	0.1
	2	60	2.3 (0.6)	0.1
	3	75	2.2 (0.8)	0.0
D120	1	70	2.3 (0.7)	−0.2
	2	60	1.9 (0.8)	−0.1
	3	75	2.5 (0.6)	0.2
S300	1	70	2.4 (0.7)	−0.5
	2	60	2.6 (0.7)	−0.7
	3	75	2.8 (0.4)	−0.2
S450	1	60	2.7 (0.4)	−0.2
	2	75	2.7 (0.6)	−1.0
	3	80	2.7 (0.5)	−0.6

source was activated, and often within the first 30 s. An exception to this behavior was noted during one mid distance test at 300 flashes per minute, when shifts in the center-of-school positions away from the source were considerably less than observed in previous trials (Figure 4). In general, avoidance was stronger and more consistent to the downstream light source than to the upstream source, and when the lights were positioned closer to the test channel (i.e., mid distance versus maximum distance). Also, avoidance responses to the upstream light, although strong, decreased slightly with each successive 10 min interval. Decreases in avoidance response to the upstream source were observed at both flash rates and source positions (mid and maximum).

In general, chinook salmon either did not respond or responded weakly to strobe light during the 60-min exposure trials conducted during daylight and early evening hours (Table 4). No distinct back and forth displacements of fish were observed during any of these trials as was observed during the 60-min nighttime trials. Weak responses were noted more frequently during evening trials than during daylight trials, and usually comprised shifts in the center-of-school position away from the active source of about one to two quadrants (0.8–1.7 m). There was no trend in the observed responses with

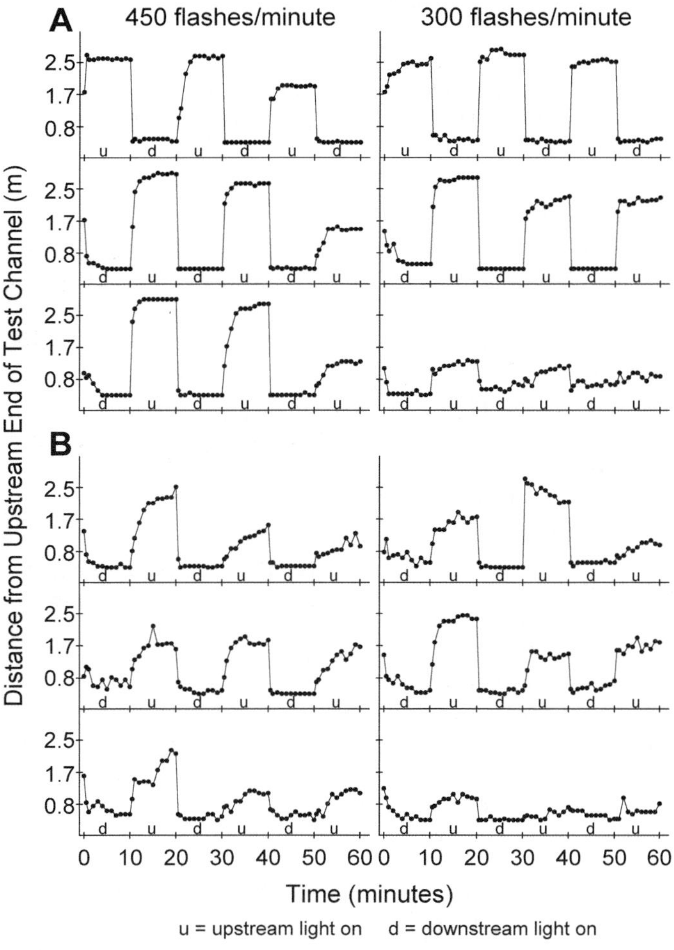

Figure 4. Center-of-school position during 60-min exposure tests with chinook salmon conducted at night and with the strobe lights at the mid-distance (A) and max-distance positions (B).The limits of the estimated center of school position are 0.4 and 2.9 m. The upstream source was activated first for the initial trials with the lights at the mid-distance position; the downstream source was activated first for all remaining tests. Operation of the two lights was alternated every 10 min. Each graph represents a single trial with a separate group of fish (i.e., fish were not used in more than one trial).

Table 4. Response of chinook salmon to two strobe light flash rates during 60-min exposure tests conducted during daytime and early evening hours. Response levels are based on changes in behavior and shifts in the center-of-school position as recorded by an observer during each test. "No response" was defined as no movement or movement of less than one quadrant away from the source; "weak" responses were classified as movements of one to two quadrants away from the source for short durations of time during a 10 min period.

Trial	Time period	Source distance position	Level of response recorded at 10 min intervals 0–10	10–20	20–30	30–40	40–50	50–60
300 flashes per minute								
1	daytime	min	no response	no response	no response	no response	no response	no response
2	daytime	min	weak	no response	weak	no response	no response	no response
3	daytime	min	no response	weak	no response	no response	no response	no response
450 flashes per minute								
1	daytime	min	no response	no response	no response	no response	no response	no response
2	daytime	min	no response	no response	no response	no response	no response	no response
3	daytime	min	weak	no response	no response	weak	weak	no response
4	dusk	min	no response	no response	no response	no response	weak	weak
5	dusk	min	no response	weak	weak	weak	weak	weak
6	dusk	min	no response	weak	weak	no response	no response	weak
7	dusk	mid	no response	no response	no response	no response	no response	no response

respect to source location, time interval, or flash rate during daytime and evening trials.

Northern pikeminnow demonstrated a moderate pattern of avoidance to 300 flashes per minute; any response that may have occurred at 450 flashes per minute was less evident (Figure 5). At 300 flashes per minute, the center of school position shifted away from light source as the operation of each light was alternated. Avoidance was more evident during the operation of the upstream light than it was for the downstream light. The center-of-school position varied considerably during the test of 450 flashes per minute and did not indicate any clear response of the fish to the strobe light.

Smallmouth bass demonstrated strong avoidance to strobe light from both directions (Figure 6). The four fish that were in the cage were moved from one end to the other with the activation of each strobe light. Similar to chinook salmon, smallmouth bass response appeared to be stronger and more consistent when the downstream light was operated compared with when the upstream light operated. That is, fish response was stronger when the stimulus came from the source located behind the fish versus when it came from the source located in front of the fish.

The small number of fish used in northern pikeminnow and smallmouth bass tests (four per test) resulted in the movements of one fish having a large impact on the calculated center-of-school position (i.e., a large shift in position of one fish resulted in a large shift in position of the center of school). Due to the small number of fish evaluated and limited replication (two tests with northern pikeminnow and one with smallmouth bass), the results of these tests should be considered preliminary in nature with further testing required to verify responses.

Infrasound Trials

Chinook salmon did not exhibit avoidance or startle reactions when exposed to any of the five infrasound frequencies generated by the PMG during daytime and nighttime trials. Northern pikeminnow evaluated during a daytime test with infrasound demonstrated moderate to strong avoidance to each frequency, except 20 Hz.

Environmental Conditions and Light Field Measurements

Water temperature and clarity did not vary much during the two-week study period. Water temperature, measured at the beginning and end of each trial, ranged from 10 to 14°C. Water turbidity ranged from 3.28 to 6.50 NTUs and Secchi depth measurements ranged from 1.7 to 2.4 m, with clear weather conditions during most of the two week test program. Water velocity at the downstream end of the test channel ranged from 0.08 m/s to

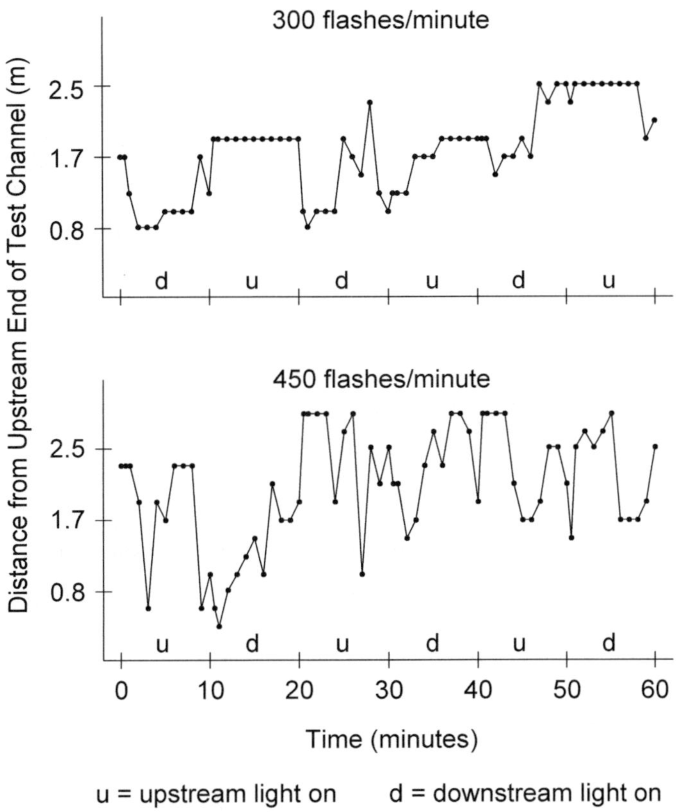

Figure 5. Center-of-school position during nighttime strobe light tests with northern pikeminnow. The limits of the center of school position are at 0.4 and 2.9 m. The downstream source was operated first for the 300 flash rate test; the upstream source was operated first for the 450 flash rate test. Operation of the two lights was alternated every 10 min.

0.15 m/s. The largest difference between test channel water velocities at the beginning and end of a trial was 0.03 m/s.

The light field produced by one strobe light in open water (i.e., no obstructions between light and points of measurement) was measured for a flash rate of 450 flashes per minute (Figure 7). As expected, the greatest intensities were observed along the centerline of the flash head. Lower intensities were observed along transects at 45 and 90 degree angles to the flash head centerline (Figure 7). These light readings were recorded at a turbidity level of about 3.3 NTU, which is at the low end of the turbidity levels that were recorded during the two-week study period.

Strobe light intensities measured in the test channel varied with time of day and light location. For both flash rates, there was a strong gradient across the length of the test channel during nighttime hours with the lights at mid-distance (Figure 8). This gradient decreased considerably with the lights at the maximum distance position. Measurements during daytime and early evening hours demonstrate an inability to detect strobe light in the test channel with high ambient light conditions (Figure 9).

The intensity of the strobe light was reduced by about 25% as it passed through the nylon sheeting of the cage. The light intensity measured outside the cage was 0.20 μE/m2/s, and the intensity directly inside the cage was 0.15 μE/m2/s. Although not measured, it is likely that another reduction of 25% in intensity also occurred as the light penetrated the nylon sheeting that enclosed the test channel within the cage.

Discussion

Although the shorter-duration tests were generally inconclusive, a rapid and prolonged avoidance

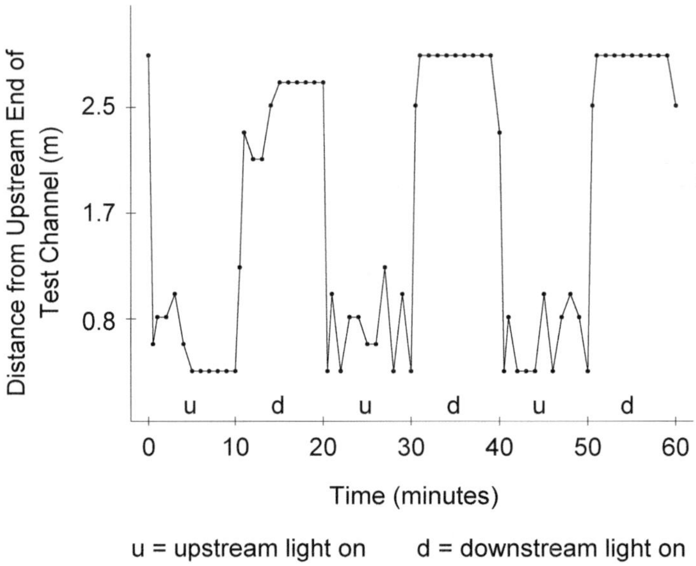

Figure 6. Center of school position during the nighttime strobe light test with smallmouth bass. The limits of the estimated center of school position are 0.4 and 2.9 m. The downstream source was activated first and the operation of the two lights was alternated every 10 min.

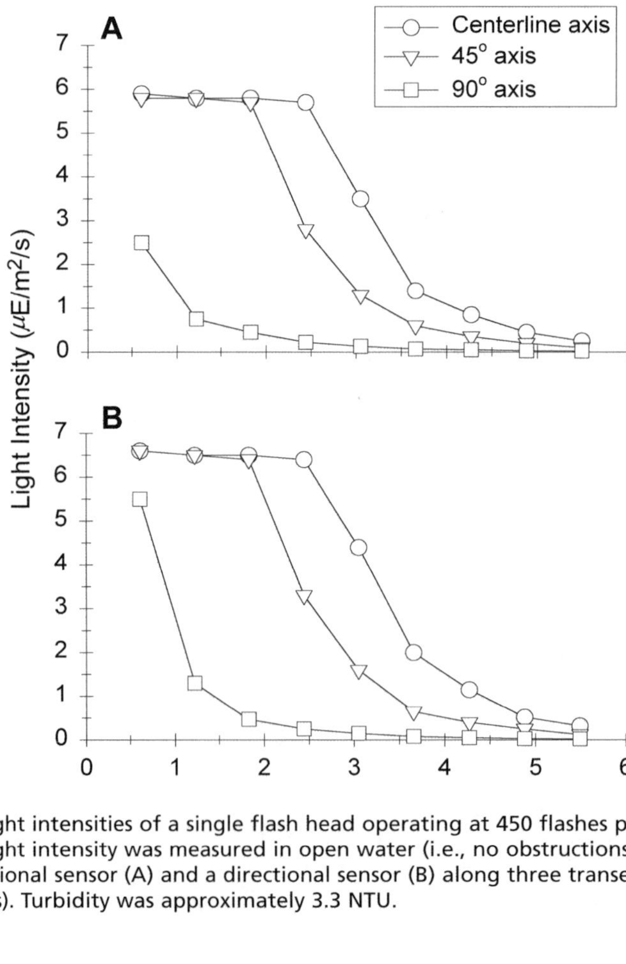

Figure 7. Strobe light intensities of a single flash head operating at 450 flashes per minute during nighttime conditions. Light intensity was measured in open water (i.e., no obstructions to light transmission) with an omnidirectional sensor (A) and a directional sensor (B) along three transects (centerline, 45 degree, and 90 degree axes). Turbidity was approximately 3.3 NTU.

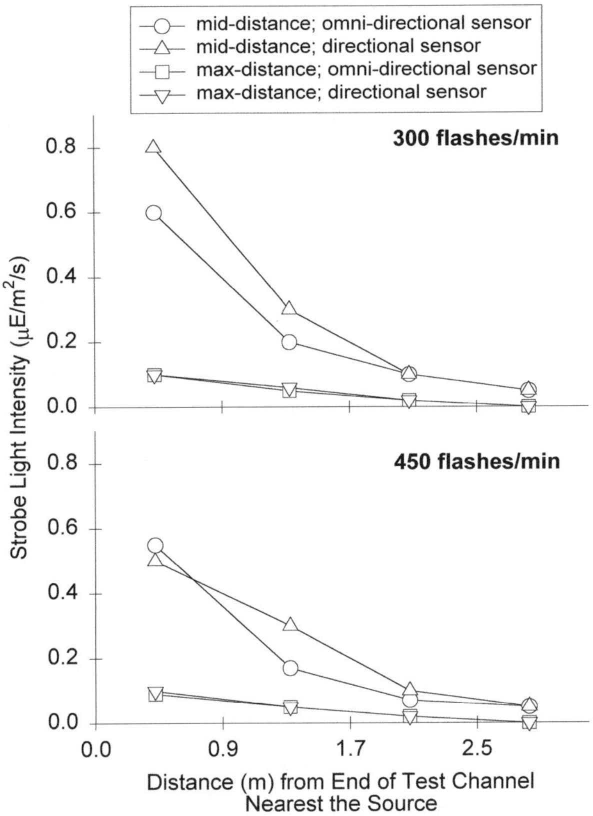

Figure 8. Strobe light intensities measured in the test channel during nighttime hours. Light intensities were recorded at the vertical and horizontal center of each test channel quadrant with the strobe light located at the mid and maximum distance positions.

response to strobe light was consistently elicited from chinook salmon smolts during the longer 60-min exposure trials conducted at night. Competing light sources (i.e., ambient and supplemental lighting) appeared to reduce the effectiveness of strobe lights. Regardless of exposure time, chinook salmon did not respond to strobe light in a substantial or consistent manner with respect to temporal and spatial movements during daytime or early evening tests, or during nighttime tests when artificial underwater lighting was used to allow for video observation of fish. Chinook salmon also exhibited a stronger response to strobe light during the 60-min nighttime tests when the light source was behind the fish (i.e., downstream source).

Based on our observations, successful application of strobe lights at Roza may be restricted to nighttime hours. However, avoidance of strobe light by juvenile chinook salmon has been demonstrated under daylight conditions in previous studies (Puckett and Anderson 1988a; Nemeth and Anderson 1992; Ploskey et al. 1998). During daytime tests in a raceway, juvenile chinook salmon were repelled by strobe light for the entire duration of 60-min tests, this included movement from one end of the raceway to the other when the operation of the active source was switched after 30 min (Puckett and Anderson 1988a; Nemeth and Anderson 1992). The average group distance from the strobe light reported by Puckett and Anderson (1988a) was about 6.7 m. Nemeth and Anderson (1992) showed that between 45 and 90% of test fish were at least 6.5 m from the active strobe light during a test. In another study, the maximum response distance for juvenile salmonids (chinook and coho salmon) evaluated during daylight conditions was estimated to be 4.5 m and was determined to be a function of strobe light intensity relative to ambient light intensity (Ploskey et al. 1998).

We would like to correlate the results from our study to the potential for fish avoidance to occur at a particular distance from an active source in open

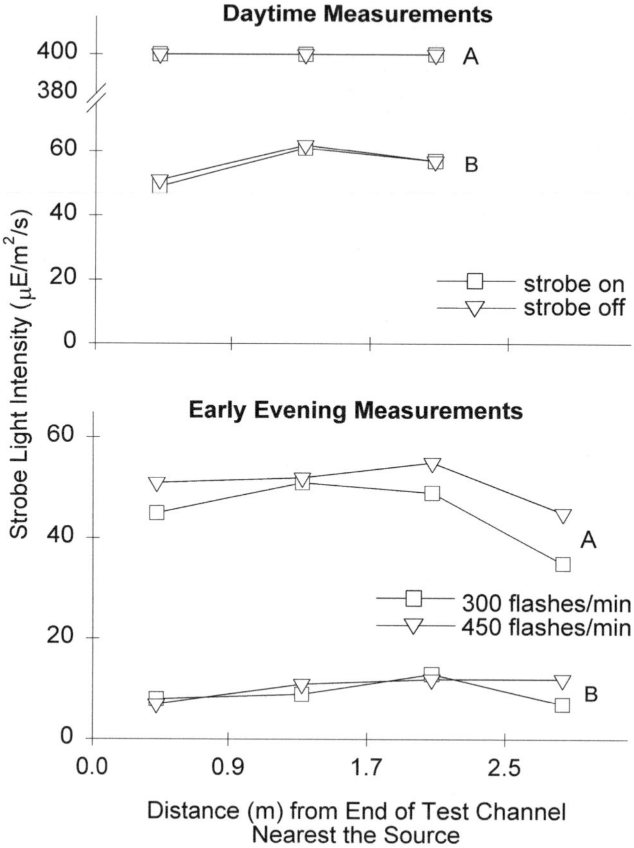

Figure 9. Light intensity measurements recorded in the test channel during daytime and early evening hours. Light intensities were measured with omnidirectional (A) and directional (B) sensors at the vertical and horizontal center of each test channel quadrant with the light positioned at the minimum distance location. Intensity measurements were recorded only for 300 flashes per minute during daytime hours.

water. However, because the two layers of nylon sheeting reduced light intensity in the test channel, direct relationships between the position of the strobe lights and the response of fish (i.e., distance fish moved from the light source) cannot be made. Nevertheless, a generalized effectiveness distance can be estimated using the light measurements in open water and from inside the test channel. The center-of-school position of fish during a test can be assessed in relation to the light intensities recorded in the test channel, which in return can be compared with the light measurements that were recorded along the three axes extending from a flash head without any impediments to light penetration. For example, if fish were repelled from the first quadrant to the third, then the light intensity recorded in the third quadrant can be used to determine the distance at which fish might respond in open water based on the distance from the flash head that this light level was recorded along the three axis in open water. Because open-water light measurements were recorded at night, conclusions regarding the distance at which downstream migrating fish would respond can only be made for nighttime conditions. Turbidity conditions affect light penetration as well, and, subsequently, also would influence the range at which fish probably would respond.

Results from the trials using different placements of the strobe lights can also be generalized to a threshold for behavioral responses. During tests with the lights at the mid-distance position, the center-of-school position shifted from the quadrant closest to the active source to the quadrant furthest away (i.e., a three quadrant shift). With the lights at the maximum distance position, the center of school position generally shifted approximately one to two quadrants away from the active source.

The light intensity levels recorded at the midpoints of each quadrant indicate that movements away from the active source decreased at levels between 0.02 and 0.05 $\mu E/m^2/s$. A decrease in chinook salmon avoidance response at strobe light intensities less than 0.05 $\mu E/m^2/s$ has been observed in previous studies (Puckett and Anderson 1988a; Nemeth and Anderson 1992; Ploskey et al. 1998).

Using the light field measured for a single light operating at 450 flashes per minute, this level of illumination may occur at centerline distances between about 5 and 6 m from a flash head at a turbidity level of 3.3 NTU. Based on measurements taken along transects 45 and 90 degrees from the centerline, the effective distance would decrease by about 1–2 m if fish approaching the light were more than 45 degrees off the centerline of the flash head. The turbidity level of 3.3 NTU that was measured during strobe light intensity readings was the lowest that was measured during the two week test program, indicating that the estimated effective range may be near the maximum that is achievable at Roza during this time of year (i.e., April).

An important parameter that probably has a strong influence on distance and light intensity threshold levels is the ambient light level to which fish are adapted (Puckett and Anderson 1988b). The contrast between the intensity of a stimulus light and ambient light (i.e., signal to noise ratio) most likely will influence how fish will respond to a light source, especially if the fish are adapted to the ambient light. Behavioral responses of juvenile chinook salmon to light have been shown to be dependent on the contrast between adaptation light intensity and the intensity of the stimulus light (Puckett and Anderson 1988b). The results from our study support this finding when strobe light is the stimulus. The strongest responses to strobe light that we observed occurred during nighttime hours when the contrast between ambient light (also the adaptation light level) and the stimulus light was greatest. We did not observe any response to strobe light when the stimulus intensity was less than the intensity of ambient light.

The hydraulic conditions (e.g., water velocity) that fish encounter when they are exposed to a light stimulus also will play an important role in the ability to elicit a behavioral reaction from approaching fish (Anderson 1991). Fish must have time to react and to find an escape path before being entrained into a water intake. The water velocities recorded in our test channel were low (0.8–0.15 m/s) and did not restrict fish movement away from the downstream light source. Because we did not evaluate the effects of water velocity on behavioral responses, we cannot speculate with respect to an upper velocity that might limit the ability of chinook salmon smolts to actively avoid strobe lights.

The experimental design of tests conducted at Roza may have contributed to the lack of response by chinook salmon to the strobe light stimuli during daytime and early evening tests. The test channel in which fish were evaluated was within about 0.6 m of the water surface. This produced high ambient light conditions, even during early evening hours after the sun had descended below the mountain ridge directly west of the test location. The strobe light intensity also was reduced as it passed through the nylon sheeting of the test cage and channel. The high ambient light level and the loss of strobe light intensity through the nylon sheeting resulted in no discernable contrast between the strobe light and ambient light within the test channel (i.e., the signal to noise ratio was zero). Under such conditions, it is possible that fish in the test channel could not detect the strobe light to a degree that it would elicit a response. Previous studies that have demonstrated daytime avoidance by juvenile salmonids to strobe light have measured strobe light intensities above ambient light levels (Puckett and Anderson 1988a; Nemeth and Anderson 1992; Ploskey et al. 1998).

If fish cannot adequately detect strobe light over ambient light conditions during daytime and early evening hours at Roza, then the use of strobe lights may not be a viable option for diverting fish there. Data collected during trapping activities at the bypass collection facility in 1996 and 1998 indicate that between 30 and 90% of out-migrating smolts travel downstream past the Roza Diversion Dam during daytime (Washington Department of fish and Wildlife, unpublished data). Additionally, radio telemetry studies in the Snake and Columbia rivers have shown that the median depth of tagged hatchery-reared chinook smolts is 1.8 m from the water surface, even when water depths are up to 30 m (John Beeman, U.S. Geological Survey, personal communication). This suggests that the daytime ambient light conditions that were evaluated during cage tests within 0.6 m of the water surface, are representative of what would be experienced by many smolts approaching a strobe light array at Roza. Should smolts travel deeper in the water column, light measurements recorded at Roza during daylight hours indicate that fish would have to be more than 6 m deep before they would encounter ambient light conditions that provide enough contrast to strobe light for avoidance to occur (i.e., similar to conditions experienced during nighttime cage tests).

For nighttime tests, stronger and more consistent avoidance of the downstream source suggests different perception of the stimulus from the two directions. This may result in different types of avoidance response or escape behavior. Another possibility for the observed differences is that the downstream light was closer to the test channel at each position (mid and max) than was the upstream light. The exact position of the two lights at each location was affected by the design of the test platform (i.e., placement of support beams from which the lights were deployed). The slightly more dramatic avoidance observed when the lights were at the mid distance position compared with the maximum distance position most likely was due to greater light intensity levels and a sharper gradient of intensity across the length of the test channel when the lights were at mid distance.

The orientation of out-migrants to the river flow also may influence the effectiveness range of strobe lights in a field application. The temporal and spatial response of chinook salmon smolts to the strobe light during nighttime tests indicated that avoidance was stronger when the downstream light source was operated and fish were facing into the flow (that is, moving downstream tail first). The effective range of strobe lights appears to be less for fish facing downstream versus fish approaching tail first.

Although both strobe flash rates that were tested produced rapid and strong directional responses, the spatial and temporal characteristics of smolt reactions appeared to be slightly stronger and more consistent during tests with 450 flashes per minute than during tests with 300 flashes per minute. Consequently, it is recommended that a strobe light array deployed at Roza should be designed to operate at 450 flashes per minute. The orientation of strobe light flash heads relative to the location to which fish are being diverted also may be an important design parameter for a field installation. However, the ability of fish to move laterally along a light field was not tested.

Northern pikeminnow and smallmouth bass exhibited varying levels of avoidance to strobe light during 60-min exposure tests conducted at night. The avoidance reaction of northern pikeminnow to the strobe light operated at 300 flashes per minute was similar to chinook salmon but not as strong. Although test fish appeared to react to the strobe light at 450 flashes per minute, the response was not as definitive as observed at 300 flashes per minute.

A strong avoidance reaction was elicited from smallmouth bass during a single test at 450 flashes per minute. During this test, the fish quickly moved to the opposite end of the cage from the active light source and remained there until the source direction was switched at the end of each 10-min interval. This movement was very similar to the movements of chinook salmon that were exposed to strobe light during nighttime tests for 60 min periods.

A previous study that examined responses of several riverine species in cage and field tests did not demonstrate potential for repelling or diverting smallmouth bass with strobe lights (EPRI 1998). During this study, conducted at the Kingsford Hydroelectric Project in Michigan, smallmouth bass that were similar in size to the fish tested at Roza did not respond to strobe light at flash rates between 200 and 600 flashes per minute. Although the same test cage and type of strobe light were used in both studies, a separate test channel with flowing water was not used in the Kingsford tests. Also, the strobe light was located in line with the long axis of the test cage, whereas at Roza the lights were offset from the test channel centerline. The larger space (entire test cage versus smaller channel), the lack of flow, and the direction of the light used in the Kingsford study may have contributed to the lack of response that was observed for smallmouth bass. However, because only one test with four fish was conducted with smallmouth bass at Roza, it would be prudent to perform a more extensive evaluation of the response of this species to strobe light before any field applications are considered. Similarly, pikeminnow also should be tested more extensively before strobe lights are ruled out as an effective behavioral deterrent.

The drop light stimuli were ineffective in eliciting any type of response regardless of background light intensity. Chinook salmon did not respond to any of the drop light stimuli (i.e., on/off intervals and continuous operation) that were tested during daytime, early evening, or nighttime trials with and without the use of additional underwater lighting.

Although avoidance of infrasound by salmonids has been demonstrated during both laboratory and field studies (Knudsen et al. 1992, 1994, 1997; Mueller et al. 1998), chinook salmon smolts exhibited no behavioral reaction to infrasound produced by the PMG during cage tests at Roza. This concurs with the results from a study that examined juvenile salmonid responses to infrasound produced by a PMG and a piston device during cage tests (Ploskey et al. 1998). Infrasound did appear to repel pikeminnow during a single test conducted with three fish. Avoidance was observed at four of the five frequencies that were evaluated. Because only one test was conducted with a small number of fish, further testing would have to be performed to verify the observations made this study.

In summary, the results of this study suggest that strobe light has considerable potential to divert chinook salmon smolts to the Roza screening facility during nighttime hours. There was no evidence produced that would indicate a strobe light system will be effective during the day at Roza. However, due to of the experimental design we used (i.e., shallow depths at which fish were tested), and because other studies have demonstrated daytime avoidance, the potential for strobe lights to effectively repel chinook salmon smolts during daytime hours should not be entirely dismissed. The use of infrasound to divert chinook salmon is not considered a viable alternative for Roza due to a lack of any distinguishable response during cage tests. Further cage testing should be conducted with pikeminnow and smallmouth bass before strobe lights are considered as an alternative for repelling either of these species in a field application. Similarly, more extensive cage tests should be conducted with pikeminnow to fully assess their responses to infrasound.

Acknowledgments

We are very thankful to the individuals and agencies who made this work possible. Bonneville Power Administration funded the research as part of the Yakima/Klickitat Fisheries Monitoring Project. The Monitoring Implementation Team (C. Busack, B. Watson, C. Knudsen, and M. Johnston) provided useful discussion about the study design and utility of behavioral guidance technologies. W. Larrick and other Bureau of Reclamation staff provided valuable support in the deployment and removal of test equipment. B. Hopley and D. Byrnes helped secure the funds for this research. G. McMichael and A. Fritts helped collect the spring chinook salmon. G. Ploskey of the U.S. Army Corps Engineers Waterways Experiment Station loaned us the drop light and R. Brown of Flash Technology supplied the strobe lights. N. Taft and R. Brown reviewed drafts of the manuscript and provided helpful comments.

References

Anderson, J. J. 1991. Fish bypass mathematical models. Pages 2082–2091 *in* D. D. Darling, editor. Waterpower '91: proceedings of the International Conference on Hydropower. American Society of Civil Engineers, New York, New York.

Bernier, K. R. 1995. 1995 report on effectiveness of the permanent downstream passage system for Atlantic salmon at Weldon Dam. Great Northern Paper Company.

Brown, R., and K. Bernier. 2000. The use of an aquatic guidance strobe lighting system to enhance the safe bypass of Atlantic salmon smolts. Pages 149–154 *in* M. Odeh, editor. Advances in fish passage technology: engineering design and biological evaluation. American Fisheries Society, Bethesda, Maryland.

Electric Power Research Institute (EPRI). 1998. Evaluation of fish behavioral barriers. Prepared by Alden Research Laboratory, EPRI Report No. TR-109483, Palo Alto, California.

Knudsen, F. R., P. S. Enger, and O. Sand. 1992. Awareness reactions and avoidance responses to sound in juvenile Atlantic salmon, *Salmo salar* L. Journal of Fish Biology 40:523–534.

Knudsen, F. R., P. S. Enger, and O. Sand. 1994. Avoidance responses to low frequency sound in downstream migrating Atlantic salmon, *Salmo salar* L. Journal of Fish Biology 45:227–233.

Knudsen, F. R., C. B. Schreck, and O. Sand. 1997. Infrasound produces flight and avoidance responses in Pacific juvenile salmonids. Journal of Fish Biology 51:824.

Mueller, R. P., D. A. Neitzel, W. V. Mavros, and T. J. Carlson. 1998. Evaluation of low and high frequency sound for enhancing fish screening facilities to protect outmigrating salmonids. DOE/BP-62611-13, U.S. Department of Energy, Bonneville Power Administration, Portland, Oregon.

Mueller, R. P., D. A. Neitzel, and B. G. Amidan. 1999. Evaluation of infrasound and strobe lights to elicit avoidance behavior in juvenile salmon and char. DOE/BP-00029-6, U.S. Department of Energy, Bonneville Power Administration, Portland, Oregon.

Nemeth, R. S., and J. J. Anderson. 1992. Response of juvenile coho and chinook salmon to strobe and mercury vapor lights. North American Journal of Fisheries Management 12:684–692.

Nestler, J. M., G. R. Ploskey, J. Pickens, J. Menezes, and C. Schilt. 1992. Responses of blueback herring to high-frequency sound and implications for reducing entrainment at hydropower dams. North American Journal of Fisheries Management 12:667–683.

Ploskey, G. R., P. N. Johnson, M. G. Burczynski, J. M. Nestler, and T. J. Carlson. 1998. Effectiveness of strobe lights, infrasound devices, and a sound transducer for eliciting avoidance by juvenile salmonids. Technical Report of the U.S. Army Engineer District, Portland, Oregon.

Puckett, K. J., and J. J. Anderson. 1988a. Behavioral responses of juvenile salmonids to strobe and mercury lights. FRI-UW-8717, Fisheries Research Institute, University of Washington, Seattle.

Puckett, K. J., and J. J. Anderson. 1988b. Conditions under which light attracts juvenile salmon. Pages 4-65–4-73 *in* W. C. Micheletti, editor. Proceedings: fish protection at steam and hydroelectric power plants. CS/EA/AP-5663-SR, Electric Power Research Institute, Palo Alto, California.

Taft, E. P., N. A. Brown, T. C. Cook, J. P. Ronafalvy, and M. W. Haberland. 1996. Developments in the use of infrasound for protecting fish at water intakes. North American Water and Environment Congress '96, ASCE, New York.

Zar, J. H. 1984. Biostatistical analysis. Prentice-Hall, Inc., Englewood, Cliffs, New Jersey.

American Fisheries Society Symposium 26:145–160, 2001

Biological Evaluation of the Behavioral Guidance Structure at Lower Granite Dam on the Snake River, Washington in 1998

NOAH S. ADAMS[1]
U. S. Geological Survey, Western Fisheries Research Center, Columbia River Research Laboratory, Cook, Washington 98605, USA

GARY E. JOHNSON[2]
Battelle Pacific Northwest Division
Mail Stop K6-85, Post Office Box 999, Richland, Washington 97352, USA

DENNIS W. RONDORF
U.S. Geological Survey, Columbia River Research Laboratory
Laboratory, Cook, Washington 98605, USA

STEVE M. ANGLEA
Battelle Pacific Northwest Division
Mail Stop K6-85, Post Office Box 999, Richland, Washington 97352, USA

TIM WIK
U.S. Army Corp of Engineers
Walla Walla District, 201 North Third Avenue, Walla Walla, Washington 99362, USA

Abstract.—In 1998, a behavioral guidance structure (BGS; a steel wall 330 m long and 17–24 m deep) was installed in the forebay of Lower Granite Dam on the Snake River, Washington. The purpose of the BGS was to change the horizontal distribution of downstream migrants approaching the south half of the powerhouse (Turbines 1–3) by guiding them toward the surface bypass and collector (SBC) attached to the dam upstream of the north half of the powerhouse (Turbines 4–6). The effectiveness of the BGS was evaluated using biotelemetry and hydroacoustic techniques. The BGS was designed to be movable, thereby allowing a comparison between the horizontal distribution of the fish when the BGS was deployed as a diversion device (i.e., BGS in) and when the BGS was moved 800 m upstream of the dam and no longer influenced fish movements immediately upstream of the powerhouse (i.e., BGS out). Radio telemetry and hydroacoustic techniques showed that about 80% of the fish migrating towards Turbines 1–3 were successfully diverted north. Radio telemetry data revealed that the mean residence times of chinook salmon *Oncorhynchus tshawytscha*, hatchery steelhead *Oncorhynchus mykiss*, and wild steelhead *Oncorhynchus mykiss* were 1.6, 1.7, and 2.4 times longer, respectively, when the BGS was out compared with when it was in. And, overall, fish passage efficiency (percentage of fish passing through nonturbine routes) was significantly (P = 0.026) higher when the BGS was in (93.7%) than out (91.2%). The BGS concept appears to be valid. However, only two of the three turbines behind the BGS were functioning during the test. Additional flow behind the BGS might change its effectiveness.

For years research has been focused on the effects of dams and reservoirs on juvenile salmonids as they migrate from their natal waters to the ocean. Raymond (1968, 1979) and Park (1969) showed that dams increased migration times and suggested that this may be detrimental to juvenile salmon survival. Since the completion of Lower Granite Dam in 1975 on the Snake River, Washington (Figure 1), delays of more than 20 d have been observed (Buettner and Nelson 1989, 1990; Buettner 1991; Buettner and Brimmer 1993; Rondorf and Banach 1996), and steelhead fingerlings *Oncorhynchus mykiss* have residualized in Lower Granite Reservoir during low flow years (Bennett 1992).

Research conducted in 1994 (Rondorf and Banach 1996) and 1995 (Adams et al. 1998b) showed

[1]Corresponding Author: noah_adams@usgs.gov, tel (509) 538-2299, fax (509) 538-2843

[2]Present Address: BioAnalysts, Inc. 105 W. Main Street, Battleground, Washington 98604, USA

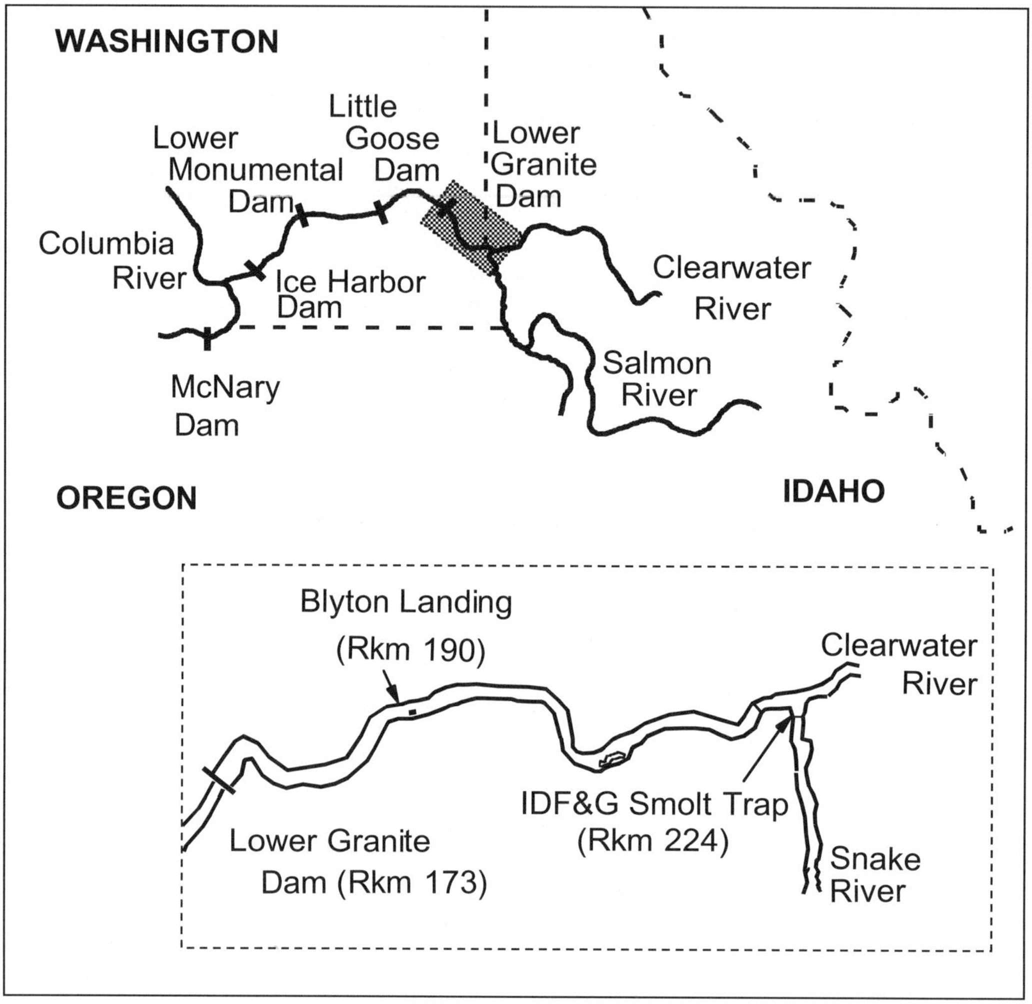

Figure 1. Map of Snake River drainage showing Lower Granite Reservoir study site, collection sites (Idaho Department of Fish and Game Smolt Trap and confluence of Snake and Clearwater rivers), and release site (Blyton Landing) of radio-tagged Juvenile salmonids.

that radio-tagged juvenile chinook salmon *Oncorhynchus tshawytscha* and steelhead traveled through the Snake River and Lower Granite Reservoir quickly compared with long residence times in the forebay of Lower Granite Dam. This observation suggested that juvenile salmonids have more difficulty passing through the dam than traveling through the reservoir.

Reservoir drawdown, flow augmentation, spill, improved fish collection, and transportation systems have been identified as some of the management actions that may improve juvenile salmonid passage and survival, and thereby assist in the recovery of endangered salmon stocks in the Snake River (NMFS 1995). One option now being evaluated is surface bypass. Under this option, surface-oriented juvenile salmonids emigrating to the ocean are intercepted before entering turbine intakes and diverted to alternative passage routes (Johnson et al. 2000). A prototype surface bypass and collector (SBC) was installed at Lower Granite Dam in 1996 and evaluated in 1996, 1997, and 1998 (Figure 2; Adams and Rondorf 2001; Johnson et al. 2000). Normandeau et al. (1997) documented safe passage at the SBC for juvenile salmonids. In 1996, fish could pass through the north half of the powerhouse by either entering the SBC or turbine intakes. About 40–45% of all fish passing the dam through the north half of the powerhouse (Turbines 4–6) did so through the SBC (Adams et al. 1997; Johnson et al. 1997). Relative to the dam

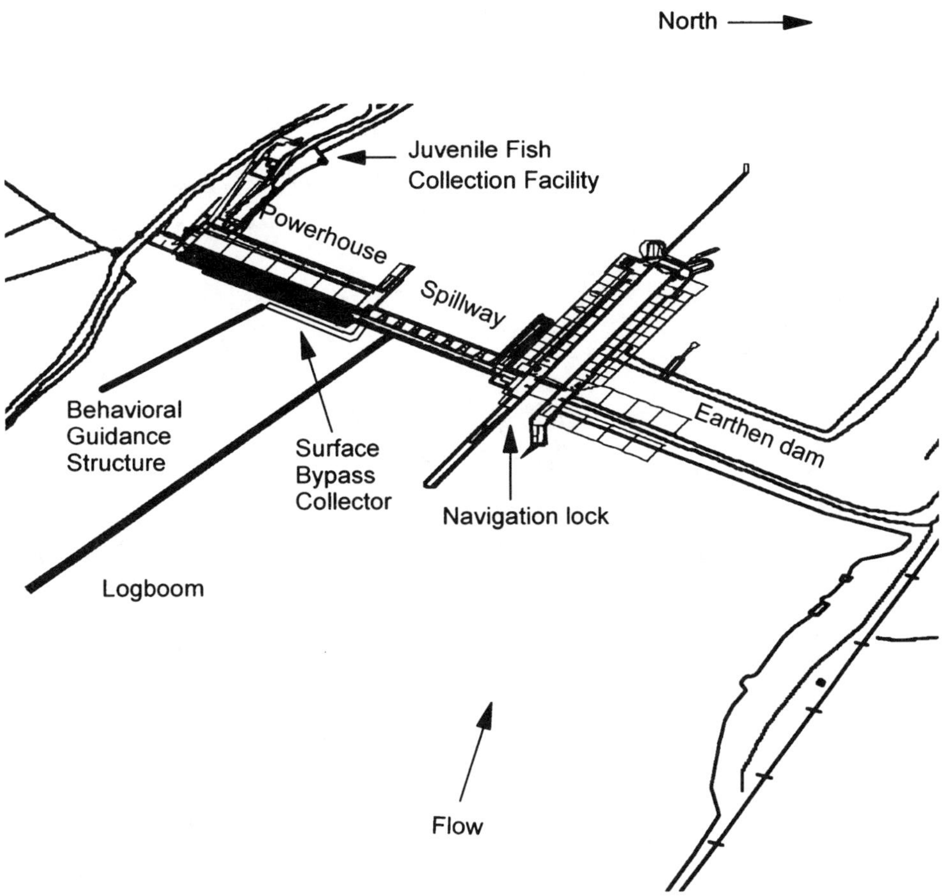

Figure 2. Perspective view of Lower Granite Dam, Snake River, Washington.

as a whole, fish could pass through the SBC, turbine intakes, or the spillway. In 1996, about 10% of all fish passing the project passed through the SBC. This measure of SBC efficiency was low partly because many fish passed over the spillway without encountering the SBC. When river flow is high, as it was in 1996, excess water is discharged through the spillway to control flooding upstream of the dam. River flows remained high in 1997 and similar results were obtained (Adams et al. 1998a; Johnson et al. 1998). Although the 1996 and 1997 radio telemetry and hydroacoustic results indicated that the surface collection concept worked at Lower Granite Dam, further improvement in SBC efficiency is necessary for the SBC to be a viable, long-term bypass alternative for juvenile salmonids (Johnson et al. 2000).

In spring 1998, the Corps installed two new components to the SBC to increase its efficiency. The first structure, called a Simulated Wells Intake, was a steel shield retrofitted to the bottom of the SBC (Figure 3). Its purpose was to decrease smolt entrainment into turbines and increase their availability to the SBC by decreasing the velocity and downward direction of water going into turbine intakes below the SBC. The second structure, called a Behavioral Guidance Structure (BGS), was a floating steel wall connected to the south end of the SBC (Figure 2). The purpose of the BGS was to divert smolts from the south half of the powerhouse (Turbines 1–3) into the SBC (upstream and above Turbines 4–6) or spillway.

We used biotelemetry and hydroacoustic technologies to determine how effective the BGS was at diverting fish. Biotelemetry is an effective means of obtaining fish movement information (Adams and Rondorf 2001). Recent miniaturization of radio tags has made it possible to monitor juvenile

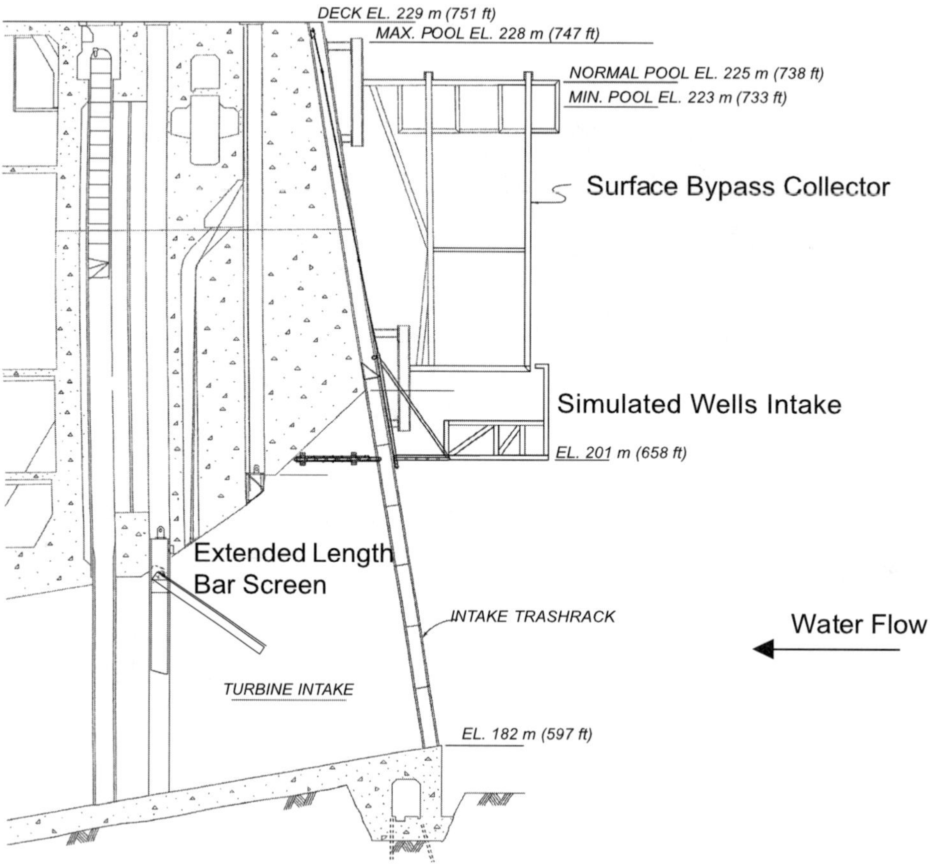

Figure 3. Cross-sectional view of Lower Granite Dam at a turbine intake showing locations of Surface Bypass/Collector (SBC), Simulated Wells Intake (SWI), and Extended length Bar Screens (ESBS).

salmonids as small as 120 mm fork length (Adams et al. 1998a). Although biotelemetry enables investigators to obtain information on individual fish that would otherwise be unattainable, sample size may be restricted by high cost of tags and limits on the number of radio-tagged fish that can be tracked concurrently. Hydroacoustic fish stock assessment is an effective means of obtaining information on numerous fish, but species-specific information is not attainable (Johnson et al. 1999). We used both biotelemetry and hydroacoustic technologies to overcome the limitations of each method and to assess juvenile salmon behavioral responses to the BGS at Lower Granite Dam. The objectives of our evaluation were to estimate the effect of the BGS on horizontal distribution of fish passage at Turbines 1–3, Turbines 4–6, the SBC, and the spillway. Because the BGS was moveable, a BGS absent/present (hereafter referred to as BGS out and BGS in, respectively) study design was used to estimate BGS diversion efficiency. Lastly, we statistically compared SBC efficiency, fish guidance efficiency (FGE; a measure of performance for intake screens), fish passage efficiency (FPE; a measure of nonturbine passage performance for the entire dam), and combined bypass efficiency (CBE; a measure of combined nonturbine passage performance for Turbines 4–6 and the SBC) for BGS in and out conditions.

Methods

Study Area

The study was conducted at Lower Granite Dam starting on 14 April 1998 and ending on 28 May 1998. Lower Granite Dam is located on the Snake River, Washington, at river kilometer 173 (Figure 1). It is a run-of-river project with an effective height of 30 m and under normal operations water levels in the pool above the dam fluctuated about 1.5 m. The dam has four primary structures: powerhouse, spillway, lock, and earthen embankment (Figure 2). The powerhouse has six turbines with a total generating capacity of 810 MW. The total hydraulic capacity of the powerhouse is about 3,681 m^3/s (130 kcfs). Each turbine has three extended-length submersible bar screens (Figure 3). The screens are intended to divert fish that enter the turbine intakes into a bypass channel (hereafter referred to as the fish collection channel) that leads to a juvenile fish collection facility downstream of the dam (Figure 2). During spring and summer, the juvenile salmonid migration seasons, water level in the pool above the dam is managed near the minimum operating pool elevation (MOP) of 223 m (733 ft) above mean sea level (MSL). The spillway comprises eight spill bays with tainter gates and is about 156 m long. A trash boom deflects debris away from the powerhouse toward the spillway (Figure 2).

Surface Bypass and Collector

The Lower Granite Dam SBC is a steel box 18 m high, 6 m deep, and 100 m long. It uses large flotation chambers to move vertically with forebay elevation. There are four entrances to the SBC, designated BGS, South (S), Middle (M), and North (N; Figure 4). Each entrance, about 5 × 15 m, is made

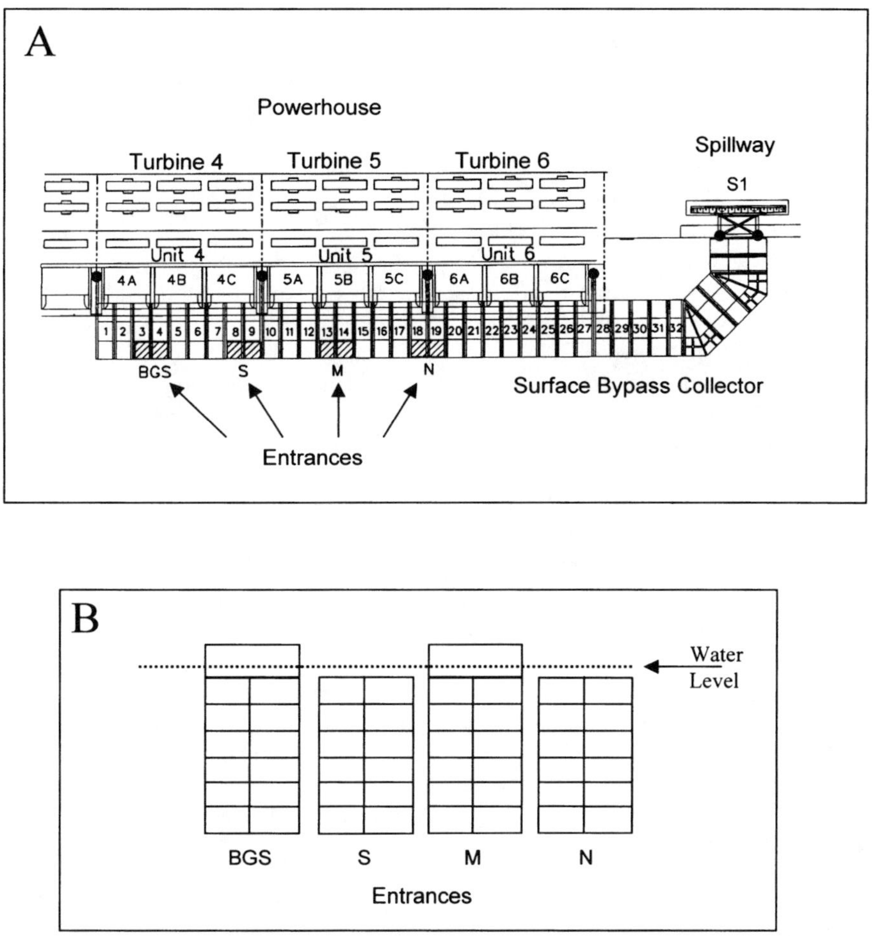

Figure 4. Plan view (A) of the SBC showing locations of SBC entrances (BGS, S, M, and N). Elevation view (B) of SBC entrances showing the 12 openings making up each entrance and the overflow at the top of the BGS and Middle entrances.

up of 12 closeable openings (Figure 4). These 12 sliding doors were used to control the size of the openings so that different entrance sizes and shapes could be tested. The BGS entrance is located over the junction of Turbine Intakes 4A and 4B, the South entrance is over Intake 4C, the Middle entrance is over Intake 5B, and the North entrance is over the junction between Intakes 5C and 6A (Figure 4). The BGS and Middle entrances extend to the surface (i.e., they have overflow). The tops of the South and North entrances are 2.3 m below the water surface. The gate at Spill Bay 1 regulates flow into the SBC and is also the SBC outfall. During the 1998 test, SBC discharge was estimated to be 113 m^3/s (4,000 cfs). The prototype SBC was constructed to determine the feasibility of providing a surface oriented passage route for out-migrating salmonids. A decision was made not to test the feasibility of dewatering capabilities within the SBC; therefore, all water and fish entering the SBC passed through the first spillbay.

Behavioral Guidance Structure

The purpose of the BGS was to change the horizontal distribution of downstream migrating salmonids as they approached the south half of the powerhouse (Turbines 1–3) and guide them towards the north where the SBC (above Turbines 4–6) and spillway are located. The BGS is a steel wall 335 m long, comprised of a series of 30 m wide steel plates that are supported by floats in a "catamaran" design. Because the BGS was moveable, a BGS absent/present (hereafter referred to as BGS out and BGS in, respectively) study design was used to estimate its diversion efficiency. When the BGS was in, it extended from the SBC upstream to the south shoreline where it was anchored about 30 m offshore to allow adult migrants to pass between the shoreline and the BGS (Figure 2). The BGS was 24 m deep where it intersected the SBC and tapered to a depth of 17 m at its upstream end. Over the length of the BGS, the distance between the bottom of the BGS and the bottom of the forebay was about 5 m. When the BGS was out, it was anchored about 800 m upstream of the dam along the south shore of the forebay.

Experimental Design

The 1998 evaluation of the SBC and BGS at Lower Granite Dam consisted of a split-plot experimental design with a 2 × 2 factorial treatment design. We analyzed the hydroacoustic data as a split-plot experiment and extracted information on SBC and BGS main effects and interactions. Specifically, we evaluated two factors, SBC entrance configuration and BGS in/out (i.e., presence/absence). The two BGS positions together with two SBC entrance configurations yielded four treatments tested in the experiment. This paper focuses on the factors designed to evaluate the performance of the BGS.

We tested the four treatments using a six-day experimental block to accommodate the different test conditions along with two additional days to move the BGS in or out. (Note that the BGS did not always have to be moved between treatment blocks.) These six-day blocks were repeated eight times during the 1998 spring out-migration (13 April to 28 May). The SBC configuration during the day when the BGS was moved was the same as that for the next test day. We continued to collect data during the time when the BGS was being moved to gather information on how this activity might affect fish behavior. Data collected on the "move" days was not used to estimate BGS diversion efficiency.

Radio-Telemetry Tagging Procedures

We radio-tagged hatchery spring chinook salmon, hatchery steelhead, and wild steelhead that were obtained from Idaho Department of Fish and Game's Snake River smolt monitoring trap at the confluence of the Snake and Clearwater rivers. Additionally, 193 hatchery spring chinook salmon were collected by beach seine about 3 km above the confluence of the Snake and Clearwater rivers. Fish were tagged and released at Blyton Landing, about 20 km upstream of the dam. This allowed time for fish to redistribute both horizontally and vertically in the water column before approaching Lower Granite Dam. All fish were surgically implanted with coded radio transmitters (Lotek Engineering Inc., Newmarket, Ontario, Canada) using procedures described by Adams et al. (1998a). Fish were transferred to holding containers (127 L) and held for 24 h in the river before being released. Releases occurred in the morning (about 0800 hours) and throughout the afternoon (between 1200 and 1600 hours) to stagger fish arrival time at the dam.

We used a coded tag technology that allowed multiple tags to broadcast on the same frequency. Multiple tags on each frequency allowed us to minimize receiver scan cycle and greatly reduced the probability of not detecting a tagged fish. To further enhance our detection capabilities, we combined receivers with digital spectrum processor units (DSP). Because a DSP listens to all frequencies and codes simultaneously, there is essentially no scan cycle and all tagged fish within range of an antenna can be detected instantaneously. The tags we implanted in chinook salmon weighed about 1.4 g in air (an average of 5.4% of the fish's body weight), measured 7.3 × 18.0 mm, and had a life expectancy of 9 d. The tags we implanted in steelhead weighed about 1.75 g in air (an average of 1.8% of the fish's body weight), measured 8.2 × 18.9 mm, and had a life expectancy of 19 d.

Radio Telemetry Antenna Locations

Sixty-one fixed aerial antennas linked to 14 data logging receivers (Lotek Engineering Inc., Newmarket, Ontario, Canada) and 215 underwater antennas (designed by Grant Systems Engineering, Newmarket, Ontario, Canada) linked to 20 data logging receivers were used to monitor fish movements within the forebay of Lower Granite Dam 24 h per day, 7 d per week. Aerial arrays were installed at nine locations: Granite Point (located 10 km upstream of Lower Granite Dam), forebay barge sites (located 0.8 km upstream of Lower Granite Dam), earthen dam, logboom, BGS, spillway, powerhouse, juvenile fish collection facility, and exit sites (located 0.8 km downstream of Lower Granite Dam). Each aerial array consisted of multiple antennas linked to a Lotek SRX 400-W16 receiver. The receivers were programmed to scan eight frequencies for 3 s each, resulting in a 24 s scan cycle. Underwater arrays were installed in seven locations: spillway, powerhouse, SBC, BGS, fish gallery (i.e., the juvenile fish collection channel inside the powerhouse), floating barge in front of the SBC, and on the back of all 18 extended-length submersible bar screens (ESBS). Each underwater array consisted of multiple antennas linked to a Lotek SRX 400-W16 receiver and DSP.

Tags deployed from a towline attached to a boat in the forebay showed that the range of detection for aerial antennas varied from 130 to 300 m, depending on the depth of a tag in the water column. Tags towed in the top 3 m of the water column could be detected from 200 to 300 m, whereas tags towed deeper than 3 m could be detected from 130 to 240 m. Due to the characteristics of underwater signal propagation, the range of detection for underwater antennas was 5–10 m. This range was much shorter than ranges for aerial antennas and provided the spatial resolution needed in areas such as the SBC, BGS, fish collection channel, and turbine intakes.

Aerial antennas were positioned to provide complete coverage of the forebay. Detection areas of individual antennas overlapped slightly to minimize the probability of fish moving into the forebay undetected by the monitoring equipment. The strength of the signal received by individual antennas was used to determine which antenna the fish was closest to. Similarly, underwater antennas on the face and inside the SBC had overlapping ranges and strength of the signal was used to determine which antenna the fish was closest to. Underwater antennas in locations other than the SBC typically did not have overlapping ranges. Because spatial resolution was not as critical in these other areas, it was not necessary for the coverage areas to overlap.

Hydroacoustic Transducer Locations and Orientations

In 1998, a total of 54 single-beam transducers (420 kHz) were used to collect fish passage data at all four SBC entrances, nine of 15 turbine intake screens, five of 18 powerhouse pier noses, and at all seven available spill bays. All systems were calibrated before and after the study. Echo strength (threshold = –52db) was used to limit the collection of data to fish that were about 100 mm fork length: the smallest expected size of juvenile salmonids migrating through the dam.

At the SBC, 12 squinted-pairs of 6° transducers, pulsing at a rate of 20 pings per second (pps), sampled fish passage. A squinted-pair transducer is essentially equivalent to two transducers mounted side-by-side and aimed in the same direction. We used this configuration to obtain information on the general direction fish were moving by analyzing which of the two transducers detected the fish first. Squinted-pairs were aimed up from the bottom of the middle of each of the entrances to the SBC (BGS, South, Middle, and North entrance; Figure 4). In addition, squinted-pairs were aimed down from the surface at the middle of the BGS and Middle entrances; these entrances had overflow openings. The up- and down-looking transducer combinations provided complete vertical coverage at the BGS and Middle entrances. Single-axis rotators were used to aim the transducer beams as close to the SBC entrances as physically possible (SBC squinted acoustic beams sampled essentially from the opening upstream about 3 m at maximum range).

To determine how effective the BGS was at guiding fish away from the turbine intakes, we deployed 18 transducers inside the turbine intakes. Each turbine intake is divided into three sections referred to as A, B, and C. We installed transducers at 1B, 3A, 3B, 3C, 4A, 4B, 4C, 5B, and 6B. These 10° transducers sampled in front of and behind the ESBS located inside the intakes just upstream of the turbine blades. Fish sampled upstream of the ESBS were guided by the screen into a bypass facility. Fish sampled behind the ESBS were not guided and continued on through the turbine blades. The transducers pulsed at 20 pps. Fish guided by an intake screen were enumerated with a transducer mounted on the tip of the screen and aimed along the upstream face of the intake screen toward the intake ceiling. The transducers were aimed 25° off the face of the screen. Unguided fish were sampled with a transducer mounted to the top and back of the intake screen frame. This transducer was aimed down the back of the screen toward the intake floor. Unguided fish were counted as they passed below the tip of the intake screen.

At pier noses of Turbines 1, 3, 4, 5, and 6 (Turbine 2 was off-line during the study), 15° transducers were mounted on sleds, lowered to the bottom of the forebay, aimed up about 10° from vertical, and canted toward the center of an adjacent intake. The transducers pulsed at 10 pps. One of three intakes from each operating turbine was evaluated. The intakes sampled were randomized for each half of the powerhouse separately with the constraint that an equal number of A, B, and C intakes be sampled. The intakes sampled were 1C, 3A, 4C, 5B, and 6A. At Spill Bays 2, 3, 4, 5, 6, 7, and 8, we deployed 15° transducers just below the water surface in the center of the each bay. The transducers were aimed down about 10° upstream and pulsed at 10 pps. These transducers allowed us to estimate the total number of fish passing through the spillway. These data were used to assess spill efficiency and effectiveness and fish passage efficiency.

Response Variables

The response variables used in the single-beam hydroacoustic analysis included fish guidance efficiency of the intake screens (FGE), fish passage efficiency for the dam as a whole (FPE), BGS diversion efficiency (P_B), SBC efficiency for the powerhouse (R_{1-6}), and SBC efficiency for the whole dam (R_{all}; Table 1). Results from BGS move-days are not included in the BGS performance or

Table 1. Definitions for response variables used in the hydroacoustic analysis of fish passage.

Parameter	Descriptor	Estimator
FGE	Fish guidance efficiency depicts intake screen performance (Figure 1).	FGE = G/(G+U), where G = fish guided by turbine screens, U = fish unguided.
FPE	Fish passage efficiency depicts non-turbine passage at the entire project.	FPE = (X+G+S)/TOT, where X = total passage into the SBC, S = spill passage, and TOT = X+G+S+U.
P_B	BGS diversion efficiency shows fraction of fish intended for Turbines 1–3 that are diverted to the north.	$P_B = 1 - (P_{in}/P_{out})$, where P_{in} = proportion of total passage at Turbines 1–3 with the BGS in and P_{out} = proportion at Turbines 1–3 with the BGS out.
P_B-corr	P_B corrected using radio telemetry data for passage through the upstream gap.	P_B-corr = P_B+(1–P_B)* (number gap/(number gap+number under)), number under = 3 radio-tagged fish, number gap = 3 radio-tagged fish.
G_{BGS}	BGS guidance efficiency shows the number of fish within 30 m that were guided to the north.	G_{BGS} = (number within 30 m of BGS – number within 30 m that passed under or around)/number within 30 m of BGS.
$G_{BGScorr}$	GBGS corrected using radio telemetry data for passage through the upstream gap.	$G_{BGScorr}$ = G_{BGS} + (1–G_{BGS})* (number gap/[number gap+number under]).
R_{1-6}	SBC efficiency that expresses SBC passage relative to passage at Turbines 1–6.	$R_{1-6} = X/(X+T_{1-6})$, where T_{1-6} = Turbine 1–6 passage.
R_{all}	SBC efficiency that expresses SBC passage relative to total project passage.	R_{all} = X/TOT.

statistical analysis results. The estimates of variances for the single-beam data include temporal and some spatial variation associated with the way data were collected. Sources of error from seasonal and spatial variation across sample locations (e.g., a spill bay) are not included. Confidence intervals (brackets in bar graphs) are at the 95% level. The main variable for the BGS analysis is BGS diversion efficiency (P_B-corr); P_B represents the proportion of fish intended for Turbines 1–3 that were diverted to other passage routes. P_B-corr is P_B corrected for passage of radio-tagged fish through the upstream gap between the BGS and the shoreline. This correction was necessary because any final BGS design would eliminate downstream passage through this gap.

Biotelemetry techniques allowed us to calculate many of the same parameters measured using hyroacoustics as well as some unique parameters such as BGS guidance efficiency (G_{BGS}). This parameter measures the guidance efficiency of the BGS for all radio-tagged fish that came within 30 m of the upstream face of the BGS. A range of 30 m was selected because it was theorized that this was the range at which fish movement would likely be influenced by the presence of the BGS. The BGS guidance efficiency was also estimated (i.e., corrected) assuming that the upstream gap could effectively be closed off to downstream migrants thereby preventing fish from passing through the gap between the BGS and the south shoreline. This corrected performance of the BGS is represented by $G_{BGScorrected}$.

Statistical Analysis of Hydroacoustic Data

Analysis of variance (ANOVA) was used on the hydroacoustic data to statistically evaluate BGS performance. For all metrics except $\hat{P}_B$ the quantities were computed daily for input into the split-plot analysis. Data for move days were excluded from the ANOVA, but were evaluated in providing study-wide estimates of smolt passage. For the analysis of diversion probabilities ($\hat{P}_B$), values were calculated under similar SBC entrance configurations providing two values for $\hat{P}_B$ per block. Paired t-tests compared $\hat{P}_B$'s between SBC entrance configurations. In addition, one-sample t-tests were used to test the following null hypothesis: $H_o{:}P_B = 0$, against $H_A{:}P_B > 0$, under each SBC configuration. All tests of significance were performed at $\alpha = 0.10$. Day was defined as 6:00 a.m. to 8:00 p.m. and night as 8:00 p.m. to 6:00 a.m.

Radio TelemetryData Aanalysis

Radio telemetry data were examined to determine passage routes and individual fish movements during BGS in and out treatments. All comparisons are descriptive because the experimental design, which included numerous treatment blocks of relatively short duration, was tailored more for the hydroacoustic evaluation.

Results

Hydraulic Characteristics

River discharge at Lower Granite Dam in spring 1998 was 101% of the historical average for that period. Daily discharge was 1,557 m^3/s (55 kcfs) at the beginning of the study on 13 April 1998. Daily discharge peaked on 27 May 1998 at 6,060 m^3/s (214 kcfs). From 13 April 1998 to 20 April 1998 there was no spill, except for SBC discharge at Bays 1 and 2. After this period, spill increased as river discharge increased so that flow through the powerhouse could be kept as constant as possible during the test. When spill occurred, it was usually 24 h/d. About 26% of mean daily total discharge was spilled at Bays 2–8 during this study. The proportion of SBC discharge out of total SBC and turbine discharge was 4.8% and 7.8% at Turbines 1–6 and Turbines 4–6, respectively. SBC discharge was 3% of total project discharge.

Migration Characteristics

Species composition of untagged migrants during the spring study period was dominated by steelhead. Sixty-seven percent of the total index used to estimated passage at the dam was hatchery steelhead. Wild and hatchery yearling chinook salmon, wild steelhead, coho *Oncorhynchus kisutch*, and sockeye *Oncorhynchus nerka* salmon also migrated through Lower Granite Dam during the study. Total juvenile fish passage peaked on 3 May 1998.

Radio Tag Detections

We tagged and released a total of 795 radio-tagged fish, of which 399 were spring chinook salmon, 214 were hatchery steelhead, and 182 were wild steelhead. Mean fish length was 134 mm for spring chinook salmon, 216 mm for hatchery steelhead, and 181 mm for wild steelhead. Total mortality of radio-tagged fish included two chinook salmon and no steelhead resulting in an overall mortality rate of 0.25%. Ninety-seven percent (386 of 399) hatchery spring chinook salmon, 99% (213 of 214) hatchery steelhead, 100% (182 of 182) wild steelhead were detected two or more times by fixed-site receiving stations in the study area. Ninety-two percent (365 of 399) of the spring chinook salmon, 99% (213 of 214) of the hatchery steelhead, and 99% (181 of 182) of the wild steelhead were detected at the upstream face of Lower Granite Dam. When the BGS was in, more than half of the spring chinook salmon were first detected at the powerhouse, and more than half of the hatchery and wild steelhead were first detected at the spillway. When the BGS was out, less spring chinook salmon were first detected at the powerhouse and more hatchery and wild steelhead were first detected at the spillway than when the BGS was in.

Passage Through Lower Granite Dam

Passage routes through the dam for radio-tagged fish were determined for 94% (361 of 386) of the spring chinook salmon, 98% (208 of 213) of the hatchery steelhead, and 99% (180 of 182) of the wild steelhead. Regardless of species, more fish entered turbine intakes and were diverted by the turbine screens into the fish collection system when the BGS was out compared with when it was in. Passage into the turbine intakes and then into the fish collection system occurred predominantly at the north half of the powerhouse through Turbines 4 and 5. When the BGS was in, it attached to the powerhouse between Turbines 3 and 4 and was designed to divert fish away from the south half of the powerhouse (Turbines 1, 2, and 3). The BGS did not appear to affect the time of day that radio-tagged fish entered the fish collection channel. Regardless of the BGS position, passage into the fish collection channel occurred between 0700 hours and 0600 hours for 76% of spring chinook salmon and wild steelhead and 83% of hatchery steelhead.

A higher proportion of radio-tagged steelhead entered the SBC when the BGS was in compared with when it was out (Table 2). Passage efficiencies for spring chinook salmon and wild steelhead, relative to BGS position, were slightly higher with the BGS out (Table 3). Fish passage efficiency (FPE), combined bypass efficiency (CBE), fish guidance

Table 2. Route of passage for all radio-tagged fish (shown in number of fish) passing through Lower Granite Dam relative to BGS position.

	Species					
	Hatchery chinook		Hatchery steelhead		Wild steelhead	
Passage route	BGS in	BGS out	BGS in	BGS out	BGS in	BGS out
Fish collection channel	85	93	26	82	24	77
Spillway	39	30	8	22	12	24
SBC	27	22	25	43	6	21
Turbine	39	26	0	3	7	9
Totals	190	171	59	150	49	131

Table 3. Passage efficiency response variables (%) for radio-tagged fish when the BGS was in, out, and overall, spring 1998. Fish that passed the dam when the BGS was moving are included in overall values but are not included in BGS in and BGS out values.

	Spring chinook			Hatchery steelhead			Wild steelhead		
Response variable	BGS in	BGS out	Overall	BGS in	BGS out	Overall	BGS in	BGS out	Overall
FPE	79	84	82	100	98	99	86	93	91
CBE_{4-6}	82	83	83	100	98	99	85	90	90
FGE_{1-6}	69	78	73	100	96	97	83	90	86
FGE_{1-3}	59	81	73	100	97	98	88	93	90
FGE_{4-6}	76	76	76	100	95	97	81	87	85
R_{all} ($SBC_{overall}$)	14	12	14	42	30	32	12	15	15
R_{1-6}	18	15	17	49	35	38	16	18	19
R_{4-6}	28	31	29	53	49	49	22	27	28
Spill	21	16	19	14	15	14	24	17	20

efficiency (FGE), and SBC efficiency related to Units 4–6 (R_{4-6}) were higher when the BGS was out than when it was in for spring chinook salmon. The same was true for wild steelhead. In contrast, passage efficiencies for hatchery steelhead were higher when the BGS was in.

Based on hydroacoustic data, the proportion of total passage at Turbines 1–3 was 16% lower with the BGS in than with it out for day and night combined (Figure 5). The proportion of fish passing through Turbines 4–6, the SBC, and spillway was higher with the BGS in than when it was out. Data for daytime had the same pattern as the combined data except for the SBC, where mean daily passage was slightly higher during day with the BGS out (Figure 6). During the night, however, mean daily passage at Turbines 4–6, SBC, and spillway were all higher with BGS in than out.

We compared hydroacoustic estimates of SBC efficiency (R_{1-6}), FGE, and FPE for BGS in and out

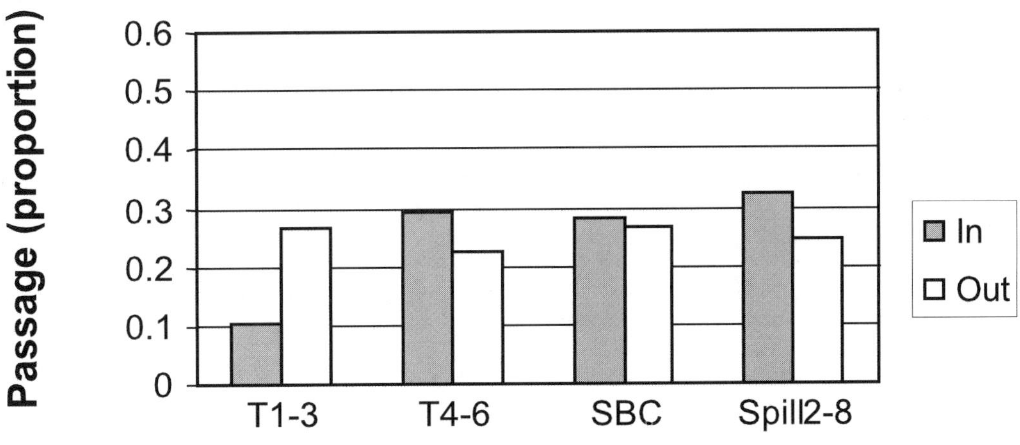

Figure 5. Hydroacoustic estimates of fish passage proportions for BGS in and out conditions. T1 – 3 = Turbine units 1, 2, and 3; T4 – 6 = Turbine units 4, 5, and 6; SBC = Surface Bypass Collector; Spill2 – 8 = spillways 2, 3, 4, 5, 6, 7, and 8.

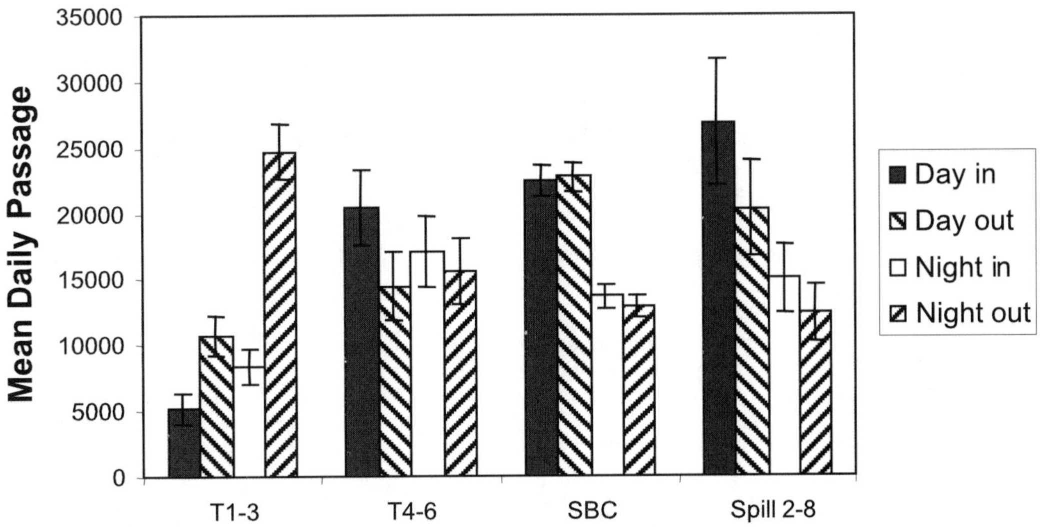

Figure 6. Hydroacoustic estimates of mean daily passage by day and night for BGS in and out conditions. T1 – 3 = Turbine units 1, 2, and 3; T4 – 6 = Turbine units 4, 5, and 6; SBC = Surface Bypass Collector; Spill2 – 8 = spillways 2, 3, 4, 5, 6, 7, and 8.

conditions. Because hydroacoustic techniques could not differentiate between species, estimates represent the population of juvenile salmonids as a whole. R_{1-6} was higher with the BGS in than out (Figure 7); this result was significant (P = 0.066) during night, but not during day or day/night combined. For all turbines combined, FGE was 2.1% higher with the BGS in than out (we did not assess the statistical significance of these FGE estimates). FPE was 2.5% (P = 0.026) higher with the BGS in than out (Figure 7); this result was statistically significant.

BGS Performance

Forebay residence times for radio-tagged fish were shortest when the BGS was in (Table 4). The mean residence times of chinook salmon, hatchery steelhead, and wild steelhead were 1.6, 1.7, and 2.4 times longer, respectively, when the BGS was out compared with when it was in. Most fish arrived at and passed through the dam during a single BGS test condition (BGS in or out). Fish that arrived at or passed the dam while the BGS was being moved were not included in the comparison of residence time relative to BGS position. After moving into the forebay for the first time, some fish traveled 3 km or more upstream before entering the forebay for a second time. Upstream movement was more common when the BGS was out than when it was in. Two spring chinook salmon, 13 hatchery steelhead, and six wild steelhead traveled upstream a minimum of 3 km after detection in the forebay when the BGS was out. Only two hatchery steelhead traveled upstream when the BGS was in. The mean time from first detection at the SBC until passage through the dam was 1.6, 0.9, and 1.7 times higher for chinook salmon, hatchery steelhead, and wild steelhead, respectively, when the BGS was out.

Fish that approached the BGS were either diverted away from the south half of the powerhouse as intended, or they traveled under the BGS or through the upstream gap between the BGS and south shore. Of the fish detected at Lower Granite Dam, 21% (76 of 365) of spring chinook salmon, 11% (24 of 213) of hatchery steelhead, and 10%

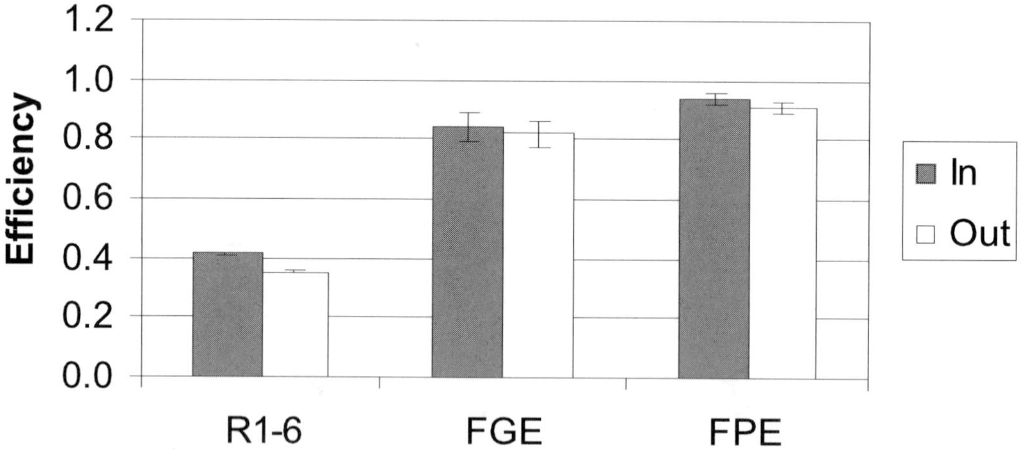

Figure 7. Hydroacoustic estimates of SBC efficiency (R_{1-6}), FGE, and FPE for BGS in and out conditions.

Table 4. Forebay residence time (h) of radio-tagged fish relative to the proportion of time the BGS was deployed in the "in" position during a fish's residence in the forebay.

Species	BGS position	N	Mean	Median	Minimum	Maximum
Hatchery chinook	Out	141	4.93	1.70	0.01	53.31
Hatchery chinook	In	143	3.17	1.06	0.04	33.29
Hatchery steelhead	Out	147	7.68	3.40	0.01	49.97
Hatchery steelhead	In	41	4.58	1.36	0.04	29.77
Wild steelhead	Out	119	6.80	3.61	0.06	57.00
Wild steelhead	In	36	2.83	0.60	0.02	17.16

(18 of 181) of wild steelhead were detected within 30 m of the upstream face of the BGS. One quarter of hatchery steelhead (6 of 24) and about half of the spring chinook salmon (40 of 76) and wild steelhead (10 of 18) detected within 30 m of the BGS traveled under it or through the upstream gap. Of those fish, 75% of hatchery spring chinook salmon (30 of 40), 33% of hatchery steelhead (2 of 6), and 60% of wild steelhead (6 of 10) passed under the BGS; the remainder passed through the upstream gap.

BGS diversion efficiency for radio-tagged fish (P_B) ranged widely, with diversion efficiencies of chinook salmon at 56%, hatchery steelhead at 72%, and wild steelhead at 37%. The BGS performance was also estimated assuming that the upstream gap could effectively be closed off to downstream migrants. This corrected performance of the BGS, represented by P_B-corr, was highest for hatchery steelhead (86%) and similar for spring chinook salmon (69%) and wild steelhead (65%; Table 5). Due to the relatively low number of fish going under or around the BGS, especially for hatchery and wild steelhead, large confidence intervals may be associated with our estimation of P_B-corr.

Based on hydroacoustic data, BGS diversion efficiency (P_B-corr, see Table 1 for definition) for the study as a whole was 78 ± 18% (Figure 8). This means that for every 100 fish destined for Turbines 1–3, 78 fish were diverted to the north (i.e., either to the SBC, Turbines 4–6, or the spillway). BGS diversion efficiency was slightly higher at night (2.3%) than during the day (Figure 8). On an experimental block basis, BGS diversion efficiency was greater than or equal to 73%, except for the last block.

To further quantify BGS performance and obtain an additional measure of BGS performance, we used radio telemetry data to measure the guidance efficiency of the BGS (G_{BGS}). This parameter used data obtained at the face of the BGS when it was in. As was the case for P_B, we also corrected

Table 5. BGS guidance efficiency for radio-tagged juvenile salmon and steelhead in 1998. Fish detected by any underwater antenna on the BGS and/or fish detected by aerial antennas on the north side of the BGS at signal strengths greater than 150 were determined to be within 30 m of BGS. Note: Fish diverted away from powerhouse units 1–3 by the BGS were not necessarily guided along the entire BGS and toward the SBC.

Species	Number in Forebay	Number within 30 m of BGS	Number through gap	Number Under	Number (%) diverted by BGS (G_{BGS})	% diverted (corrected for gap) ($G_{BGScorrected}$)	P_B-corr
Spring chinook	366	76	10	30	36 (47%)	61%	69%
Hatchery steelhead	213	24	4	2	18 (75%)	92%	86%
Wild steelhead	181	18	4	6	8 (44%)	67%	65%
Subyearling fall chinook	261	188	47	41	100 (53%)	78%	n/a

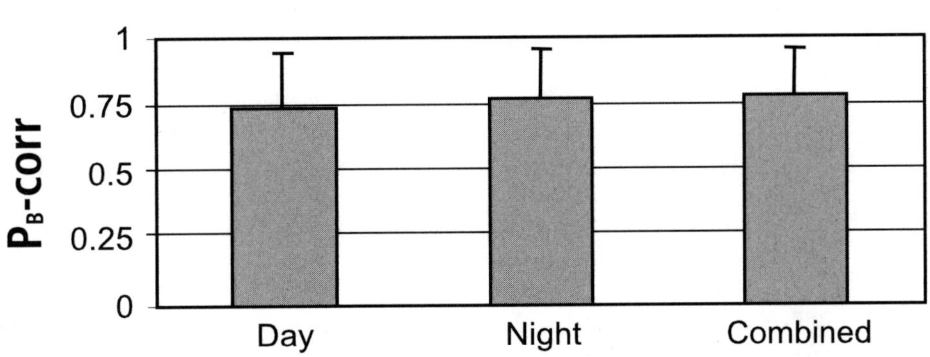

Figure 8. Hydroacoustic estimates of BGS diversion efficiency (P_B-corr) for day, night, and combined.

G_{BGS} for fish that passed through the gap between the BGS and south shoreline. Both measures of BGS performance were relatively similar (Table 5). The BGS was most efficient at guiding hatchery steelhead (86–92%), whereas about the same percentage of spring chinook salmon (61–69%) and wild steelhead (65–67%) were guided.

Discussion

Radio telemetry and hydroacoustic results indicated the BGS diverted about 80% of the smolts destined for Turbines 1–3 to the north. Mean daily passage rates and proportions were generally lower at Turbines 1–3 and higher at Turbines 4–6, the SBC, and the spillway with the BGS in rather than out. Thus, the BGS apparently influenced fish migration through much of the forebay. The comparable results of BGS diversion efficiency obtained from hydroacoustics and radio telemetry is supported by the independent measure of BGS guidance efficiency obtained from radio telemetry data. Although some fish passed under the BGS or through the upstream gap, the majority (61–92%) of fish that came within 30 m of the BGS were guided to the south. This displacement increased the number of spring migrants that encountered the SBC in 1998 compared with 1997. In 1997, 13% of the radio-tagged fish detected at the dam passed within 10 m of the openings to the SBC. In 1998, 39% of the radio-tagged fish detected at the dam passed within 10 m of the openings to the SBC. The BGS was also effective in reducing residence time and upstream movement of spring migrants. Increased residence time when the BGS was out can be attributed to the fish's inability to discover a passage route through the dam.

Fish were exposed to many different conditions at the dam besides those at their ultimate passage route. Although the BGS successfully diverted fish, not all fish that were diverted passed into the SBC. For example, with the BGS in, enhanced passage at the spillway improved FPE such that there was a significant (P = 0.026) difference in FPE between BGS in and out conditions. The horizontal extent of the effect of the BGS was apparently tens of meters, not meters.

The BGS performed especially well at night when downstream migrants will sound if necessary to pass a dam (J. Williams, National Marine Fisheries Service, personal communication). At night, mean daily passage rates at Turbines 1–3 were about one-third lower with the BGS in than out. Most importantly, SBC efficiency (R_{1-6}) was significantly (P = 0.066) higher at night with the BGS in than out. Even though fish migrate deeper during night than day (Thorne and Johnson 1993), day and night BGS diversion probabilities were comparable. Thus, the BGS was deep enough to intercept most fish headed toward Turbines 1–3 during both day and night.

In general, the BGS approach to fish guidance seems valid, although the first-year test in 1998 was not sufficient for long-term decision making. A major caveat to the 1998 test was that Turbine 2 was off-line during the entire study. This means there was one-third less flow under the BGS than there would have been with Turbines 1–3 at full load. This is an appreciable amount of water (~566 m^3/s or ~20 kcfs) that could influence the number of fish moving under the BGS. The BGS should be studied with Turbines 1–3 at full load (~1,699 m^3/s or ~60 kcfs) to examine how well it diverts fish with maximum discharge through the turbines behind it.

Besides the prototype BGS at Lower Granite Dam, other guidance walls have shown a propensity to divert juvenile salmonids. Powerhouses at Rocky Reach and The Dalles dams are oriented parallel to the main river flow. Juvenile fish have been observed guiding along the concrete face of these dams above the trashracks (Stevenson et al. 1997 for Rocky Reach and Sheer et al. 1997 for The Dalles). The only other flat-wall diversion structure for downstream migrant anadromous fish that we are aware of is at Bellows Falls Dam on the Connecticut River in Vermont. The flows (~283 m^3/s or ~10 kcfs) there are relatively low. The structure diverted 94% of radio-tagged juvenile Atlantic salmon *Salmo salar* smolts (CH2M-Hill 1998). Walls in rivers can efficiently divert fish as long as the migrants are surface oriented, water velocity under the wall is not extremely high (<0.6 m/s or 2 fps at Lower Granite Dam), and safe structures can be reliably deployed. In conclusion, we recommend that the BGS be considered as a long-term smolt protection measure at Lower Granite Dam, and that feasibility of the concept be tested at other selected mainstem dams.

Acknowledgments

Walla Walla District, U. S. Army Corps of Engineers funded these studies under the Anadromous Fish Evaluation Program, Pacific Northwest Division

(DACW68–96-D-0002 to Battelle and E-86930151 to the U.S. Geological Survey). We thank Corps staff at Lower Granite Dam and Walla Walla, Washington for logistic and programmatic assistance. We appreciate statistical guidance from L. Calvin, C. Pereira, and J. Skalski. Staff at Battelle, BioSonics, Grant Systems Engineering, Lotek, and Precision Acoustic Systems helped us immensely. We thank S. Evans, R. Perry, J. Plumb, M. Novick, and our colleagues at the Columbia River Research Laboratory, USGS. Use of trade names does not imply endorsement of commercial products.

References

Adams, N. S., and D. W. Rondorf. 2001. Migrational characteristics of juvenile spring and fall chinook salmon and steelhead in the forebay of Lower Granite Dam relative to the 1998 surface bypass collector tests. Report of U.S. Geological Survey (Contract E-86930151) to U.S. Army Corps of Engineers, Walla Walla, Washington.

Adams, N. S., D. W. Rondorf, S. D. Evans, and J. E. Kelly. 1998a. Effects of surgically and gastrically implanted radio transmitters on growth and feeding behavior of juvenile chinook salmon. Transactions of the American Fisheries Society 127:128–136.

Adams, N. S., D. W. Rondorf, E. E. Kofoot, M. J. Banach, and M. A. Tuell. 1997. Migrational characteristics of juvenile spring chinook salmon and steelhead in the forebay of Lower Granite Dam relative to the 1996 surface bypass collection tests. Report of U.S. Geological Survey (Contract E86930151) to U.S. Army Corps of Engineers, Walla Walla, Washington.

Adams, N. S., D. W. Rondorf, and M. A. Tuell. 1998b. Migrational characteristics of juvenile spring and fall chinook salmon and steelhead in the forebay of Lower Granite Dam relative to the 1997 surface bypass collector tests. Report of U.S. Geological Survey (Contract E-86930151) to U.S. Army Corps of Engineers, Walla Walla, Washington.

Bennett, D. H. 1992. Residualism of salmonid fishes in Lower Granite Reservoir, Idaho-Washington. Pages 81–86 *in* Passage and survival of juvenile chinook salmon migrating from the Snake River basin. Proceedings of a technical workshop 26-28 February, 1992. University of Idaho, Moscow, Idaho.

Buettner, E. W. 1991. Smolt monitoring at the head of Lower Granite Reservoir and Lower Granite Dam. Report for 1991 operations to Bonneville Power Administration, (Contract DE-BI79-83bp11631) Portland, Oregon.

Buettner, E. W., and V. L. Nelson. 1989. Smolt condition and timing of arrival at Lower Granite Reservoir. Report for 1988 operations to Bonneville Power Administration, (Contract DE-AI79-83BP11631) Portland, Oregon.

Buettner, E. W., and V. L. Nelson. 1990. Smolt monitoring at the head of Lower Granite Reservoir and Lower Granite Dam. Report for 1989 operations to Bonneville Power Administration, (Contract DE-AI79-83BP11631) Portland, Oregon.

Buettner, E. W., and A. F. Brimmer. 1993. Smolt monitoring at the head of Lower Granite Reservoir and Lower Granite Dam. Report for 1992 operations to Bonneville Power Administration, (Contract DE-BI79-83BP11631) Portland, Oregon.

CH2M-Hill/Montgomery Watson, BioAnalysts, and Pacific Northwest Laboratory. 1998. Bonneville Dam second powerhouse physical guidance device. 100% Letter Report, August 1998, submitted to U.S. Army Corps of Engineers, Portland, Oregon.

Johnson, G. E., N. S. Adams, R. L. Johnson, D. W. Rondorf, D. D. Dauble, and T. Y. Barila. 2000. Evaluation of the prototype surface bypass for salmonid smolts in spring 1996 and 1997 at Lower Granite Dam on the Snake River, Washington. Transactions of the American Fisheries Society 129:381–397.

Johnson, R., S. Anglea, S. Blanton, M. Simmons, R. Moursund, G. Johnson, E. Kudera, J. Skalski, and J. Thomas. 1999. Hydroacoustic evaluation of fish passage and behavior at Lower Granite Dam in spring 1998. Report of Battelle's Pacific Northwest Division (Contract DACW68-96-D-0002) to U.S. Army Corps of Engineers, Walla Walla, Washington.

Johnson, G. E., R. L. Johnson, S. M. Anglea, C. S. Abernethy, S. Blanton, M. Simmons, E. A. Kudera, C. M. Sullivan, and J. R. Skalski. 1998. Fixed-location hydroacoustic evaluation of the prototype surface bypass and collector, spill efficiency, and fish guidance efficiency at Lower Granite Dam in spring and summer 1997. Report of Battelle's Pacific Northwest Division (Contract DACW68-96-D-0002) to U.S. Army Corps of Engineers, Walla Walla, Washington.

Johnson, G. E., R. L. Johnson, E. A. Kudera, and C. M. Sullivan. 1997. Fixed-location hydroacoustic evaluation of the prototype surface bypass and collector at Lower Granite Dam in 1996. Report of Battelle's Pacific Northwest Division (Contract DACW68-96-D-0002) to U.S. Army Corps of Engineers, Walla Walla, Washington.

(NMFS) National Marine Fisheries Service. 1995. Responsible and prudent measure 11 of the NMFS Biological Opinion on 1994-1998 operation of the Federal Columbia River power system and juvenile transportation program (March 2, 1995). NMFS, Seattle.

Normandeau Associates, Inc., J. R. Skalski, and Mid Columbia Consulting, Inc. 1997. Evaluation of chinook salmon smolt (*Oncorhynchus tshawytscha*) survival and condition associated with passage through the surface bypass collector at Lower Granite Dam, Snake River. Report of Normandeau Associates to U.S. Army Corps of Engineers, Walla Walla, Washington.

Park, D. L. 1969. Seasonal changes in downstream migration of age-group 0 chinook salmon in the upper Columbia River. Transactions of the American Fisheries Society 98:315–317.

Raymond, H. L. 1968. Migration rates of yearling chinook salmon in relation to flows and impoundments in the Columbia and Snake rivers. Transactions of the American Fisheries Society 97:356–359.

Raymond, H. L. 1979. Effects of dams and impoundments on migrations of juvenile chinook salmon and steelhead from the Snake River, 1966 to 1975. Transactions of the American Fisheries Society 108:505–529.

Rondorf, D. W., and M. J. Banach. 1996. Migrational characteristics of juvenile spring chinook salmon and steelhead in Lower Granite Reservoir and tributaries, Snake River. Report of U.S. Geological Survey (Contract E 86930151) to U.S. Army Corps of Engineers, Walla Walla, Washington.

Sheer, M. B., G. S. Holmberg, R. S. Shively, H. C. Hansel, T. L. Martinelli, T. P. King, C. H. Frost, T. P. Poe, J. C. Snelling, and C. B. Schreck. 1997. Movement and behavior of radio-tagged juvenile chinook salmon in John Day and The Dalles forebays, 1995. Report of U.S. Geological Survey to U.S. Army Corps of Engineers, Portland, Oregon.

Stevenson, J., A. Giorgi, W. Koski, K. English, and C. Grant. 1997. Evaluation of juvenile spring chinook and steelhead migratory patterns at Rocky Reach and Rock Island dams using radio telemetry techniques, 1996. Report of BioAnalysts for Chelan County PUD, Tacoma, Washington.

Thorne, R. E., and G. E. Johnson. 1993. A review of hydroacoustic studies for estimation of salmonid downriver migration past hydroelectric facilities on the Columbia and Snake rivers in the 1980s. Reviews in Fisheries Science 1:27–56.

American Fisheries Society Symposium 26:161–170, 2001

A New Multibeam Sonar Technique for Evaluating Fine-Scale Fish Behavior near Hydroelectric Dam Guidance Structures

Robert L. Johnson
Battelle Pacific Northwest Division, Post Office Box 999 K6-85, Richland, Washington 98352, USA

Mary Ann Simmons
Battelle Pacific Northwest Division, Post Office Box 999 K6-85, Richland, Washington 98352, USA

Carver S. Simmons
Battelle Pacific Northwest Division, Post Office Box 999 K9-33, Richland, Washington 99352, USA

Susan L. Blanton
Battelle Marine Sciences Laboratory, 1529 West Sequim Bay Road, Sequim, Washington 98382, USA

Abstract.—A Dual-Head Multibeam sonar (DHMS) system was developed for and deployed at two dam sites on the Snake and Columbia rivers in Washington State to evaluate the fine-scale (<1 m) movement behavior of salmon smolt as they approach prototype surface bypass and guidance structures intended to facilitate downstream fish passage. The DHMS system was mounted beneath a moored barge in front of the fish guidance structures to record the smolt's directional behavior. A system of locating devices was placed on the barge to translate target coordinates back to a stationary reference frame for locating fish relative to the dam. The multibeam sonar system is superior to more commonly used splitbeam sonar for ensonifying an adequate region (volume of water) in the proximity of a bypass opening. Multibeam sonar techniques allow the collection of relatively longer and continuous fish trajectories over larger regions, compared with earlier narrow-angle splitbeam hydroacoustic systems. In conjunction with analyzing fish track trajectories in three-spatial dimensions as produced by the DHMS, tracking software and computer visualization programs were developed to study the nature of fish behavior. The tracking software integrates data from the two multibeam systems and accommodates both automatic and manual processing of the data. Visualization software provides a reconstruction of the fish tracks to study entrainment behavior. Fish swimming straight into a bypass or guidance opening, or instead milling about, can also be distinguished with a simple quantitative metric called the tortuosity index. Additional metrics for quantifying the nature of fish trajectories are being developed to describe the detailed tracking data provided by the DHMS.

This chapter describes a new technique for obtaining fine-scale fish behavior data near prototype surface bypass and behavioral guidance structures at hydropower projects. Although the focus here is on hydropower projects, the same techniques may be applied to other situations where fish behavior monitoring and evaluation is needed to establish critical uncertainties or responses to behavioral technologies.

In 1995, the National Marine Fisheries Service, after evaluating the federal Columbia River power system and juvenile salmonid transportation, issued a biological opinion that directed the U.S. Army Corps of Engineers Pacific Northwest Division (Corps) to investigate the effectiveness of surface bypass and collection on juvenile salmonid out-migration (NMFS 1995, 1998). In response to the opinion, both Portland and Walla Walla Corps Districts began an aggressive development of surface flow bypass concepts for design, construction, and testing (USACE 1995). Surface bypass structures modify flow fields near dams in an attempt to attract salmonids to the bypass structure and thereby provide an alternative to entrainment in the turbines. Understanding the behavior of fish in front of these prototype structures will play a major role in assisting the Corps in designing permanent structures. Recent studies indicate that surface bypass structures are only moderately effective at passing fish (NMFS 2000). Therefore, more understanding of how smolts react to these entrances is required to provide the design criteria necessary for effective smolt passage. In this chapter, we describe a new multibeam sonar technique that uses remote sensing to evaluate smolt behavior in front of prototype bypass and guidance structures.

Hydroacoustic techniques have been adopted as a preferred means for evaluating fish behavior because traditional methods using nets or other direct capture techniques are no longer acceptable for fish stocks listed as threatened or endangered under the Endangered Species Act. Additionally, intrusive techniques might affect the behavior of the smolt being studied. Thus, studies of fine-scale fish behavior (i.e., movement in the scale of 1 m [3.3 ft] or less) in the last 3 years have focused on hydroacoustic evaluations coupled with radio telemetry studies using unlisted or hatchery stocks. Radio telemetry studies, while providing species-specific data on individual fish, have been limited by low sample size and the potential for attached transmitters to alter the smolt's behavior.

Traditional hydroacoustic sampling tools have a limited sample volume, while the new multibeam technique offers the ability to cover an expanded volume. In previous studies, a single splitbeam transducer was automatically rotated through as many as 20 sample locations each hour, producing many short tracks (typically less than 2 m [6.6 ft] in length at a maximum range of 20 m [65.6 ft]) (Johnson and Johnson 2000). The short tracks were then pieced together to create a comprehensive picture of fish behavior (Johnson et al. 1997; 1998; Steig and Johnston 1996). With the advent of real-time digital beam forming, it was possible, using a multibeam approach, to sample volumes that were 10 or more times those of traditional splitbeam sampling using a typical 6° transducer. Multibeam sonar has also been recommended by researchers to address biases associated with estimating school size in the marine environment using the small sample volume of traditional echosounding techniques (Gerlotto et al. 1999). Multibeam techniques permitted us to collect relatively long (5–6 m [16.4–19.7 ft] in low noise environments) contiguous fish tracks that better described the behavior of individual fish. The longer fish tracks provided an improved indicator of fish behavior because fish were observed over longer distances. It should be noted, however, that the advantage of longer tracks may not be realized when background noise levels are high as is the case with most amplitude-based hydroacoustic sampling techniques.

In the late 1990s, both the Walla Walla and Portland Corps Districts opted to deploy this new multibeam technology at Lower Granite and Bonneville dams (Figure 1). The new approach, called Dual-Head Multibeam sonar (DHMS), allows three-dimensional tracking of individual fish within a relatively large volume of water (approximately 248 m^3 [8,758 cubic feet] at a 20-m [65.6-ft] range). The use of DHMS has resulted in unprecedented insight into fine-scale salmon smolt behavior in a nonintrusive manner (Johnson and Moursund 2000).

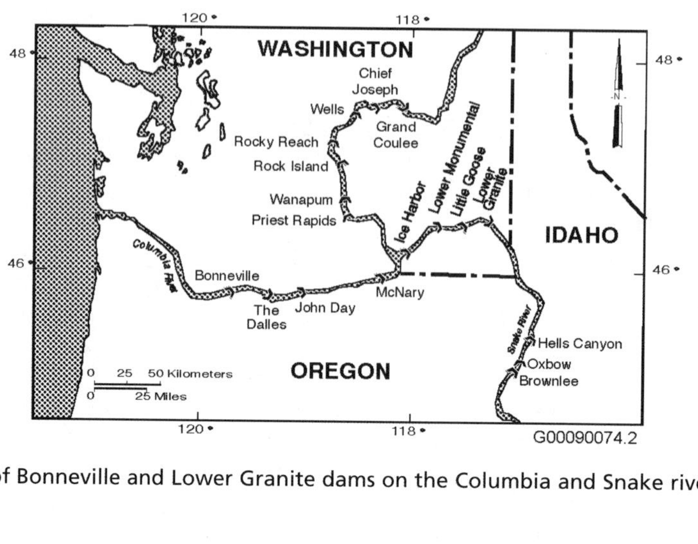

Figure 1. Location of Bonneville and Lower Granite dams on the Columbia and Snake rivers in Washington, respectively.

Until 1996, when the first behavioral study was undertaken using splitbeam hydroacoustics at Lower Granite Dam, little was known about the fine-scale behavior of smolts as they approached a hydropower project. Past research was devoted primarily to large-scale smolt behavior evaluations covering distances of 32–64 km (20–40 mi) (Giorgi and Stevenson 1995; Steig and Johnston 1996). But, the fine-scale behavior of smolts in the immediate forebay region of the structure will likely determine the success or failure of a bypass system. As Pitcher (1993) states, "*The behavior of fishes intimately reflects unique and efficient solutions to the problems raised by their three-dimensional environment.*" Migratory fish approaching a dam are undoubtedly reacting to a myriad of stimuli as they determine which path to take. Among these might be light, sound, hydraulics, and electromagnetic fields. Understanding fish reactions to these stimuli and the mechanisms behind them will likely be the key to designing better and more efficient collection and bypass systems. It follows that successfully understanding the environmental factors that affect fish behavior as they approach dams also will contribute to the design of surface flow bypass structures for effective and safe passage of smolts past dams or other areas of difficult or hazardous passage (Johnson et al. 1999; Coutant 1999).

Multibeam Sonar—A Description

The hydrographic industry has used narrow-beam echo sounding since the 1960s, and it has developed as a tool of preference within that industry (Vilming 1998). Multibeam swath bathymetry has emerged in recent years. It has the advantage of large swath coverage with relatively fine resolution for bottom mapping and seabed classification. It has only been within the last few years that multibeam sonar has been considered a viable tool for fishery research (Hamitouche et al. 1998; Gerlotto et al. 1999). The primary reasons for this are increased sample volume and target position resolution beyond that achievable with traditional hydroacoustic techniques. Scientific echo sounders of the future will likely have a multibeam configuration and may incorporate the functionality of a splitbeam configuration within beam target positioning as well.

Data Collection Using the DHMS

Multibeam technology was used first at Lower Granite Dam in 1997 and subsequently at Bonneville Dam in 1998. The DHMS used to collect behavior data used a unique application of the multibeam concept by using two sonar heads positioned on a fixed mount. The sonar heads were oriented at a 90° angle to one another, referenced to the primary-axis of the two beams, producing horizontal and vertical components within a "hot zone." The hot zone, or region of overlap of the two beams, was the volume sampled for fish tracking. The vertical aiming angle of the pole mount was controlled remotely via a motor-driven actuator arm.

The sonar heads were deployed from a barge moored in the forebay in front of the surface bypass collector (Figures 2 and 3). The platform was secured in place by two large upstream anchors attached to buoys and two tensioning lines attached

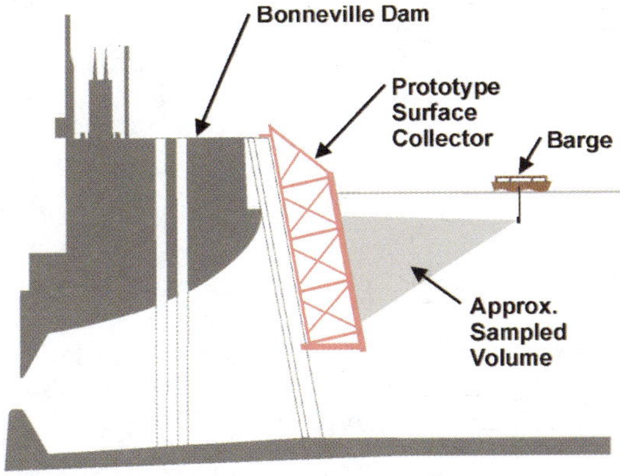

Figure 2. Location of the DHMS barge relative to Bonneville Dam, and surface collector deployed in 1998.

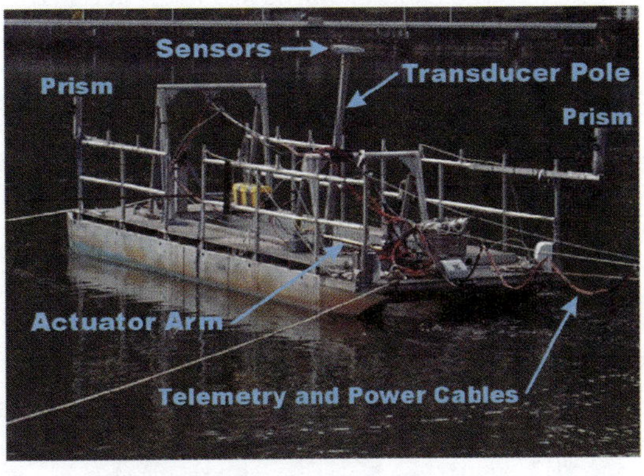

Figure 3. Photograph of the barge deployed at Bonneville Dam in 1998.

to the bypass structure. Two Simrad[1] SM2000 sonar heads (multiple transducers and associated underwater electronics) were mounted near the end of a 6.0-m (19.7-ft) (3.5-m [11.5-ft] submerged) pole. The orientation of the pole was monitored continuously by a KVH[2] C100 fluxgate compass and a Crossbow[3] CXTILT dual-axis tilt sensor attached to the top of the pole. The position of the floating platform relative to the bypass structure was monitored at 1-s intervals using an MDL[4] Laser Radar Fanbeam MKIII. Prisms were placed at both the upstream and downstream extents of the barge to provide additional orientation data. Power, data telemetry, and actuator control lines were attached to a 0.635-cm (0.25-in) insulated cable umbilical belayed to the bypass structure.

The barge was positioned approximately 20 m (65.6 ft) upstream of the bypass structure to preclude bottom interference and to avoid interference from turbine noise. The DHMS collected data 24 h/day. Both horizontal and vertical coverage was approximately 10 m (32.8 ft) at the maximum range (Figure 4). That translates to a horizontal coverage of almost an entire turbine intake.

The control and logging equipment for the DHMS were housed in a trailer positioned on the deck of the dam. The system comprised two Simrad SM2000 surface processor units, a control computer running Seismic Positioning System Software for the MDL Laser Radar Fanbeam MKIII, and a third computer running Battelle's Multibeam Data Acquisition Software (MBAQ) for logging the data (Figure 5). Additionally, a Garmon Global Positioning System was mounted on the roof of the powerhouse, and the MDL Laser Radar Fanbeam MKIII was positioned at the northern-most end of the bypass structure. An additional prism was attached to the handrail of the upstream side of the bypass structure to provide a quality control reference point for platform positioning. An actuator control switch was positioned near the surface units for remote aiming of the sonar heads. The operator could thereby view the sonar displays while adjusting the aiming angle for optimal positioning or exploration.

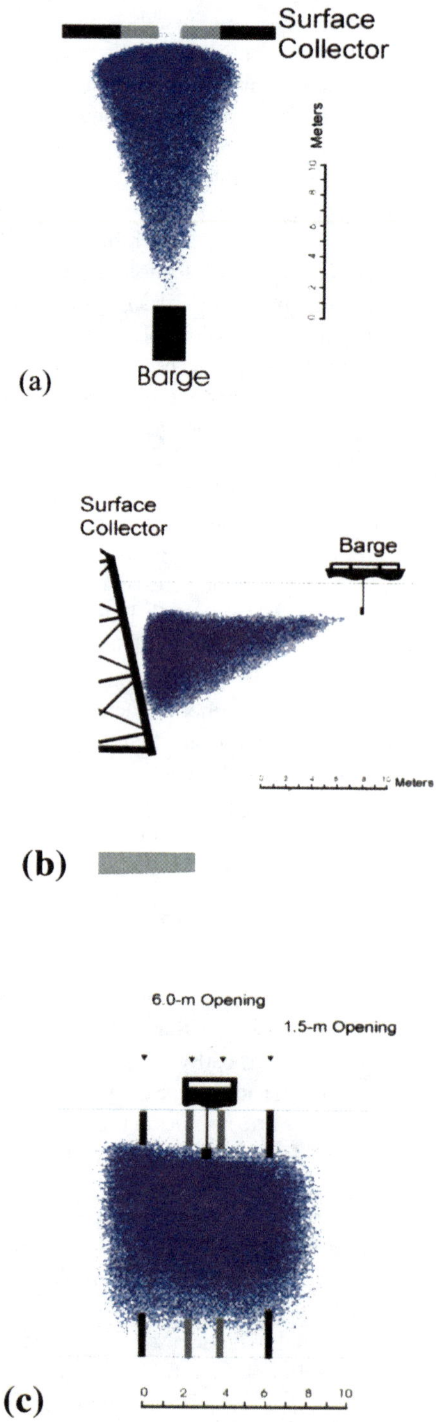

Figure 4. Tracked target maps in relation to surface bypass structures at Bonneville Dam in 1998: (a) top view; (b) side view; (c) front (forebay) view. Gray areas in (a) refer to two opening configurations shown in (c).

[1]Kongsberg Simrad Mesotech Ltd., Port Coquitlam, British Columbia.

[2]KVH Industries Inc., Middletown, Rhode Island.

[3]Crossbow Technology Inc., San Jose, California.

[4]MDL Technologies Inc., Houston.

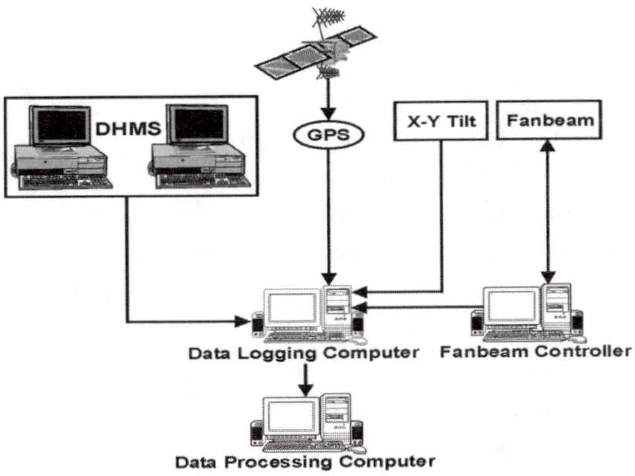

Figure 5. DHMS system diagram.

MTrack Fish Tracking Software

MTrack software developed at Battelle was used to filter out the fish target signals from permanent dam structures and to group targets together based on their proximity in space, time, and angle units through a visual user interface (Figure 6). Once the MTrack software had grouped tracks together in a meaningful way, technicians read the visual track echograms and separated fish targets from the noise according to expected track characteristics in three dimensions, range versus time, angle versus range, and angle versus time. Tracks that had continuity in time and space were deemed legitimate targets. If a signal remained in one place for too long, or was sporadic in nature, it was deemed noise. The system was calibrated periodically by suspending a target of known target strength in the sampled region. A calibrated copper sphere with a measured target strength of −45 decibels was used. This corresponds to the on-axis dorsal target strength expected from a large migratory smolt (Love 1971). Calibration files were obtained each day at the field site for both stationary and transecting targets providing an in-season check on the operational parameters of the DHMS. The calibration files were processed by the MTrack software to confirm that the track of the ball was clearly visible and selected as a potential fish track by the software. Additional algorithms in the MTrack software were used to retain only fish targets that overlapped in time and space within the horizontal and vertical transducer beams. The resulting fish track data were stored as computer files.

The text files resulting from fish tracking were further processed by a series of Fortran programs that calculated fish positions within the forebay (in the x, y, z planes). These programs derived fish location from transducer location, tilt, and aspect. The data were then loaded into a database where

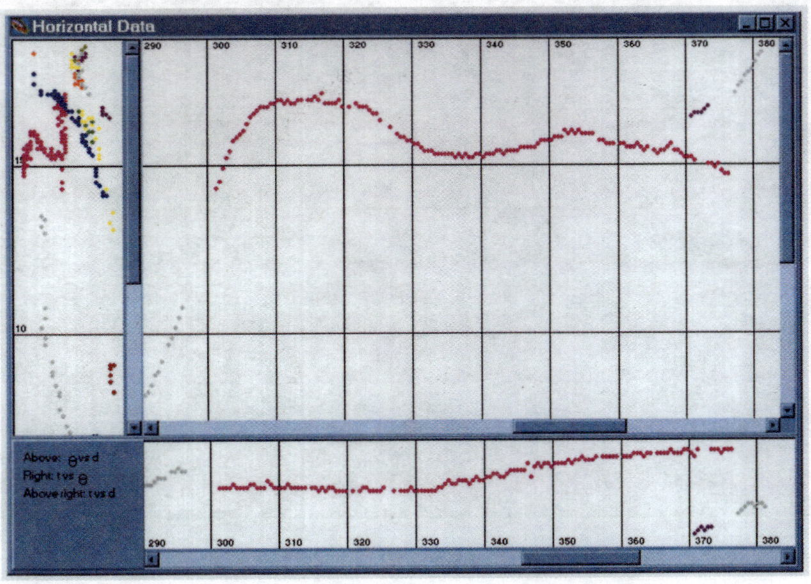

Figure 6. An example of fish tracks resulting from sonar echoes as displayed by the Mtrack visual user interface. The top left panel is a display of range versus angle. The top right panel is range versus time. The bottom panel represents angle versus time.

velocity and acceleration were calculated for each target. Average positions and average speed were also calculated for individual fish targets. The data were then queried to obtain the subsets of data used to generate the figures and example results.

3-D Visualization

An animated, three-dimensional, visualization tool was developed to view observed fish movement in the context of the dam (Figure 7). The visualization includes user controls for manipulating the scene graphic, speed of data playback, and display of track histories. These visualizations were examined both to increase understanding of fish behavior and to spot trends and anomalies in the dataset. The presence of the latter prompted further analysis of the tracked targets or quality control procedures. In Figure 7, the barge with transducer (red cylinder) is shown at upper left; the upstream face of the surface bypass is at right with the opening shown as the dark rectangular area. Fish tracks are shown as irregular lines (yellow).

Figure 7 illustrates a feature commonly noted as fish approach the bypass: transition from low tortuosity (straight) movement away from the surface collector to high tortuosity (turning) near it. High tortuosity may indicate that fish are reluctant to enter a bypass entrance. Providing better directional cues, such as attracting flows in a bypass (Coutant 2001, this volume), could improve fish entrainment as displayed by visualization.

Two analytical methods are commonly used to model animal movement. One is based on population dynamics (as described by a population distribution aggregated in space and time) and the other on the movement of individuals (trajectories). Because multibeam sonar allowed us to collect data for individual fish, we focused our development of new metrics on the latter individual-based, or Lagrangian, approach. We adapted techniques from particle-diffusion theory (Parrish and Hamner 1997; Turchin 1998), multiple target tracking (Bar-Shalom and Li 1995), spherical statistics (Zar 1984), geology (Davis 1986), and animal behavior (Parrish and Hamner 1997; Turchin 1998) to develop our own suite of descriptive information that we believe will be useful for applications at hydropower dams.

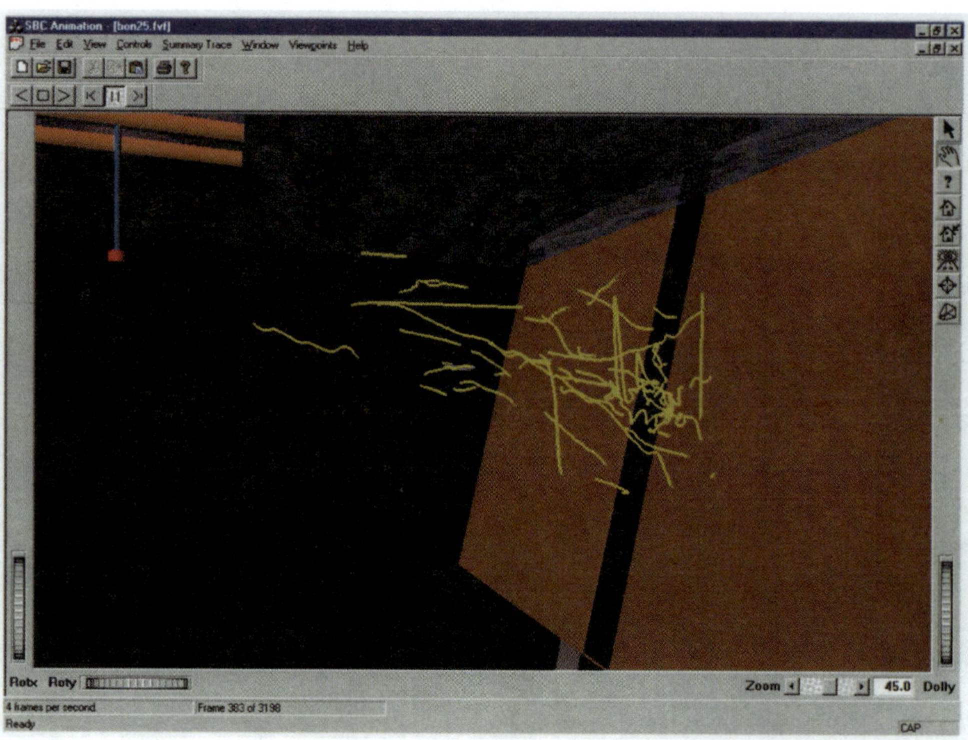

Figure 7. Illustration (single frame) of the three-dimensional interactive visualization developed for displaying fish track data.

Track Descriptors and Metrics

Because it is impractical and nearly impossible to visualize all the individual movements of a large sample of smolt tracks, simplified metrics that summarize behavior information are essential quantitative tools. Metrics allow the multitude of swimming behavior patterns to be put into a probability context that enables the sample population of fish to be evaluated as a whole. Some of the descriptors and metrics we have used include: tortuosity, polar histograms and circular statistics, vector plots, and frequency distributions. The collection of a substantial archive of fish track data (in millions) for many locations since these early endeavors has motivated devising metrics to condense behavior information and to facilitate interpretation.

A particularly useful metric is tortuosity, which describes how much a fish track is turning or departing from a linear path.

The tortuosity index (τ) is a term to quantify linearity (linearity was used historically in single-beam hydroacoustics), where:

$$\tau = \left(\frac{\text{Distance Traveled}}{\text{Straight Line Distance}} \right)$$

Distance traveled is the sum of the small distances a target moves between each echo along a track. The straight-line distance is the difference between the target's first and last recorded positions. Using this definition, a fish traveling in a straight line will have an index of tortuosity equal to one. Fish traveling more circuitous paths have index values greater than one. Examples of tracks having various tortuosity index values are plotted in Figure 8.

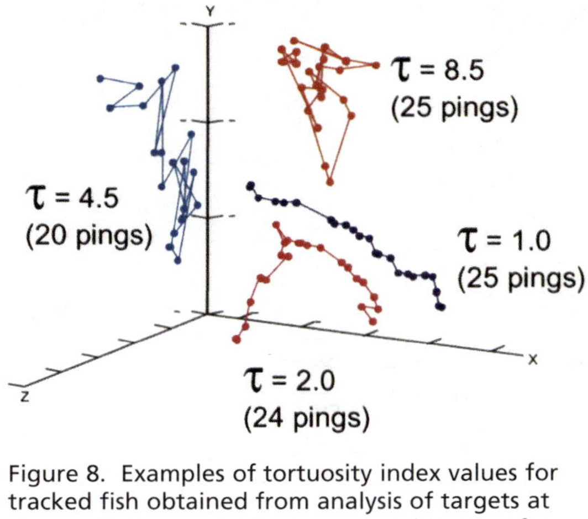

Figure 8. Examples of tortuosity index values for tracked fish obtained from analysis of targets at Bonneville Dam in 1998. Axes are arbitrary reference points.

Direction of Travel

The direction traveled by tracked fish is presented as a polar histogram for the three planar axes described for the sonar coverage (Figure 9). This information is important because it indicates the direction fish were moving in the water and the distribution of movement within a particular region. Statistics generated can be used to test for differences in movement between regions. The distribution also shows the variability in heading. It is apparent from this example that while the predominant direction, in this region, is down and toward the dam, not all the fish targets in the sampled population are headed in this direction. This range in the direction of movement is another common feature of fish behavior found close to the guidance structures. Even in areas of high flows, not all fish targets are headed in the direction of flow. The mean direction shown is an angular mean based on circular statistics and should not be confused with a usual Cartesian coordinate mean.

Fish swimming effort is related to the direction of fish travel and the flow conditions at each location. In Figure 10, the green (observed) vector component represents the direction of fish movement relative to the dam structure as "observed" with the DHMS system. The blue (flow) vector is the average speed and direction of "flow" for a specified water volume as measured by a physical hydraulic model located at the Corps' Waterways Experiment Station in Vicksburg, Mississippi, or by computational fluid dynamics models. The red (effort) vector represents the speed and direction fish are exerting swimming "effort" to maintain their position in the current. Effort, thus, is the difference between the observed and water flow vectors.

An example of a fish swimming field map is shown in Figure 11. The figure displays binned data from Bonneville Dam DHMS sampling in 1998. The swimming field map illustrates the observed, flow, and fish swimming effort upstream of a narrow fish entrance slot (centered at the zero location on the abscissa) in a prototype surface collector.

Depth Distribution

Frequency distributions of the depth of fish targets relative to distance from a guidance or collection structure provide a picture of vertical location during the sample period (Figure 12). This pattern of distribution is useful for establishing whether fish are moving toward the shallow surface collector or the deeper turbine intake.

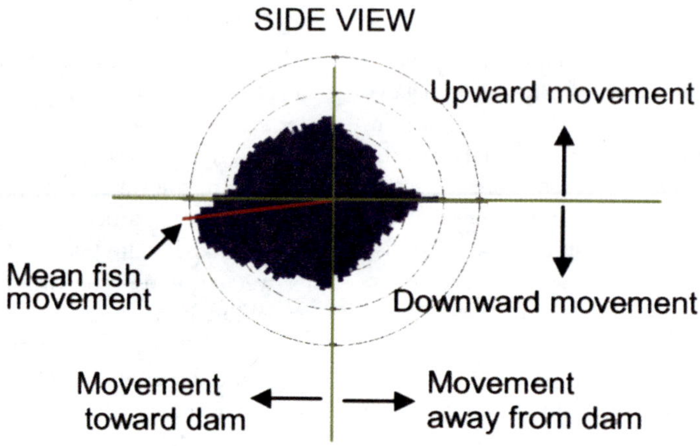

Figure 9. Example side view of a polar histogram to illustrate the distribution of movement by a tracked fish population within a defined area in front of the bypass system.

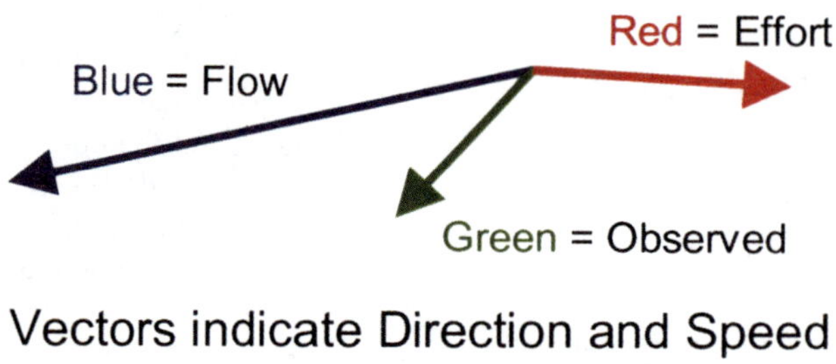

Figure 10. Water flow, observed fish movement, and fish swimming effort vectors at a location.

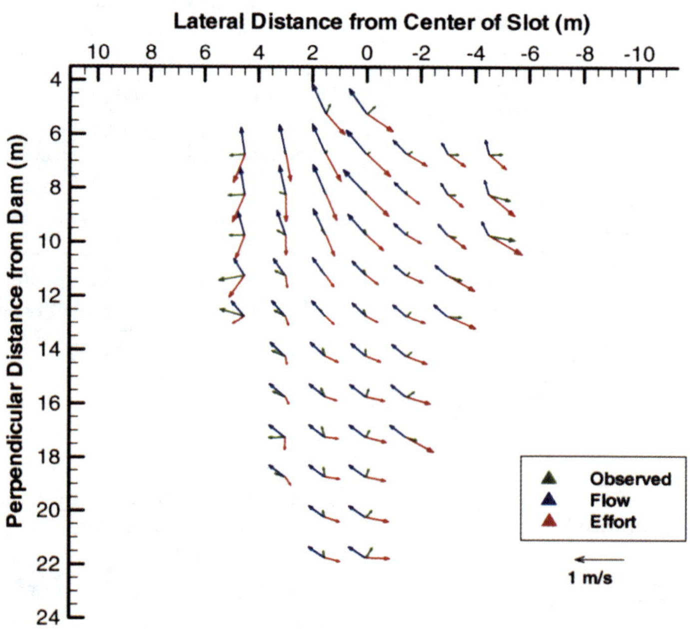

Figure 11. Example of observed, flow, and effort fish swimming vectors (plan view) from Bonneville Dam in 1998.

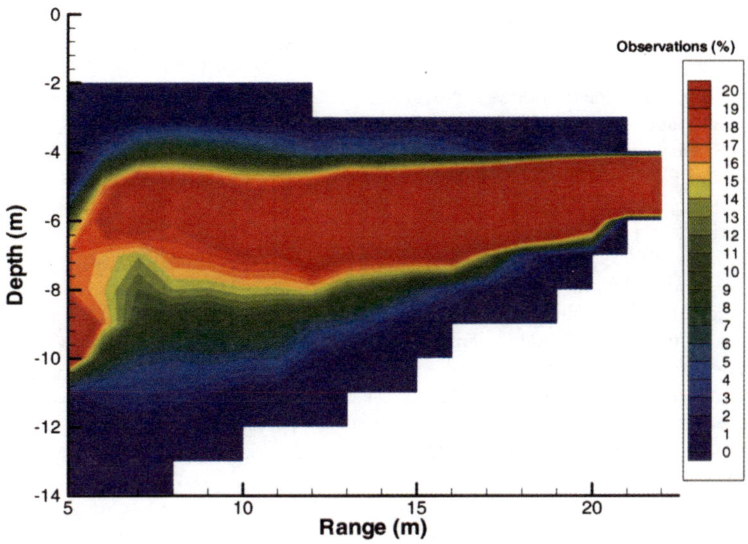

Figure 12. Depth frequency distribution (% fish observations) with range from the multibeam sonar heads. Example from Bonneville Dam (1998) for one configuration of the bypass entrance (1.5 m [4.9 ft] wide in this case).

Kinematics

The kinematic, or motion, analysis was divided into two parts. First, descriptive statistics were developed to illustrate the overall distribution of fish velocity relative to any test conditions at the bypass structure, and for both the observed and effort (flow-corrected) data. Second, velocity component ratios were plotted for each axis for each plane of the sample volume. Velocity component ratios are the ratios of the velocity in each axis (horizontal, vertical, and depth) to the total velocity. Such ratios are a compact form to display the relative contribution of the velocity in each axis to the total velocity, and results in another basis for defining behavior regions.

Conclusion

The DHMS is an emerging technology for fisheries applications that addresses the current nonintrusive information needs of the scientific community for salmon smolt behavior at hydropower facilities. It has two principal advantages over traditional scientific hydroacoustic techniques. It 1) allows for simultaneous, large-volume coverage of a region of interest, and 2) affords the researcher with three-dimensional tracking capability within a large volume of water.

The current DHMS is a modular system composed of two synchronized sub-systems including two separate sonar heads. The data from each are stored in a common location but requires postprocessing software to generate three-dimensional fish tracks. Future systems will likely be fully integrated, both at the surface and underwater, with integrated real-time tracking built into the system. The DHMS system may be used to track targets relative to physical capture and guidance structures or other behavior modifying technologies such as light (Maiolie et al. 2001, this volume; Johnson et al. 2001, this volume), turbulent flow (Coutant 2001, this volume), electrical/magnetic fields, or low-frequency sound and vibration (Goetz et al. 2001, this volume). The DHMS may also be applied to general aquatic behavioral studies in both freshwater and marine environments.

Acknowledgments

Support for the projects that resulted in the DHMS was supplied by the Walla Walla and Portland Districts, U.S. Army Corps of Engineers (Contracts No. DACW68–96-D-0002 and Project 28777, respectively). We especially thank Tim Wik and Blaine Ebberts for their support of this work. Our gratitude to C. Coutant, G. Johnson, and one anonymous reviewer who provided helpful review of this chapter.

References

Bar-Shalom, Y., and X. Li. 1995. Multitarget-multisensor tracking: principles and techniques. Artech House, Norwood, Massachusetts.

Coutant, C. C. 1999. Think Like a Fish! Emphasizing the 'Behavior' in Behavioral Guidance Systems. Hydro Review Volume XVIII:(3)18–25.

Coutant, C. C. 2001. Turbulent attraction flows for juvenile salmonid passage at dams. Pages 57–77 *in* C. C. Coutant, editor. Behavioral technologies for fish guidance. American Fisheries Society, Symposium 26, Bethesda, Maryland.

Davis, J. C. 1986. Statistics and data analysis in geology. Wiley, New York, New York.

Gerlotto, F., M. Soria, and P. Fréon. 1999. From two dimensions to three: the use of multibeam sonar for a new approach in fisheries acoustics. Canadian Journal of Fisheries of Aquatic Science 56:6–12.

Giorgi, A., and J. Stevenson. 1995. A review of biological investigations describing smolt passage behavior at Portland District Corps of Engineers projects: implications to surface collection systems. Report to the U.S. Army Corps of Engineers, Portland District, Portland, Oregon.

Goetz, F. A., J. J. Dawson, T. Shaw, and J. Dillon. 2001. Evaluation of low-frequency sound emitters for guiding salmon smolts away from a navigation lock. Pages 57–77 *in* C. C. Coutant, editor. Behavioral technologies for fish guidance. American Fisheries Society, Symposium 26, Bethesda, Maryland.

Hamitouche, C., V. Fracasso, and C. Scalabrin. 1998. Multibeam sonar image processing and three-dimensional analysis of fish schools. Journal of the Acoustical Society of America 103:2939.

Johnson, G. E., R. L. Johnson, E. A. Kudera, and C. Sullivan. 1997. Fixed-location hydroacoustic evaluation of the prototype surface bypass and collector at Lower Granite Dam in 1996. Final report to the Walla Walla District, U.S. Army Corps of Engineers, Walla, Walla, Washington.

Johnson, G. E, R. L Johnson, S. M. Anglea, C. S. Abernethy, S. L. Blanton, M. A. Simmons, E. A. Kudera, C. Sullivan, J. R. Skalski, and J. Thomas. 1998. Fixed-location hydroacoustic evaluation of the prototype surface bypass and collector, spill efficiency, and fish guidance efficiency at Lower Granite Dam in spring and summer 1997. Report to Walla Walla District, U.S. Army Corps of Engineers, Walla Walla, Washington.

Johnson, R. L., R. A. Moursund, and M. A. Simmons. 1999. Fish behavior in front of the prototype surface collector at Bonneville Dam in 1998. Report to U.S. Army Corps of Engineers, Portland District, Portland, Oregon.

Johnson, R. L., and G. E. Johnson. 2000. Fish behavior relative to a prototype surface bypass, and collector (SBC) at Lower Granite Dam, using innovative hydroacoustic techniques. Pages 57–67 *in* M. Odeh, editor. Advances in fish passage technology: engineering design, and biological evaluation. American Fisheries Society, Bethesda, Maryland.

Johnson, R. L., and R. A. Moursund. 2000. Evaluation of juvenile salmon behavior at Bonneville Dam, Columbia River, using a multibeam technique. Aquatic Living Resources 13:(2000)1–6.

Johnson, P. N., F. A. Goetz, and G. R. Ploskey. 2001. Evaluation of strobe lights for vertically displacing juvenile salmon near a filling culvert intake at the Hiram M. Chittenden Locks, Seattle, Washington. Pages 57–77 *in* C. C. Coutant, editor. Behavioral technologies for fish guidance. American Fisheries Society, Symposium 26, Bethesda, Maryland.

Love, R. H. 1971. Dorsal-aspect target strength of an individual fish. Journal of the Acoustical Society of America 49:816–823.

Maiolie, M. A., B. Harryman, and B. Ament. 2001. Response of free-ranging kokanee to strobe lights. In Behavioral technologies for fish guidance. Pages 57–77 *in* C. C. Coutant, editor. Behavioral technologies for fish guidance. American Fisheries Society, Symposium 26, Bethesda, Maryland.

NMFS (National Marine Fisheries Service). 1995. Biological opinion—reinitiation of consultation on 1994-1998 operation of the Federal Columbia River Power System and juvenile transportation program in 1995 and future years. NMFS, Hydro Program, Portland, Oregon.

NMFS (National Marine Fisheries Service). 1998. Supplemental biological opinion—operation of the Federal Columbia River Power System including the smolt monitoring program and the juvenile fish transportation program: a supplement to the biological opinion signed on March 2, 1995, for the same projects. NMFS, Hydro Program, Portland, Oregon.

NMFS (National Marine Fisheries Service). 2000. Biological opinion—operation of the Federal Columbia River Power System including the juvenile fish transportation program, and the Bureau of Reclamation's projects, including the entire Columbia Basin Project. NMFS, Hydro Program, Portland, Oregon.

Parrish, J. K., and W. M. Hamner, editors. 1997. Animal groups in three dimensions. Cambridge University Press, Cambridge.

Pitcher, T. J., editor. 1993. Behavior of teleost fishes. 2nd edition. Chapman and Hall, London.

Steig, T. W., and S. V. Johnston. 1996. Monitoring fish movement patterns in a reservoir using horizontally scanning split-beam techniques. International Council for the Exploration of the Sea, Journal of Marine Science 53:435–441.

Turchin, P. 1998. Quantitative analysis of movement. Sinauer Associates, Sunderland, Massachusetts.

U.S. Army Corps of Engineers (USACE). 1995. Lower Snake and Columbia Rivers surface bypass and collection systems prototype development program. USACE, Portland, Oregon.

Vilming, Stale. 1998. The development of the multibeam echosounder: an historical account. Journal of the Acoustical Society of America. 103:2935.

Zar, J. H. 1984. Biostatistical analysis. Prentice-Hall, Englewood Cliffs, New Jersey.

American Fisheries Society Symposium 26:171–182, 2001

Use of Electrical Barriers to Deter Movement of Round Goby

Jacqueline F. Savino
Great Lakes Science Center, U.S. Geological Survey,
1451 Green Road, Ann Arbor, Michigan 48105, USA

David J. Jude
Center for Great Lakes and Aquatic Sciences, University of Michigan,
501 East University, Ann Arbor, Michigan 48109, USA

Melissa J. Kostich
Great Lakes Science Center, U.S. Geological Survey
1451 Green Road, Ann Arbor, Michigan 48105, USA

Abstract.—An electrical barrier was chosen as a possible means to deter movement of round goby *Neogobius melanostomus*. Feasibility studies in a 2.1-m donut-shaped tank determined the electrical parameters necessary to inhibit round goby from crossing the 1-m stretch of the benthic, electrical barrier. Increasing electrical pulse duration and voltage increased effectiveness of the barrier in deterring round goby movement through the barrier. Differences in activity of round goby during daytime and nocturnal tests did not change the effectiveness of the barrier. In field verification studies, an electrical barrier was placed between two blocking nets in the Shiawassee River, Michigan. The barrier consisted of a 6-m wide canvas on which were laid four cables carrying the electrical current. Seven experiments were conducted, wherein 25 latex paint-marked round goby were introduced upstream of the electrical barrier and recovered 24 h later upstream, on, and downstream of the barrier. During control studies, round goby moved across the barrier within 20 min from release upstream. With the barrier on and using the prescribed electrical settings shown to inhibit passage in the laboratory, the only marked round goby found below the barrier were dead. At reduced pulse durations, a few round goby (mean one/test) were found alive, but debilitated, below the barrier. The electrical barrier could be incorporated as part of a program in reducing movement of adult round goby through artificial connections between watersheds.

The round goby *Neogobius melanostomus* is a Great Lakes nonindigenous fish that was first recorded in the St. Clair River in 1990 (Crossman et al. 1992; Jude et al. 1992). Round goby are small, benthic fish that generally prefer cobble/rock or macrophyte-dominated substrate (Jude and DeBoe 1996). Round goby are primarily benthivores, usually eating molluscs, aquatic insects, and other invertebrates (Jude et al. 1995, French and Jude in press). The zebra mussel *Dreissena polymorpha*, another nonnative species that has colonized throughout the Great Lakes, provides a ready and plentiful food source. Consequently, round goby have moved quickly throughout the Great Lakes region (Charlebois et al. 1997). They have now entered the Illinois Water System, Chicago, Illinois, an artificial connection between Lake Michigan and the Mississippi River drainage, and are moving inland from Lake Michigan (Charlebois et al. 1997).

Some evidence suggests that round goby could compete with or displace native fish (Jude et al. 1995; Janssen and Jude in press). Round goby observed in the laboratory were extremely aggressive and can directly displace mottled sculpin *Cottus bairdi* (Dubs and Corkum 1996). Populations of mottled sculpin and logperch *Percina caprodes* have declined dramatically in areas colonized by round goby (Jude et al. 1995), possibly from competition for space or food or disruption of spawning. Entry of round goby into the Illinois Waterway System could threaten many other native species particularly as this introduced species enters the Mississippi River drainage (Charlebois et al. 1997). Zebra mussels entered the Mississippi River through this connection with major economic and ecological consequences (Miller and Payne 1997; Morton 1997), and therefore, management agencies are interested in deterring further introductions of nonindigenous species through this connection (Charlebois et al. 1997).

Although most nonindigenous introductions of fishes have occurred through other methods (primarily stocking, bait release, and aquarium release), the spread of these fish into neighboring waterways has been enhanced via artificial connections (irrigation and boat canals) (Fuller et al. 1999). In some cases, fish native to the United States (such as blacktail shiner *Cyprinella venusta*, brassy minnow *Hybognathus hankinsoni*, comely shiner *Notropis amoenus*, brindled madtom *Noturus miurus*, flathead catfish *Pylodictis olivaris*, Atlantic needlefish *Strongylura marina*, and white perch *Morone americana*) have expanded their range through canal systems (Fuller et al. 1999). A number of clupeids (blueback herring *Alosa aestivalis*, skipjack herring *Alosa chrysochloris*, alewife *Alosa pseudoharengus*, and gizzard shad *Dorosoma cepedianum*) have entered the Great Lakes through a variety of canals (Fuller et al. 1999). To prevent further expansions of fish populations through canal systems, methods need to be developed to prevent movement of fish (native or nonnative to the United States) in either direction within these unnatural connections between watersheds.

Among possible control alternatives in the Chicago Waterway, the U.S. Army Corps of Engineers and U.S. Fish and Wildlife Service concluded that an electrical barrier would be most feasible and effective in the first part of an integrative approach to preventing fish movement through an artificial connection between watersheds (Keppner and Theriot 1997). Unfortunately, round goby have now moved beyond the original barrier site in this case, but managers are still interested in developing a test barrier to prevent additional fish movement through the site and will use information on round goby in determining their overall management scheme. Researchers and managers acknowledge that any one program or method is unlikely to be 100% effective in preventing fish movement through a canal, but even slowing down the migration of nonindigenous fish could have strong economic benefits. A first step in the management program would be to develop a barrier that would prevent a large majority of fish from crossing. Through time, additional methods to complement the original barrier would increase the likelihood of preventing fish movement through the canal. Such barriers would have global applications wherever canals have connected previously separate watersheds.

Electrical barriers have been used to guide fish into specific areas. For example, chinook salmon *Oncorhynchus tshawytscha* were guided into traps for assessments (Palmisano and Burger 1988). Electrical barriers can prevent fish (such as sea lamprey *Petromyzon marinus*) from moving upstream against the current—they can be stunned and float downstream from the area of concern (Swink 1999). However, little is known about round goby migration and behavioral responses to electrical barriers (Charlebois et al. 1997). As round goby are a current nonindigenous species of interest, our objective was to determine whether a benthic, electrical barrier could be effective in deterring passage of round goby through a waterway. We used a series of laboratory tests to quickly determine appropriate electrical settings that affected round goby behavior and a set of field tests to verify laboratory results over a longer time period and in a larger and more open setting.

Methods

Laboratory Study

Round goby were collected from the St. Clair River using hook and line. They were held in well aerated, oval (190 L) tanks with a flow-through, well-water system (12°C) on a 12 h light:12 h dark cycle. Conductivity was 1030 μS. Round goby were fed an alternating diet of zebra mussels and rainbow trout *Oncorhynchus mykiss* eggs. Shelters such as PVC pipe were placed in each tank to provide cover; tanks were shaded.

In pilot studies, round goby did not move in shallow, long tanks or in still water as they apparently reacted to movement by researchers setting up or observing a test. Therefore, a deeper tank with a more disturbed water surface was used for testing. Laboratory studies were conducted in a donut-shaped tank (2.1-m outer diameter, 0.9-m inner diameter, 0.9-m deep), lined with black plastic (Figure 1). The test tank had a water flow-through system to maintain water temperatures (12°C). An electric outboard motor placed in the tank was used to generate low (about 0.1–0.3 m/s) flow rates over the electrodes. A screen placed in front of the motor prevented round goby from moving in an upstream direction.

A benthic, electrical barrier developed by Smith-Root, Inc. was placed opposite the motor in the tank. Three electrodes were positioned parallel to each other; each electrode was placed along the

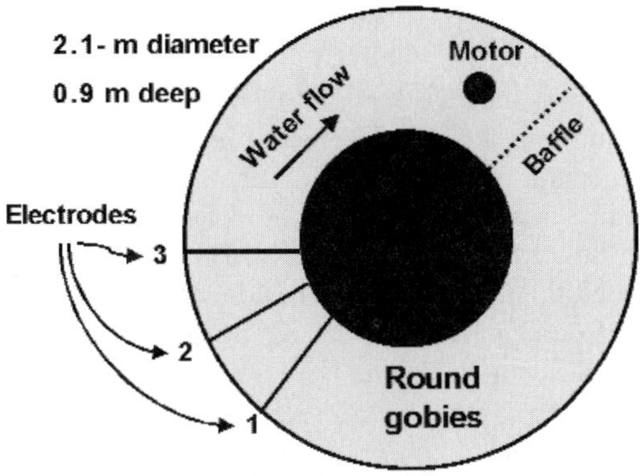

Figure 1. Diagram of laboratory test tank with location of electrodes, electric outboard motor, baffle, and initial placement of round goby.

inside walls and bottom of the tank. Each electrode could be controlled independently to provide a gradual or a sharp gradient of electricity. The gradient of the electrical field was mapped using voltage meters (EFP-2 Electric Field Probe by Smith-Root, Inc.; reference to trade names or manufacturer's does not imply U.S. Government endorsement). This barrier provided electrical current to the total water column. We did not test voltages greater than 100 V because members of the Chicago Sanitary and Ship Canal Aquatic Nuisance Species Dispersal Barrier Advisory Panel recommended that voltages applied to the barrier remain as low as possible and should not exceed 100 V to minimize potential for human injury in instances of accidental immersion near the barrier.

In canal systems, we would like to prevent movement of species downstream as well as upstream. As downstream barriers may be more difficult to achieve, we used the more conservative approach and released round goby upstream of the barrier. To start a test, a group of five round goby was measured (nearest mm, total length, TL) then released in front of the screen (in front of the motor). A video camera positioned above the electrodes filmed the behavior of round goby for the 120-min long test. Generally, if fish had adapted to test conditions in pilot studies they moved within the first 2 h of viewing. No food was available in the test chamber. Tanks were checked about three times an hour; round goby that were stunned on an electrode were removed. At the end of 2 h, round goby were removed and placed in another holding tank. At least two sets of five round goby were tested for each set of electrical parameters and controls (no electrical current). Naive (untested) round goby were used unless otherwise noted.

Behaviors were analyzed by observing the videotapes. The number of attempts and successes by round goby in crossing each active electrode was recorded. Repels were defined as a round goby approaching (entering camera's field-of-view) or crossing an electrode, turning, and moving back out of the field-of-view. Round goby that successfully crossed the electrical barrier were termed 'pass throughs.' Stuns were round goby that stayed in the electrical field until they rolled over and floated downstream or were removed. We removed all stunned fish during their occurrence in the test rather than allow them to remain in the electrical field until death.

To determine if changes in nocturnal activity changed repel rate of round goby, a set of tests was run under dark conditions. The lighting schedule of round goby in holding tanks was altered for at least 2 weeks before testing such that the 12-h dark cycle occurred from 6:00 a.m. to 6:00 p.m. A night vision scope (American Eagle Miniature Pocketscope, model 603 C, Night Vision Equipment Company) was attached to the video camera. This device allowed viewing of round goby under low-light conditions (<0.001 μmol s^{-1} m^{-2}), but also reduced the field of view of the camera by about one-third. As we had observed that most round goby traversed the barrier along the outside wall, we focused the video camera on the area between the first and second electrode along the outer wall.

To efficiently determine what factors were most important in causing a response in round goby, we set up a series of tests. Because round goby often appear to demonstrate erratic behavior, we needed to first determine if we could observe differences in behavior with and without an electrical field present. We also wanted to know if fish retained knowledge of the electrical field and therefore compared responses of naïve controls to experienced controls (tested previously in electrical fields). Other variables tested included water velocity (0.1–0.3 m/s), pulse frequency (0, 2–30 Hz), pulse duration (0, 0.05–5 ms), voltage (0, 70, 100 V), and day or night responses. A complete matrix of tests with all possible combinations of variables would have yielded an unreasonable number of tests to complete. Therefore we grouped variables in a series of tests and provided separate analyses for each of these series.

The number of repels-to-attempts ratio was analyzed with the odds ratio statistic generated in a binary logistic regression model (Proc Logistic, SAS 1995). The odds ratio was computed by exponentiating the slope of the regression model. The odds ratio in this study described the change in odds of repelling based on a test variable. If significant, the odds of repelling increase by the generated odds ratio for each unit increase in the test variable. Because we could not determine which attempts were conducted by different individuals in a group of round goby, we measured the total number of repels and the total number of attempts per test (or group of fish). Generally, the logistic analyses sum attempts among replicates. However, to avoid biasing our results with replicates with more active fish, we determined the average number (rather than the total number) of repels and attempts among replicates in each treatment for use in the model. This procedure provides a more accurate measure of the ratio of repels-to-attempts but reduces sample size of attempts used in the model. Analysis of variance (ANOVA) was used to compare differences in number of attempts to cross the electrical barrier and the total lengths of round goby used in each set of tests.

Field Study

After developing appropriate settings for the electrical barriers, small-scale field studies were developed to test the effectiveness of the electrical barrier under field conditions. A test site was selected on the Shiawassee River, near Argentine, Michigan. The stream section used was about 20 m across and 0.5–1.0 m deep with a sand bottom and occasional macrophyte beds. The bank was well defined in a straight section more than 20 m long. Tests were conducted during 20 August 1998–9 September 1998 when water temperatures ranged from 21 to 25°C, dissolved oxygen ranged 10.2–13.7 mg/L, and conductivity was 570 μS. Water depths over the barrier were initially about 45 cm (20 August 1998 through 31 August 1998) then dropped rapidly to about 15 cm depth (2 September 1998 through 9 September 1998). Water velocity (0.3–0.5 cm/s, range) was slow directly upstream and downstream of the barrier.

The electrical barrier consisted of four braided-wire cables as electrodes, secured to a canvass sheet (about 6 m wide), evenly spaced and parallel to each other along the length of the canvass (at least 22 m long). The canvass was placed across the width of the stream and up each bank. One end was attached to electrical power. Warning signs and float lines were placed upstream and downstream from the site. To prevent movement of fish into or out of the test area, block nets (7-mm bar mesh, 2 m high, two layers) were placed 9 m above and below the barrier, tied to stakes, and the bottom edge buried in sediment and rock.

Round goby were collected by seine from locations near the field site. They were held temporarily in a large, aerated tank, placed in the stream to maintain stream water temperatures. Two hours before testing, round goby were anesthetized (MS-222), measured (nearest mm), and held in a small, aerated tank until the test. At this time, each group of round goby was uniquely tagged by latex paint injections of different color combinations under dorsal scales (Kelly 1967; Hill and Grossman 1987). To begin a test, 25 round goby were released about 9 m above the electrical barrier. Round goby were recovered after a test 24 h later to determine the number above and below the barrier. Recovery methods for finding round goby after release included (1) repeated 'kick-seining' with a small bag seine (3-m x 1-m dimensions and 7-mm bar mesh, Jude and DeBoe 1996)—two people pulling the seine and one person in front of the seine kicking the substrate such that the round goby dart up and into net, and (2) walking along and on the white canvas barrier looking for stunned or dead animals before and after kick-seining.

In a control test, we first documented if round goby would typically cross the barrier within 24 h of release upstream. After the control test, another group of round goby was tagged with a different color and again placed above the barrier. Movement downstream and through the electrical barrier was monitored with the electrical barrier energized. Presence of dead or stunned round goby was monitored by seining 24 h after release and by visual inspection throughout the test. Location and movement of marked goby from previous tests were noted in subsequent trials as well. The electrical field during tests was measured with a voltage meter (EFP-2 Electric Field Probe by Smith-Root, Inc.).

Results

Laboratory

Round goby reacted to the electrical barrier but not to the range in water velocities tested (Figure 2). The odds ratio did demonstrate that signifi-

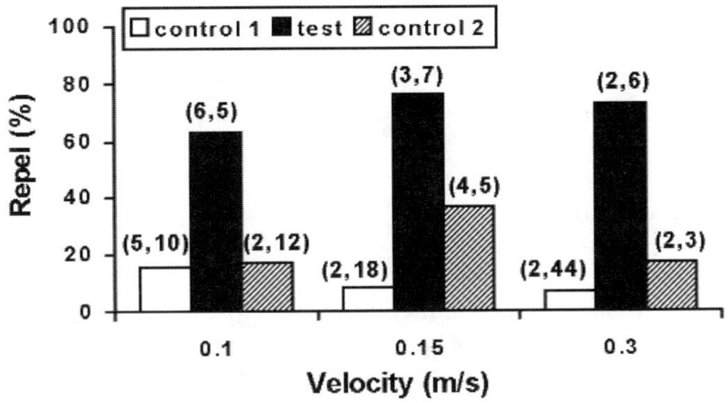

Figure 2. The percentage of round goby that were repelled from crossing the electrical barrier under laboratory conditions at each water velocity (m/s) and test condition. The number of tests and the average number of attempts per test are shown respectively in parentheses. Test conditions were: *control 1* refers to naive round goby controls (0 V) that had never gone through testing; *test* refers to round goby that were tested at 70 V (5-ms-pulse durations and 2–3-Hz-pulse frequencies); and *control 2* refers to experienced round goby controls (0 V) that had been previously tested once.

cantly more attempts to cross the barrier were repelled in tests with the barrier operating at 70 V (5 ms pulse durations and 2–3-Hz pulse frequencies) than those 'repelled' in controls (Test versus Controls, Table 1). Water velocity had no significant effect on round goby behavior in either tests or controls. Naive controls (shown as *control 1* in Figure 2) that had not been tested previously did not differ in repel rates from experienced controls (*control 2*) that had been tested previously (Odds ratio overall model: $\chi^2 = 2.70$, 2 df, $P = 0.26$). Fish in five tests (one in *control 1*, three in tests at 70 V, one in *control 2*) showed no attempts to cross the barrier and were not used in the analyses. The total number of attempts to cross the barrier did change with test condition (ANOVA, $F_{8,22} = 3.93$, $P = 0.005$). The number of attempts to cross the barrier was significantly higher with naive round

Table 1. Odds-ratio analyses using binary logistic models. Chi-squares for overall models based on -2 Log Likelihood statistic; χ^2 for individual tests based on Wald χ^2 (SAS 1995). An asterisk (*) designates significance at $P < 0.05$.

Test	Slope (SE)	Odds ratio	χ^2	*P*
		Test vs. controls (Figure 2)		
Overall model			27.47 (2 df)	0.0001*
Test vs. controls	0.04 (0.01)	1.04	21.70 (1 df)	0.0001*
Water velocity	−3.11 (3.22)	0.04	0.93 (1 df)	0.33
		Pulse frequency and duration (Figure 3)		
Overall model			7.47 (2 df)	0.02*
Duration	0.38 (0.17)	1.46	4.90 (1 df)	0.03*
Frequency	0.05 (0.05)	1.05	0.97 (1 df)	0.32
		Voltage during day (Figure 4)		
Overall model			4.02 (1 df)	0.04*
Voltage	0.02 (0.01)	1.02	3.45 (1 df)	0.06
		Voltage at night (Figure 4)		
Overall model			17.82 (1 df)	0.0001*
Voltage	0.04 (0.01)	1.04	15.86 (1 df)	0.0001*

goby (*control 1*) than with either round goby in the tests with the barrier operating at 70 V or in *control 2* (those tested previously). Average length of round goby used did not differ with test condition in this or any of the following laboratory comparisons (ANOVA, $P > 0.10$).

The effects of pulse duration and pulse frequency on round goby repel rates were explored at 70 V in the next set of tests (Figure 3). Three sets of controls were used in the analyses to provide the balance required in the statistical analyses (as shown in Figure 3). Only pulse duration was significant in the model (Table 1). In general, more round goby were repelled as pulse duration increased from 0.05 to 5 ms. We observed one aggressive interaction in this set of tests (0.5 ms, 2 Hz) in which a round goby nipped another and pushed it into the electrical field. Fish in three tests (one control, two tests) did not attempt to cross the barrier and were not used in the analysis. Round goby that were not repelled after an attempt were either stunned or passed through the barrier. No significant differences were found for the number of stuns ($\chi^2 = 2.41$, 2 df, $P = 0.30$) with changes in pulse duration and pulse frequency; an average of 12% of fish were stunned (1–23%, 95% confidence interval) when the electrical barrier was on. The percent of fish passing through the barrier was inversely related to the percent of fish repelled and also showed significant differences in the pulse duration and pulse frequency model ($\chi^2 = 10.66$, 2 df, $P = 0.005$) with pulse duration significant ($\chi^2 = 4.27$, 1 df, $P = 0.04$) but not pulse frequency ($\chi^2 = 3.48$, 1 df, $P = 0.06$).

Changes in repel rate increased with increasing voltage and were similar between day and night (Figure 4). As the videotaped observation area differed between day and night tests, we did not compare results directly. However, night tests produced similar odds ratios for voltage when compared with the day tests (Table 1). Generally, the repel rate increased with increasing voltage; this trend was more pronounced and became significant at $P = 0.05$ during night observations. Round goby were 10 times more active at night than during the day in the control. Activity, as demonstrated by the number of attempts to cross the barrier, was significantly reduced once the barrier was activated (ANOVA, $F_{2,9} = 9.24$, $P = 0.01$).

As expected, the electrical current field was measurably different when 70 and 100 V were applied to the barrier. When 70 V were applied to the three electrodes, the measured voltage in the test tank peaked at 2.6 V/cm near the middle electrode (Figure 5A). When 100 V were applied, the measured voltage again peaked near the middle electrode at 4.9 V/cm, almost twice the measured voltage when 70 V were applied (Figure 5B).

Field

Control tests were conducted to determine whether marked round goby would move quickly across the barrier if no electricity was applied. Marked round goby were released above the barri-

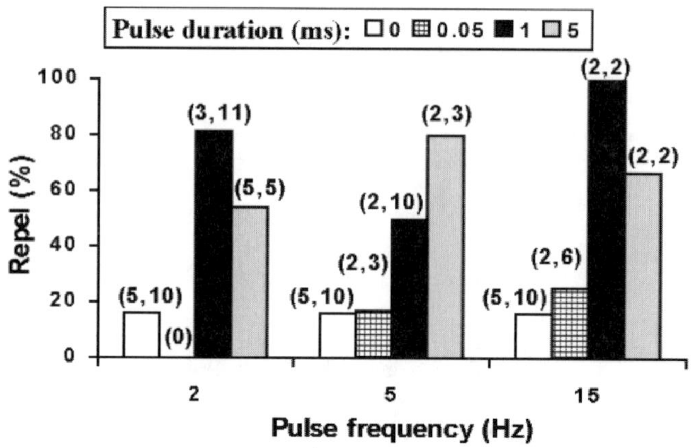

Figure 3. The percentage of round goby that were repelled from crossing the electrical barrier under laboratory conditions with 70-V pulsed direct current and different pulse frequencies (Hz) and pulse durations (ms). The number of tests and the average number of attempts per test are shown respectively in parentheses. The control data (0-ms-pulse duration or 0 V) are shown at each pulse frequency for comparison.

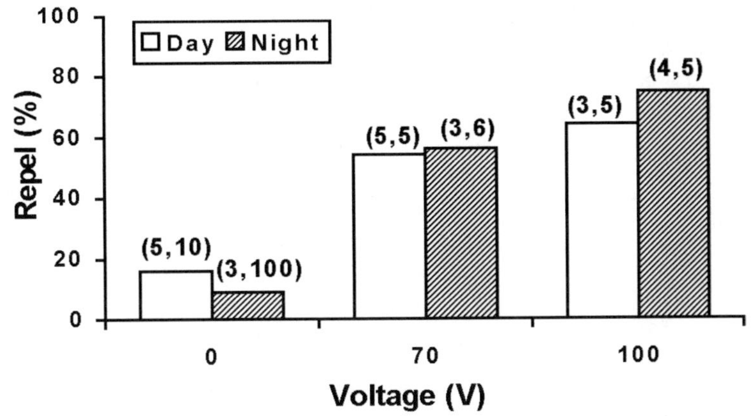

Figure 4. The percentage of round goby that were repelled from crossing the electrical barrier under laboratory conditions at different voltages of pulsed direct current with a 2-Hz-pulse frequency and at 5-ms-pulse duration during day and night observations. The number of tests and the average number of attempts per test are respectively shown in parentheses.

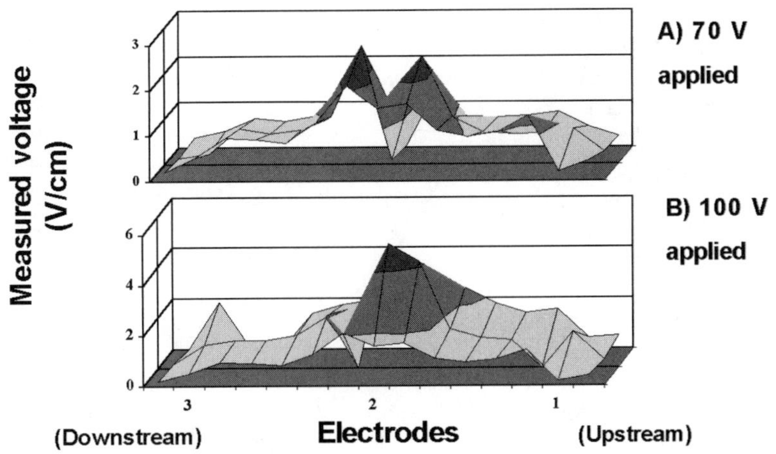

Figure 5. Voltage (V/cm) measured over the three electrodes in the laboratory test pool when (A) 70 V and (B) 100 V were applied to the electrodes. Placement of electrodes 1–3 is shown in Figure 1.

er on 20 August 1998. Observers stationed across the stream width at the barrier observed 5 of 18 round goby crossing the barrier within 20 min of release. After 1 d, four marked round goby were recaptured from seining below the barrier. After conducting the control study, further work was spent on better securing block nets and developing effective recovery methods.

The field electrical barrier was tested at two electrical settings (differing in electrical pulse duration) with three tests at each setting (Table 2). The electrical barrier at the stream site produced a measurable electrical field (Figure 6). As in the laboratory, when 100 V were applied to the electrodes, the electrical field measured peaked at 4.9 V/cm at the center of the barrier. Changes in pulse durations from 5 to 3 ms did not alter the field appreciably.

In the field tests, most round goby were recovered within 1 d of release (Table 2). Only live round goby were found above the barrier. Although not quantified during recovery efforts, round goby were often concentrated in deep areas (or holes) and in vegetation upstream of the barrier. Both live and dead round goby were found on the barrier, but the live round goby were stunned and easily recovered by hand nets. In the first set of tests (trial 1,

Table 2. Location (above, on, or below the electrical barrier) of round goby recovered after each 24-h test under the electrical parameters noted above each set of experiments. Numbers of round goby alive or dead were recorded in each location. In addition, the number recovered after 1 day and the total recovered throughout the study for each test (25 round goby released for each test) are given.

	Location of recovered round gobies			Round gobies recovered
Test	Above barrier	On barrier	Below barrier	Day 1/total
		100 V, 5 ms, 2 Hz		
1	2 live	0 live, 18 dead	2 dead	19/22
2	14 live	3 live, 5 dead	0 dead	17/22
3	18 live	2 live, 4 dead	0 dead	18/24
		100 V, 3 ms, 2 Hz		
4	12 live	3 live, 5 dead	1 live, 0 dead	19/21
5	7 live	4 live, 6 dead	2 live, 5 dead	23/24
6	8 live	0 live, 3 dead	0 live, 3 dead	14/14

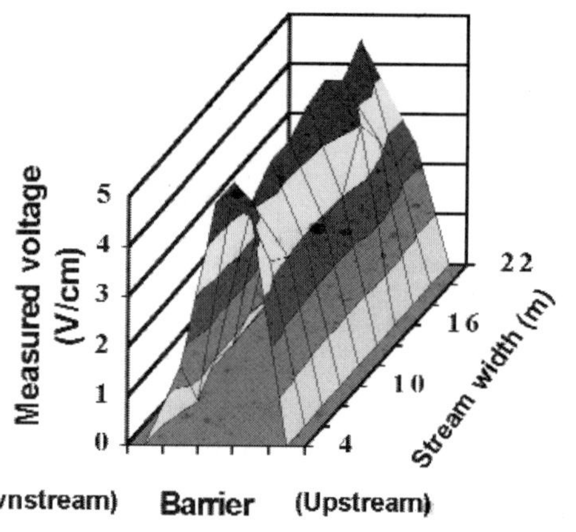

Figure 6. Voltage (V/cm) measured over the electrical barrier at the stream test site when 100 V has applied to the four electrodes.

three replicates), the only marked round goby found below the barrier (100 V, 5 ms pulse duration, and 2 Hz pulse frequency) were dead. At reduced pulse durations (100 V, 3 ms, 2 Hz) in the second set of tests (trial 2, three replicates), a few round goby (mean one/test) were found alive below the barrier, but were in poor condition and easily captured by hand nets.

Round goby tested averaged 95 mm TL (range: 60–120 mm). Sizes of round goby did not differ between those released and recaptured in trial 1 ($F_{1,141}$ = 0.23, P = 0.64) or trial 2 ($F_{1,131}$ = 0.04, P = 0.84). Among those recaptured, sizes of round goby did not differ between those found live or dead in trial 1 ($F_{1,66}$ = 0.43, P = 0.52) or in trial 2 ($F_{1,56}$ = 0.79, P = 0.38). Sizes also did not differ in location of those found above, on, or below the barrier in trial 1 ($F_{2,65}$ = 0.01, P = 0.99) or in trial 2 ($F_{2,55}$ = 0.78, P = 0.46).

A few small resident, unmarked round goby were captured in the stream during tests while seining: the mean size of two round goby per test (range: 0–4) averaged 50 mm TL (SD = 9.4 mm). Other stream fishes collected during seining at the field site (within the block nets) included: bowfin *Amia calva*, common carp *Cyprinus carpio*, golden shiner *Notemigonus crysoleucas*, common shiner *Luxilis cornutus*, spottail shiner *Notropis hudsonius*, spotfin shiner *Cyprinella spilopterus*, bluntnose minnow *Pimephales notatus*, madtoms *Noturus* spp., largemouth bass *Micropterus salmoides*, bluegill *Lepomis macrochirus*, pumpkinseed *Lepomis gibbosus*, and darters *Etheostoma* spp. Crayfish, frogs, snapping turtles, and water snakes were also plentiful at the field site, but not identified to species. A few nontarget species were found dead on the barrier during the six trials; these included one large snapping turtle; one large bullfrog, two bluntnose minnows; one common carp (about 80 cm TL), one resident round goby (unmarked), and several bluegills and pumpkinseeds.

Discussion

Round goby, once introduced into the Great Lakes, have dispersed through three main mechanisms: ballast-water transfer, bait-bucket dumping, or natural swimming movement and dispersal. This laboratory and field study addresses unaided migratory movements and was conducted to determine the feasibility of using electrical barriers to deter movement of round goby through canal systems. The barrier was about 80% effective in repelling round goby during short-term tests in the laboratory. Differences in day-time and nocturnal activity of round goby did not change effectiveness of the barrier. However, the scale-up study in a small river was almost 100% effective in deterring round goby movement downstream. This difference in effectiveness could be the result of several factors. A primary difference in methods was that round goby in laboratory studies were subjected to the electrical field for only 2 h, and fish were removed from the electrical field if stunned. However, round goby in the field study that moved within the electrical field tended to dive and remained near or under electrodes until they died. Anodic attraction is one of the possible fish responses to electrical fields (Kolz and Reynolds 1989). About 12% of the round goby in the laboratory study were stunned while 20–40% of the round goby were stunned or killed on the barrier in the field study. The low flow (0.3–0.5 cm/s) at the field site carried few dead or stunned round goby through the barrier. On a larger scale, the barrier design in the Illinois Waterway System includes the use of railroad rails for large electrodes. Round goby that are stunned near these rails could remain trapped against the protruding upstream side even under regular flow conditions until killed.

Other changes in scale between the laboratory and field studies could account for further differences in round goby responses. The electrical barrier in laboratory studies had a shorter crossing distance (<1 m) and round goby were concentrated in a narrow channel (0.6 m wide). Round goby are an aggressive fish (Dubs and Corkum 1996). In one instance, we observed a round goby forcing another through the barrier. Even though this aggression was rarely observed directly in the laboratory, concentrating round goby in a narrow channel could enhance aggressive interactions and possibly increase the number of passages through the barrier. In the field, the electrical barrier was over a 6-m stretch downstream and covered a 20-m stream width. Round goby could swim through the relatively short length of electrical current in the laboratory, but appeared unable to complete the passage in the field.

In each group of round goby, some individuals appeared determined to cross the barrier. In the laboratory about 12% of those attempting to cross the barrier, actually made it through; in the field they were stunned or died. Other round goby appeared to learn from encountering the barrier. In the laboratory, the number of attempts decreased after experience with the barrier. The number of attempts decreased significantly from controls under both day and night conditions by activating the barrier. Learning has been demonstrated in other fish behaviors such as foraging (Savino and Hudson 1995; Savino et al. 1993). Nevertheless, attempts to cross the barrier continued to be made in both the laboratory and field. In practical applications, it is extremely important that the barrier remain active. Immediately after an electrical barrier was shut down, sea lampreys moved through the stream section (Swink 1999). Through constant exploration, Atlantic salmon *Salmo salar* determined in less than an hour that barriers were off in laboratory studies and moved across an electrical barrier (Stewart 1981).

Our electrical parameters varied from others used in the literature. Our peak electrical field intensities were somewhat higher (5 V/cm) than those used in another study (3.0 V/cm) to effectively block the movement of a variety of species (gizzard shad *Dorosoma cepedianum*, golden shiner, rainbow trout, brown trout *Salmo trutta*, and largemouth bass; Barwick and Miller 1996). Lowering the peak to 3 V/cm in this laboratory study produced a noticeable decrease in effectiveness. Duration of electrical pulse was also important in determining the effectiveness of the field in both this laboratory and the field study—longer pulse duration better deterred movement downstream. Swink (1999) also found that longer pulse durations provided a more effective barrier; pulse durations of 2 ms (10 Hz) completely blocked sea lamprey migration upstream whereas 1 ms pulse duration (10 Hz) allowed 1 out of 900 marked sea lampreys to pass the barrier. In water with lower conductivity (40 μS/cm), chinook salmon were guided with 168 V, 3.7 ms duration, and 120 Hz (Palmisano and Burger 1988).

Native fish species were affected little by the barrier. Either they were repelled or they moved through the barrier. A few larger organisms died on the barrier; larger organisms are more affected by electricity as they conduct more current (Monan and Engstrom 1963; Klima 1972). However, we also saw several common carp and bowfin pass upstream through the barrier. Behavioral and physiological studies have shown that fish in many situations are affected by electric shocking, but generally return to normal parameters within 24 h (Schreck et al.

1976; Sorensen 1994; Maxfield et al. 1971; Mesa and Schreck 1989). Rainbow trout survival, growth, and fecundity were not affected by shocking (1 V/cm, 40 ms, 8 Hz). Goldfish *Carassius auratus* and brook trout *Salvelinus fontinalis* did not alter spawning behavior 1 d after being electroshocked (Sorensen 1994). Feeding and aggression of cutthroat trout *Oncorhyncus clarki* initially decreased but returned to normal within a day of shocking (Mesa and Schreck 1989). Some fish have shown spinal injuries, e.g. rainbow trout captured by electrofishing (260 V, 60 Hz; Sharber and Carothers 1988). Further tests will be needed to determine effects on native species. However, recall that what may be native to one watershed may not be native to the adjoining watershed. If the objective is to achieve complete blockage of fish passage in either direction along an artificial connection, electrical parameters may need to be enhanced or alternative barriers may need to be put in place.

Life stage is another important factor to consider in determining the effectiveness of electric barriers in deterring downstream movement. We worked with fish 60–120 mm in length, generally adults. Smaller life stages have not been studied. Godfrey (1957) showed that shocking with high voltages (550 V) caused high mortality in green eggs of Atlantic salmon and brook trout but not in eyed eggs; alevins were killed by prolonged shocking. Survival and growth of age-0 rainbow trout were not affected (Maxfield et al. 1971). Generally smaller fish are impacted less, since they conduct less electrical current (Monan and Engstrom 1963; Klima 1972). Round goby eggs are adhesive and are attached to the underside of rock or other surfaces (Corkum et al. 1998); they should not flow down river. However, larval or juvenile round goby could migrate. Recent studies suggest that young round goby migrate farther than adults (L. D. Corkum, University of Windsor, personal communication). Therefore, barriers need to be developed and tested that are effective for fish of smaller size classes.

The barrier tested as part of this study should be considered the first phase in deterring fish movement through a canal. At the suggested location for the electrical barrier in the Illinois Waterway System, water velocities are generally less than 0.3 m/s, but can reach 1.5 m/s (Keppner and Theriot 1997). The planned electrical barrier is designed to extend throughout the water column, but under conditions of high-water velocities, stunned round goby might wash downstream. The electrical barrier would prove more effective in smaller canals in areas with little or no flow.

Barriers are most effective for stopping upstream movement, as with common carp and bigmouth buffalo *Ictiobus cyprinellus* (Verrill and Berry 1995) or sea lamprey (Swink 1999). Guiding fish around obstacles, inlets, or channels and into preferred areas (mainstream, fish ladders) has had mixed success (Palmisano and Burger 1988; Kynard and O'Leary 1993; Barwick and Miller 1996). Barriers have not been shown to be entirely successful in preventing downstream movement of fishes (Kynard and O'Leary 1993; Swink 1999). Other types or additional barriers will be essential if complete blockage of fish movement through the area is desired. Considerations include acoustic bubble barriers or hydrojets. Bubble barriers proved somewhat effective in confining roundfish but not flatfish over several-hour field tests (Stewart 1981). Infrasound (10 Hz) produced avoidance behavior in chinook salmon and rainbow trout (Knudsen et al. 1997). In addition, the Illinois Waterway System demonstration project calls for a second electrical barrier. The strategy is that if round goby pass through this first barrier, they may be temporarily stopped by the second. Our studies suggest that round goby appear to learn from their first experience with an electrical barrier that results in a reduction of further downstream attempts. However, this learning does not prove 100% effective in deterring further attempts. Hence any round goby caught between two electrical barriers, should be trapped and removed from this area. In our study, round goby concentrated above the barrier in deep holes and macrophytes. This fish behavior may be able to be used as an advantage at bigger installations to concentrate round goby and other target fish and remove them by some other means (predators, chemicals, electricity). Ideally, a flat featureless area could be installed in front of the barrier with areas of structure, holes, or trenches to concentrate fishes so that some type of removal process is more efficient and effective.

In summary, the choice and verification of the efficacy of an electrical barrier was an important first phase in the demonstration barrier project. The electrical barrier proved effective in deterring adult round goby movement across a section of a small stream. However, effects of scale need to be considered in canals where the water is deeper and the flow is faster. Other studies are planned to determine effects of the electrical barrier on younger life stages of round goby and on native fauna. To form a more effective barrier and provide backup, the electrical barrier could be used in tandem with other (acoustic bubble) barriers.

Acknowledgments

David Smith, President, and Jeff Smith, General Manager, of Smith-Root, Inc. (14014 NE Salmon Creek Avenue, Vancouver, Washington 98686 USA) provided the electrical barrier equipment and logistical support. Margret and Hugh Mead graciously provided access and use of their streamside property for this study. Numerous volunteers aided in collecting round gobies and in helping carry out laboratory and field studies. We especially thank Stephanie Carman and Julie Ryan for their assistance in setting up the barrier nets, John French and Margi Chriscinski for their constant aid, and S. Jerrine Nichols for supplying zebra mussels for food. Rainbow trout eggs were provided by Wes Orr (Ennis National Fish Hatchery, Ennis, Montana 59729). This study was partially funded by the U.S. Army Corps of Engineers with Military Interdepartmental Purchase Request W81G6681317119. Dr. Philip Moy (Wisconsin Sea Grant, Manitowoc, Wisconsin), formerly of the USACE, instigated the study and provided constant feedback. This is contribution 1132 of the Great Lakes Science Center and contribution 617 of the Center for Great Lakes and Aquatic Sciences, University of Michigan.

References

Barwick, D. H., and L. E. Miller. 1996. Effectiveness of an electrical barrier in blocking fish movement. Proceedings of the Annual Conference of the Southeastern Association of Fish and Wildlife Agencies 50:139–147.

Charlebois, P. M., J. E. Marsden, R. G. Goettel, R. K. Wolfe, D. J. Jude, and S. Rudnicka. 1997. The round goby, *Neogobius melanostomus* (Pallas), a review of European and North American literature. Illinois-Indiana Sea Grant Program and Illinois Natural History Survey. INHS Special Publication No. 20.

Corkum, L. D., A. J. MacInnis, and R. G. Wickett. 1998. Reproductive habits of gobies. Great Lakes Research Review 3(2):13–20.

Crossman, E. J., E. Holm, R. Cholmondeley, and K. Tuininga. 1992. First record for Canada of the rudd, *Scardinius erythrophthalmus*, and the round goby, *Neogobius melanostomus*. Canadian Field-Naturalist 106:206–209.

Dubs, D. O. L., and L. D. Corkum. 1996. Behavioral interactions between round gobies (*Neogobius melanostomus*) and mottled sculpins (*Cottus bairdi*). Journal of Great Lakes Research 22:838–844.

French, J. R., III, and D. J. Jude. In press. Diet and diet overlap of round and tubenose gobies and associated benthic species in the St. Clair River, 1992–1993. Journal of Great Lakes Research.

Fuller, P. L., L. G. Nico, and J. D. Williams. 1999. Nonindigenous fishes introduced into inland waters of the United States. American Fisheries Society, Special Publication 27, Bethesda, Maryland.

Godfrey, H. 1957. Mortalities among developing trout and salmon ova following shock by direct-current electrical fishing gear. Journal of the Fisheries Research Board of Canada 14:153–164.

Hill, J., and G. D. Grossman. 1987. Effects of subcutaneous marking on stream fishes. Copeia 1987:492–495.

Janssen, J., and D. J. Jude. In press. Recruitment failure of mottled sculpin *Cottus bairdi* in southern Lake Michigan induced by the newly introduced round goby *Neogobius melanostomus*. Journal of Great Lakes Research.

Jude, D. J., and S. F. DeBoe. 1996. Possible impact of gobies and other introduced species on habitat restoration efforts. Canadian Journal of Fisheries and Aquatic Sciences 53(Supplement 1):136–141.

Jude, D. J., J. Janssen, and G. Crawford. 1995. Ecology, distribution, and impact of the newly introduced round and tubenose gobies on the biota of the St. Clair and Detroit Rivers. Pages 447–460 *in* M. Munawar, T. Edsall, and J. Leach, editors. The Lake Huron ecosystem: ecology, fisheries and management. SPB Academic Publishing, Amsterdam, The Netherlands.

Jude, D. J., R. H. Reider, and G. R. Smith. 1992. Establishment of Gobiidae in the Great Lakes basin. Canadian Journal of Fisheries and Aquatic Sciences 49:416–421.

Kelly, W. H. 1967. Marking freshwater and a marine fish by injected dyes. Transactions of the American Fisheries Society 96:163–175.

Keppner, S. M., and E. A. Theriot. 1997. A recommended control strategy for round goby, *Neogobius melanostomus*, in the Illinois Waterway System. Joint Report prepared for the Aquatic Nuisance Species Task Force. U.S. Fish and Wildlife Service, Lower Great Lakes Fisheries Resource Office, 405 North Amherst, New York, and U.S. Army Corps of Engineers, Waterways Experiment Station, Vicksburg, Mississippi.

Klima, E. F. 1972. Voltage and pulse rates for inducing electrotaxis in twelve coastal pelagic and bottom fishes. Journal of the Fisheries Research Board of Canada 29:1605–1614.

Knudsen, F. R., C. B. Schreck, S. M. Knapp, P. S. Enger, and O. Sand. 1997. Infrasound produces flight and avoidance responses in Pacific juvenile salmonids. Journal of Fish Biology 51:824–829.

Kolz, A. L., and J. B. Reynolds. 1989. Determination of power threshold response curves. Pages 15–24 *in* Electrofishing, a power related phenomenon. U.S. Fish and Wildlife Service Fish and Wildlife Technical Report 22.

Kynard, B., and J. O'Leary. 1993. Evaluation of a bypass system for spent American shad at Holyoke Dam, Massachusetts. North American Journal of Fisheries Management 13:782–789.

Maxfield, G. H., R. H. Lander, and K. L. Liscom. 1971. Survival, growth, and fecundity of hatchery-reared rainbow trout after exposure to pulsating direct current. Transactions of the American Fisheries Society 100:546–552.

Mesa, M. G., and C. B. Schreck. 1989. Electrofishing mark-recapture and depletion methodologies evoke behavioral and physiological changes in cutthroat trout. Transactions of the American Fisheries Society 118:644–658.

Miller, A. C., and B. S. Payne. 1997. Density and size demography of newly established populations of *Dreissena polymorpha* in the U.S. inland waterway system. Pages 99–116 *in* F. M. D'Itri, editor. Zebra mussels and aquatic nuisance species. Ann Arbor Press, Inc., Chelsea, Michigan.

Monan, G. E., and D. E. Engstrom. 1963. Development of a mathematical relationship between electric-field parameters and the electrical characteristics of fish. U.S. Fish and Wildlife Service, Fishery Bulletin 63:123–136.

Morton, B. 1997. The aquatic nuisance species problem: a global perspective and review. Pages 1–54 *in* F. M. D'Itri, editor. Zebra mussels and aquatic nuisance species. Ann Arbor Press, Inc., Chelsea, Michigan.

Palmisano, A. N., and C. V. Burger. 1988. Use of a portable electric barrier to estimate chinook salmon escapement in a turbid Alaskan river. North American Journal of Fisheries Management 8:475–480.

SAS. 1995. SAS Institute Inc., Logistic Regression Examples Using the SAS System, Version 6, 1st edition. Cary, North Carolina.

Savino, J. F., M. G. Henry, and H. L. Kincaid. 1993. Factors affecting feeding behavior and survival of juvenile lake trout in the Great Lakes. Transactions of the American Fisheries Society 122:366–377.

Savino, J. F., and P. L. Hudson. 1995. Feeding competition between larval lake whitefish and lake herring. Archives Hydrobiologia, Special Issues Advanced Limnologie 46:153–162.

Schreck, C. B., R. A. Whaley, M. L. Bass, O. E. Maughan, and M. Solazzi. 1976. Physiological responses of rainbow trout (*Salmo gairdneri*) to electroshock. Journal of the Fisheries Research Board of Canada 33:76–84.

Sharber, N. G., and S. W. Carothers. 1988. Influence of electrofishing pulse shape on spinal injuries in adult rainbow trout. North American Journal of Fisheries Management 8:117–122.

Sorensen, P. W. 1994. Effects of electroshocking on the sexual behavior of goldfish and brook trout. North American Journal of Fisheries Management 14:862–865.

Stewart, P. A. M. 1981. An investigation into the reactions of fish to electrified barriers and bubble curtains. Fisheries Research 1:3–22.

Swink, W. D. 1999. Effectiveness of an electrical barrier in blocking a sea lamprey spawning migration on the Jordan River, Michigan. North American Journal of Fisheries Management 19:397–405.

Verrill, D. D., and C. R. Berry, Jr. 1995. Effectiveness of an electrical barrier and lake drawdown for reducing common carp and bigmouth buffalo abundances. North American Journal of Fisheries Management

American Fisheries Society Symposium 26:183–193, 2001

Detection of Infrasound in Fish and Behavioral Responses to Intense Infrasound in Juvenile Salmonids and European Silver Eels: A Minireview

Olav Sand, Per S. Enger, Hans Erik Karlsen, and Frank R. Knudsen
Department of Biology, University of Oslo, P.B. 1051 Blindern, N-0316 Oslo, Norway

Abstract.—This paper reviews recent studies of infrasound detection in fish and the use if intense infrasound as a fish deterrent. The possible use of intense infrasound in acoustic fish barriers is discussed. Fish are sensitive to infrasound (<20 Hz), even down to below 1 Hz, and the otolith organs are the sensory system responsible. Particle acceleration, and not sound pressure, is the relevant sound parameter for low frequency detection in fish, and the threshold values are in the range 10^{-5}-10^{-4} ms^{-2}. The hydrodynamic noise generated by swimming fish is mainly in the infrasound range, and may be important in prey-predator interactions. An infrasound source able to generate large near-field particle acceleration has been employed to divert downstream migrating Atlantic salmon (*Salmo salar*) smolts in a small river. At the observation site the river branched into a main course and a minor channel. The infrasound source was positioned in the channel, and at intensities above 10^{-2} ms^{-2} the 10 Hz sound effectively blocked passage of smolts through this alternative route. An improved version of the original infrasound source has been employed to test the effect of intense infrasound on downstream migrating European silver eels *Anguilla anguilla*. The tests were carried out in a river where a trap that catches all the descending eels is installed near the river mouth. The trap was separated in four equal sections. During the periods with infrasound exposure, the proportion of silver eels entering the section closest to the sound source was reduced to 43% of the control value. In the section closest to the opposite riverbank, infrasound increased the proportion of trapped eels to 144% of the control values. We conclude that intense infrasound has a great potential in acoustic fish barriers.

Excluding fish from areas where they are not wanted (e.g. inlets to hydroelectric power stations, cooling water intakes of thermal-electric power stations and irrigation canals) is desirable. Physical barriers are widely used to achieve this, although such methods may reduce the available water flow and incur high maintenance costs. Attempting to overcome these disadvantages, a variety of behavioral systems has been developed for deflection of fish from specific regions (for review, see OTA 1995). Among the nonmechanical barriers that have been tested are electrical screens, bubble curtains, artificial light arrays and underwater acoustic systems. For some freshwater clupeids of the genus *Alosa*, intense ultrasound (80–150 kHz) produces significant avoidance responses (Dunning et al. 1992; Nestler et al. 1992), and full-scale field application of an acoustic barrier based on ultrasound clearly reduced the number of alewives *Alosa pseudoharengus* entering a power plant intake on Lake Ontario (Ross et al. 1993, 1996). Such high frequencies have previously been considered above the upper limit of the audible frequency range in fish (Hawkins 1981). However, controlled laboratory experiments have recently demonstrated that at least the Atlantic cod *Gadus morhua* and the American shad *Alosa sapidissima* can detect intense ultrasound (Astrup and Møhl 1993; Mann et al. 1997), although it is still not clear which sensory organ is involved in the reported sensitivity to ultrasound.

The majority of fish species are insensitive to ultrasound, and attempts to construct efficient acoustic barriers based on normal sound frequencies have generally failed (for reviews see Carlson 1994; Carlson and Popper 1997; Popper and Carlson 1998). However, during the last 15 years we have shown that fish have an acute sensitivity to extremely low sound frequencies, or infrasound, even down to below 1 Hz, and we have demonstrated that intense infrasound is an efficient deterrent for juvenile salmonids and silver eels. The present paper is a review of our recent work on infrasound detection and behavioral responses to intense infrasound in fish.

Infrasound Sensitivity in Fish

Most of the early studies on fish hearing employed sound sources unable to produce tones below 20–30 Hz, and the sound pressure audiograms obtained in these studies indicated a steadily declining sensitivity towards lower frequencies (Fay 1988). Therefore, the shape of the sound pressure audiograms made the idea that fish might detect infrasound, which is usually defined as sound with frequencies below 20 Hz, less interesting and generally ignored. However, the unaided otolith organs of the inner ear are *not* sensitive to sound pressure. Sound has a dualistic nature, with both a pressure component and a kinetic component, and it is the latter component that stimulates the otolith organs. The kinetic component is the vibrational movements of the water particles in a sound field, and may be expressed as particle displacement, particle velocity or particle acceleration.

Otolith Organs are Accelerometers

The hair cell receptor current is dependent on the hair bundle displacement, and the hair cells are thus displacement detectors. The otolith organs in fish may be treated as nearly critically damped, simple harmonic oscillators (de Vries 1950); (Figure 1A and B). The otoliths have a mass density about three times that of the surrounding tissue. At frequencies below the natural frequency of the system the deflection of the otolith relative to the sensory epithelium follows the acceleration of the nearly neutrally buoyant fish in a sound field, and the system is therefore an *accelerometer*. Figure 1C shows the response of the model for a given acceleration as a function of frequency. The model indicates a working range of otolith organs reaching from zero to the upper frequency limit of hearing.

The tradition of relating fish audiograms to sound pressure rather than to particle acceleration has led to misinterpretations concerning optimal frequency ranges and hearing capabilities, since the shape of the audiogram greatly depends on the acoustic parameter to which the thresholds are related. Figure 2 shows a hypothetical fish audiogram related to particle displacement, sound pressure (or particle velocity) and particle acceleration. When thresholds are related to particle acceleration, the apparent drop in sensitivity towards low frequencies disappears.

The Swim Bladder Transforms Sound Pressure to Particle Motion

When a volume of gas in water is exposed to pressure changes, it will display larger volume pulsations than a comparable volume of water, due to its higher compressibility. Upon exposure to sound, the surface of a gas-filled swim bladder may thus show larger motion amplitudes than the water particles in the absence of a gas-filled bladder. These amplified motions may be transmitted to the inner ear, providing an auditory gain and making the fish as a whole sensitive to sound pressure (see Popper and Carlson 1998, for a more complete discussion of the mechanism of hearing in fish). This gain is frequency dependent, and the swim bladder pulsations exceed the free field particle motions only

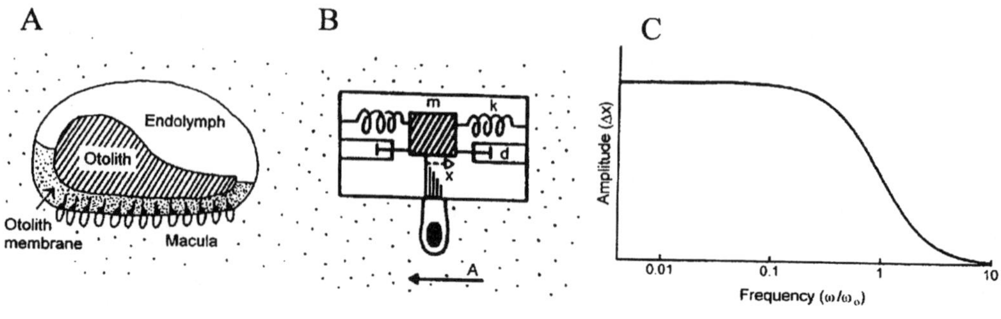

Figure 1. (A) Section of a simplified otolith organ. (B) Model of an otolith organ as a simple harmonic oscillator. m = effective otolith mass, k = spring force per unit of displacement, d = frictional force per unit of velocity, x = displacement of the otolith relative to the hair cells, A = peak acceleration of the fish. (C) Hair bundle displacement of a critically damped model system for a given acceleration as a function of frequency, presented as the ratio between the stimulus frequency (ω) and the natural frequency (ω_o) of the system (modified from Sand and Karlsen 2000).

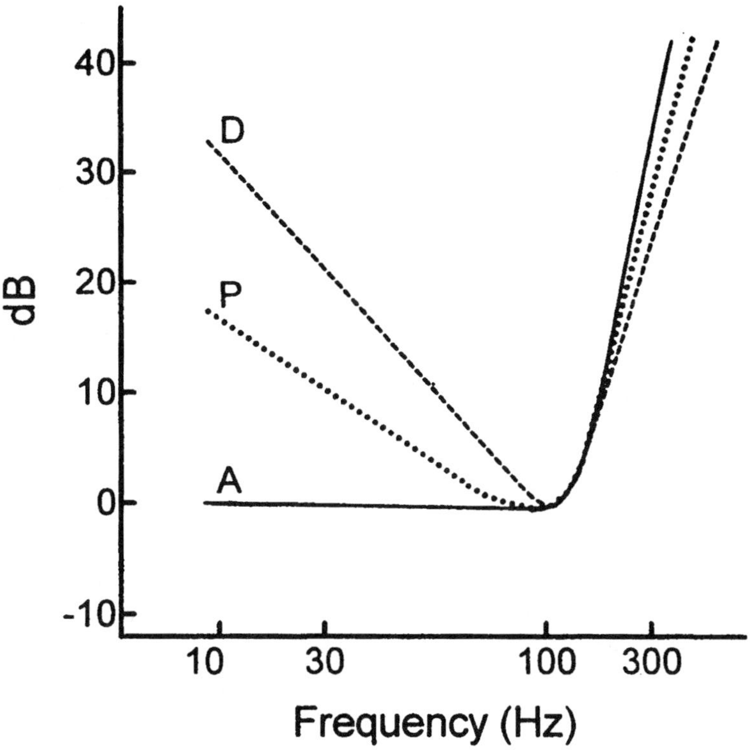

Figure 2. Hypothetical fish audiogram related to particle displacement (D), sound pressure or particle velocity (P) and particle acceleration (A) (modified from Sand and Karlsen 2000).

above a certain frequency (Figure 3; Sand and Hawkins 1973) that depends on both swim bladder volume and depth. Therefore, even fish possessing a swim bladder will be insensitive to sound pressure in the infrasound range, where the relevant stimulus parameter is particle acceleration in all species.

Detection of Infrasound

We have tested the infrasound sensitivity in fish using either an acoustic tube or a closed chamber suspended in a swing. The water mass in the tube, including the fish, is accelerated by vibrators operating 180° out of phase and connected to rubber membranes at each end (Sand and Karlsen 1986; Karlsen 1992a). In the swing, a water filled chamber enclosing the fish is suspended by steel strings, and the whole chamber, including the water and the fish, is vibrated (Karlsen 1992b). Both these methods simulate the kinetic component of infrasound stimulation in the field.

The traditional cardiac conditioning technique (Chapman and Hawkins 1973) was employed to obtain infrasound thresholds. By giving the fish a weak electric shock at the termination of the stimulus, the fish was trained to give a fright response, in the form of reduced heart rate (bradycardia), when the sound was presented. The actual thresholds were then determined by step-wise changes of the stimulus intensity (the staircase method, Dixon 1965). We have tested the infrasound sensitivity of the Atlantic cod *Gadus morhua*, the plaice *Pleuronectes platessa*, which is a flatfish, and the European freshwater perch *Perca fluviatilis* down to 0.1 Hz (Sand and Karlsen 1986; Karlsen 1992a,b). The thresholds at 0.1 Hz were 10^{-5}–10^{-4} ms^{-2}, which represents a sensitivity to linear acceleration about 10,000 times higher than in humans. The otolith organs, and not the lateral line, were responsible for the observed infrasound sensitivity. The lateral line cupulae have a mass density close to that of the surrounding water, and no relative movements deflecting the sensory hair bundles occur when the fish and the surrounding water is accelerated in a sound field (Sand 1981).

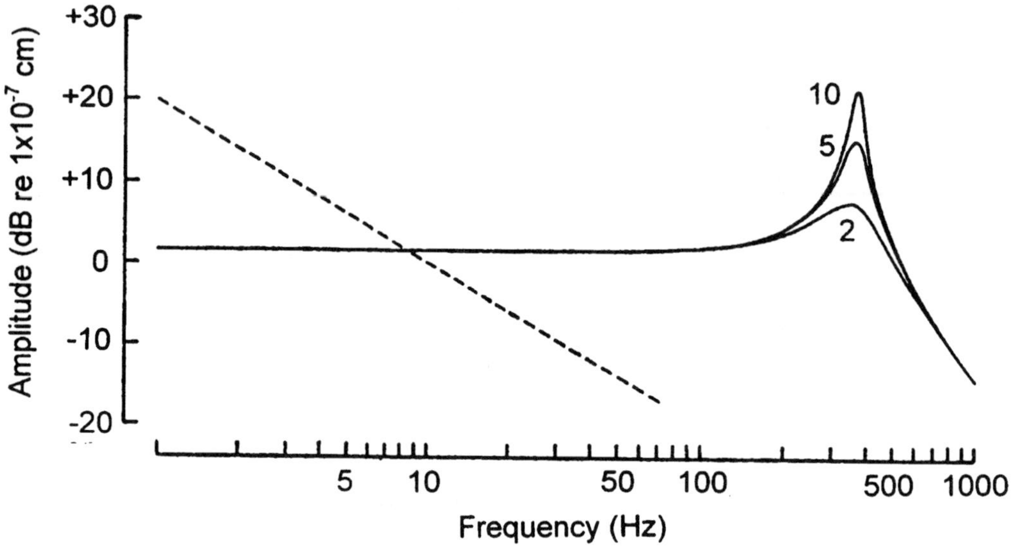

Figure 3. Radial pulsation of a damped bubble (Q values of 10, 5 and 2 are included) in a sound field. Bubble radius is 1.5 cm and the depth 20 m. The sound pressure is 10^{-1} Pa. The water displacements accompanying this sound pressure for a propagated plane wave in the free field are shown by the broken line (from Sand and Hawkins 1973).

Fish may use its high sensitivity to infrasound, or linear acceleration, to obtain a wide range of information about the environment (see Sand and Karlsen 2000), for instance the presence of moving objects in the surroundings. The hydrodynamic noise produced by swimming fish displays acceleration components mainly in the infrasound range below 10–20 Hz (Kalmijn 1989). Infrasound sensitivity in fish may thus be important in prey-predator interactions. Since avoidance responses to infrasound noise produced by swimming predators would have high survival value in prey fish, it is reasonable to suggest that avoidance reactions to infrasound may have evolved in prey fish as a protection against predators. This argument naturally leads to the suggestion to employ infrasound in acoustic barriers for juvenile fish or adult prey fish.

Physical Considerations Regarding the Use of Infrasound in Acoustic Fish Barriers

Far Field and Near Field

Under free field conditions the sound pressure is inversely related to the first power of the distance to the sound source. The pressure variations are associated with particle motions due to compression and rarefaction of water. These particle motions are proportional to the sound pressure, and the ratio between pressure and particle motion is determined by the acoustic impedance of the medium. In addition to the elasticity-dependent particle motions, which are usually denoted the far field component, the medium surrounding the sound source is also displaced due to the movements of the source. This latter component of the particle motions is frequently referred to as the local flow field or the near field component, and will greatly exceed the far field component in close vicinity of the source (Harris 1964; Kalmijn 1988). However, the drop-off with distance is much steeper for the near field component than for the elasticity-dependent particle motions, and the far field component will therefore be gradually more dominant as the distance to the sound source increases. For a monopole source, which pulsates in volume, the near field motions drop off with the square of the distance. In shallow water the situation is more complex. Due to reflections of the pressure waves from both the surface and the bottom, standard formulas for sound dispersion are not applicable in such environments. Measurements in the field may show little sound pressure attenuation across the width of a river, due to reflections and reduced geometric spread. It may thus be difficult to create a restricted area of high sound pressure at desired locations close to irrigation- or turbine inlets, leaving the main part of the river undisturbed. However, the steep drop off with distance of the near field particle motions makes it easy to generate restricted areas of high particle acceleration, which is the relevant stimulus parameter in the infrasound range.

The Working Range of Moderately Sized Infrasound Sources at Shallow Depths

The sound energy emitted as propagated sound from a monopole sound source at constant volume pulsation is proportional to the fourth power of the frequency. To keep a certain energy output at decreasing frequencies, the volume pulsations therefore has to be greatly increased. To be an efficient emitter of propagated infrasound, the sound source must thus be excessively large, and it would be a formidable task to construct a sound source able to generate high infrasound pressure in the field. Furthermore, the propagation of infrasound is greatly hampered at shallow depths due to the long wavelength (Rogers and Cox 1988). On the other hand, these considerations are less critical regarding the ability to generate large near field particle motions in close vicinity of the source. As mentioned above, a short effective range of the sound sources included in an acoustic fish fence may even be an advantage, making it possible to limit and shape the area from which the fish should be diverted. An elongated acoustic fence may for instance be designed by operating several sources in parallel. Strict phase synchronization of the sources is a prerequisite for such an array to operate as a unit.

An obvious requirement for an acoustic fish fence to be effective is incorporation of an alternative flight route constituting a natural choice for the fish. For instance, if the normal reaction to frightening stimuli is to seek deeper water, the bypass channel should be constructed accordingly. The fence should also be installed at a site where the current velocity is sufficiently modest to allow the fish to oppose its original course.

Underwater Infrasound Should be Measured as Particle Motion

It should be stressed that sound pressure measurements are of little value when monitoring the output from underwater infrasound sources. The relevant acoustic component is particle motion, which should be measured directly by using kinetic hydrophones. These may for instance be accelerometers enclosed in a watertight housing, designed to give the submerged unit neutral buoyancy. At shallow depths the kinetic sound component cannot be estimated from sound pressure measurements using the ordinary wave equations. In fact, such estimates may be rather misleading, since the acoustic impedance of water at shallow depths is much less than the theoretical value for a homogeneous water mass.

Behavioral Reactions of Juvenile Salmonids to Intense Infrasound

Initially, we studied the possibility of using intense sound as a deterrent for juvenile Atlantic salmon *Salmo salar* by recording physiological awareness reactions in the previously mentioned acoustic tube (Knudsen et al. 1992). Spontaneous awareness reactions, consisting of decreased heart rate and breathing movements (Figure 4), were compared for frequencies between 5 and 150 Hz. The lowest frequencies (5–10 Hz) tested were most effective in eliciting an awareness reaction from the fish.

The responses of freely swimming fish to 10 Hz and 150 Hz sounds were subsequently studied in an artificial pool. To produce the infrasound we constructed a low frequency source consisting of a cylinder with a 19 cm diameter piston driven by eccentric coupling to an electric motor, with a peak to peak piston movement of 4 cm. The construction was firmly bolted to the ground, with the motor in air and the piston front about 20 cm below the surface.

Juvenile salmon showed flight and avoidance responses to 10 Hz stimulation at intensities 10–15 dB above the threshold for spontaneous awareness reactions measured in the acoustic tube, corresponding to a particle acceleration of 10^{-2} ms^{-2}. The 150 Hz sound failed to evoke avoidance responses, even at a level 30 dB above the threshold for spontaneous awareness reactions. Using the same design of the infrasound generator, similar studies were later conducted on juvenile Pacific salmonids, chinook salmon *Oncorhynchus tshawytscha* and rainbow trout *Oncorhynchus mykiss*. These species also displayed flight and avoidance responses to intense 10 Hz infrasound (Knudsen et al. 1997). Mueller et al. (1998 and 2001, this volume) have also observed avoidance responses to infrasound in juvenile Pacific salmonids, particularly in wild chinook salmon. However, other researchers have failed to observe avoidance reactions in salmonids employing a different design of an infrasound

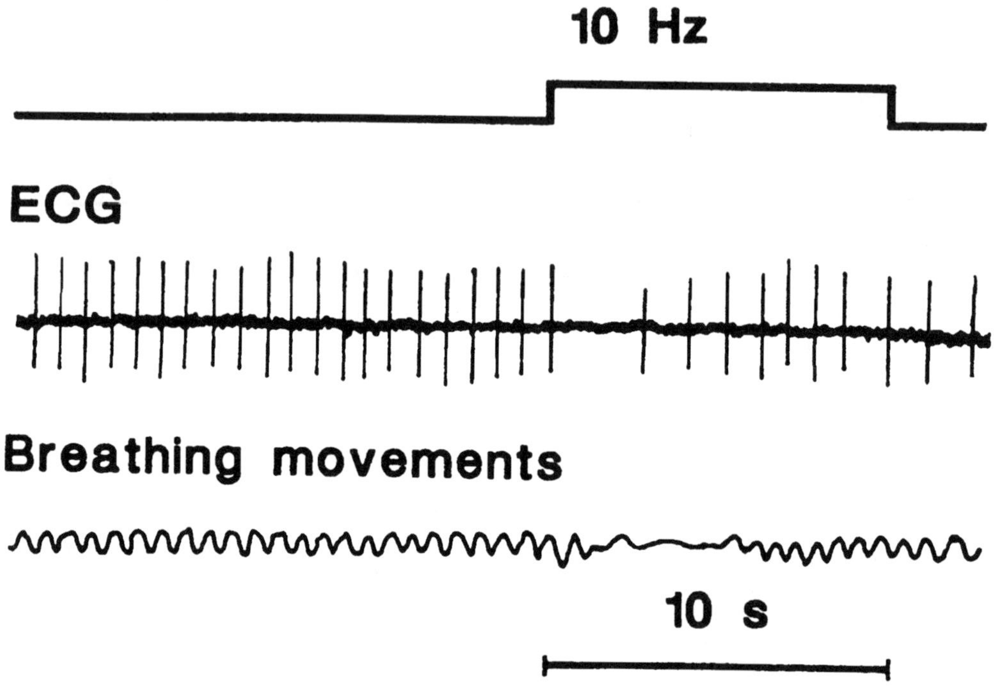

Figure 4. Spontaneous awareness or fright reaction in a juvenile Atlantic salmon caused by intense 10 Hz infrasound. The upper trace indicates the presence of the sound stimulus, the middle trace presents an electrocardiogram (ECG) from the fish, and the lower trace is a recording of breathing movements. The fish showed bradycardia and reduced breathing movements in response to the infrasound (from Knudsen et al. 1992).

source (Amaral et al. 2001, this volume; Ploskey and Johnson 2001, this volume). The reason for this discrepancy is unclear. Some of the negative tests have been performed on hatchery reared smolts, which may show a blunted reaction pattern to infrasound compared with wild fish (Knudsen et al. 1992, Mueller et al. 2001). Furthermore, the alternative source generates a low frequency of motion pulses rather than sinusoidal motions of low frequency, which shifts the frequency composition of the output towards higher frequencies (Ploskey and Johnson 2001).

Both the Atlantic salmon and the Pacific salmonids showed habituation to repeated exposures to infrasound (Knudsen et al. 1992, 1997; Mueller et al. 2001), which might reduce its effectiveness as a deterrent for local populations. However, habituation will be of less importance for the deflection of migratory fish (Knudsen et al. 1994).

We have studied the possibility of using intense sound as an acoustic barrier for downstream migrating smolts of the Atlantic salmon by observing the reactions of smolts to 10 Hz and 150 Hz sounds in a small river (Knudsen et al. 1994). At the test location the river branches into two parallel streams, a main lower course and a narrow upper course (Figure 5). The upper course forms a natural inlet to a salmon and trout hatchery. This channel is 30 m long and has a width and depth of about 4 m and 70 cm, respectively. During the observation period the flow velocity in the channel was about 0.3 ms^{-1}. The end of the channel was fitted with a metal grid, which prevented smolts from entering the hatchery. Smolts in the channel could access the main river course either through a 70 cm wide and 30 cm high slit in the side wall of the channel close to the metal grid, or by swimming back upstream to the branching point. Smolts exiting through the slit were caught in a fish trap for counting, and then released in the mainstream. The length of the migrating smolts ranged between 11 and 13 cm.

The 10 Hz sound source was of the same design as described above, and was bolted to a concrete wall behind the metal grid. The distance between the cylinder front and the center of the

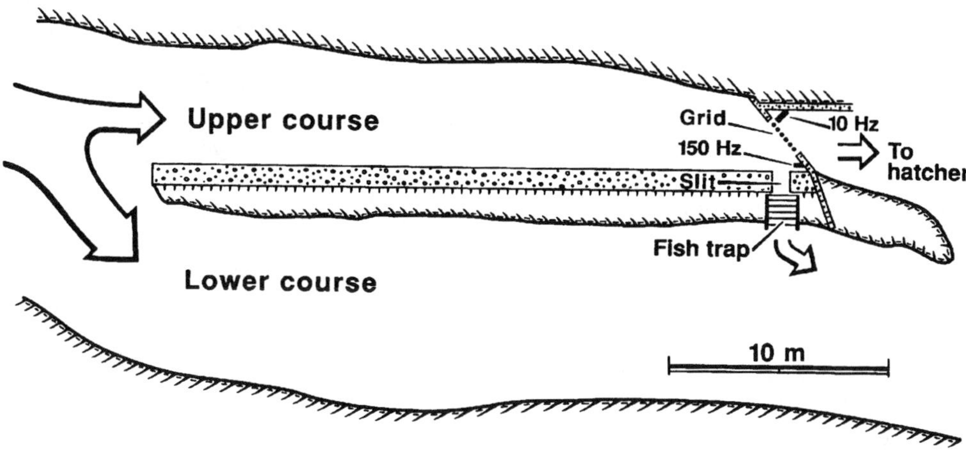

Figure 5. Sketch of the observation site for testing the effect of sound on descending salmon smolts. Arrows indicate direction of the water current. See the main text for further details (from Knudsen et al. 1994).

slit in the side wall was about 1.5 m. Installation of the 10 Hz sound source behind the metal grid eliminated possible visual components of the stimulus, such as the actual piston movements or surface ripples. The 150 Hz sound source was a traditional underwater sound projector (Type J9, Atlantic Res. Corp.), and was suspended by strings close to the slit in the sidewall. Sound stimulation was not associated with any visible motions at 150 Hz.

The kinetic sound component was measured at different positions in the channel using calibrated kinetic hydrophones. As noted, we measure the particle acceleration directly, using accelerometers enclosed in air-filled Perspex or glass housings. The mass density of the assembled units is adjusted to make the hydrophones nearly neutrally buoyant in water. The background noise in the lower end of the channel was measured in 1-octave bands with center frequencies 15 and 150 Hz. In the low frequency range the background noise was $3–10 \cdot 10^{-3}$ ms^{-2}, whereas the noise centered around 150 Hz was $5–9 \cdot 10^{-4}$ ms^{-2}.

The number of smolts reentering the mainstream at the lower end of the upper channel was recorded during alternating periods with and without sound. Intense 150 Hz sound had no observable effects on the smolts, even at intensities 114 dB above the hearing threshold at this frequency (Hawkins and Johnstone 1978). On the other hand, at intensities above 10^{-2} ms^{-2} the 10 Hz sound was an effective deterrent for the smolts, which turned and left the channel at the upstream branching point. During nine control periods without sound 338 smolts passed through the channel, whereas only 6 smolts passed during the nine similar periods with infrasound exposure.

Improved Design for an Underwater Infrasound Source

Our original infrasound source had several shortcomings. It was not completely submersible and had to be firmly attached to the ground. The unbalanced piston movements furthermore caused strong vibrations of the construction, which could lead to fractures due to material fatigue. These problems were solved in our subsequent version of the infrasound source, which is completely submersible and operating freely suspended in the water mass (Sand et al. 2000).

The source generates water movements by means of two symmetrical pistons in an air-filled cylinder. The pistons are driven by eccentric coupling to an electric motor, and are moved along the same axis. The cylinder diameter and the piston amplitude are 21 cm and 5 cm, respectively. The source is intended to approach a monopole by creating volume changes, and the piston movements are thus 180° out of phase. As a consequence, the reaction forces from the accelerating water to the pistons are opposed, leading to vibration free operation of the unit in spite of its moderate mass. The source may thus be freely suspended in the water, whereas a single piston source requires a very large mass or firm attachment to the ground. The unit is submersible, and pressure compensation is therefore necessary for operation at variable depths. To drive the unit we are using a 1.5 kW electric motor. The motor runs at 705 rotations per minute, corresponding to an emitted sound frequency of 11.8 Hz, which is within the optimal range for deterring salmon smolts.

With a source depth of 0.7–1.0 m at a water depth of 1.5–2.0 m, the near field particle acceleration is about 0.01 ms^{-2} at a distance of about 3 m along the main axis to the source. This range was reduced to about 2.5 m at an angle of 30° to the main axis. A further improved version of this sound source, fitted with electronics allowing phase synchronization of several units, will soon be commercially available (Infrasound Fish Fence, Oslo, Norway).

Deflection of Descending Silver Eels by Intense Infrasound

We have employed the sound source to test the effect of intense infrasound on migrating European silver eels (Sand et al. 2000). In Europe, the protection of migrating salmon smolts is a major concern mainly in Scandinavia and Scotland, whereas survival of eels at hydropower facilities and cooling-water intakes is a serious problem on the continent. Among the techniques that have been tested to keep eels from being entrained are the use of lights, air bubbles and electric screens (Hadderingh et al. 1988). Most of these methods were unsuccessful, although the eels tended to avoid strongly illuminated areas.

Initial tests showed that silver eels confined in an observation tank displayed startle behavior and prolonged stress reactions in response to intense infrasound (Sand et al. 2000). Subsequent field tests were carried out in the River Imsa, which had a water discharge between 7 and 14 m^3s^{-1} during the test period. The current velocity in the surface layer was 0.9–1.3 ms^{-1} in the test area. The background acceleration noise measured in the 1-octave band centered on 12 Hz was $4 \cdot 10^{-4}$–$8 \cdot 10^{-3}$ ms^{-2}. At the Ims Research Station close to the river mouth the annual runs of both juvenile and adult salmonids and eels have been closely monitored for several decades. The migration of silver eels takes place between August and December, with the peak descent in September–October. The silver eels usually start to migrate shortly after sunset and stop after midnight or when the moon appears above the horizon (Vøllestad et al. 1986). The weight of the silver eels ranged from about 100 g for the smallest males to more than 1.5 kg for the largest females, with average length and weight of about 63 cm and 440 g, respectively (Vøllestad and Jonsson 1988). The yearly silver eel run usually amounts to 1000–2000 individuals.

A trap constructed as a grid that filters all the running water (Wolf trap, Wolf 1951) is installed about 100 m from the river mouth (Figure 6). The trap has 10 mm apertures and 1:10 inclination, and catches all the descending eels. The width of the trap screen is 10 m. The collecting trough at the lower end of the screen was separated in 4 equal sections, and the number of trapped eels was recorded in each section. At the trap entrance a concrete barrier covers the entire width of the river. The upper edge of the barrier is horizontal and secures an approximately even water flow across the screen. The water flow was thus nearly the same in the 4 sections.

The sound source was suspended from a wire stretched in air between the riverbanks. The source was positioned about 4.5 m upstream of the trap. The width of the river is here about 14.5 m, and the distance between the source and the right bank was about 3 m. The source was angled about 45° to the main current direction by the use of guide ropes attached to the shore. The source was positioned at a depth of 0.7–1 m. The maximum river depth was about 3 m in the test area.

The field tests were conducted during the peak of the silver eel run. Depending on the migration activity, the duration of the alternating periods with or without infrasound exposure varied between 30 min and one hour; the shorter periods corresponding to higher density of migrating eels. A total number of 370 eels were trapped during the periods without activation of the infrasound source, and 28% of the eels where then trapped in section I, directly downstream to the sound source. Only 18% of the eels entered the trap in section IV, close to the opposite riverbank, during the periods when the infrasound source was silenced. A total number of 296 eels entered the trap during the periods when the infrasound source was activated, and the distribution of trapped eels in the different sections was then clearly shifted. The proportion of eels trapped in section I was reduced to

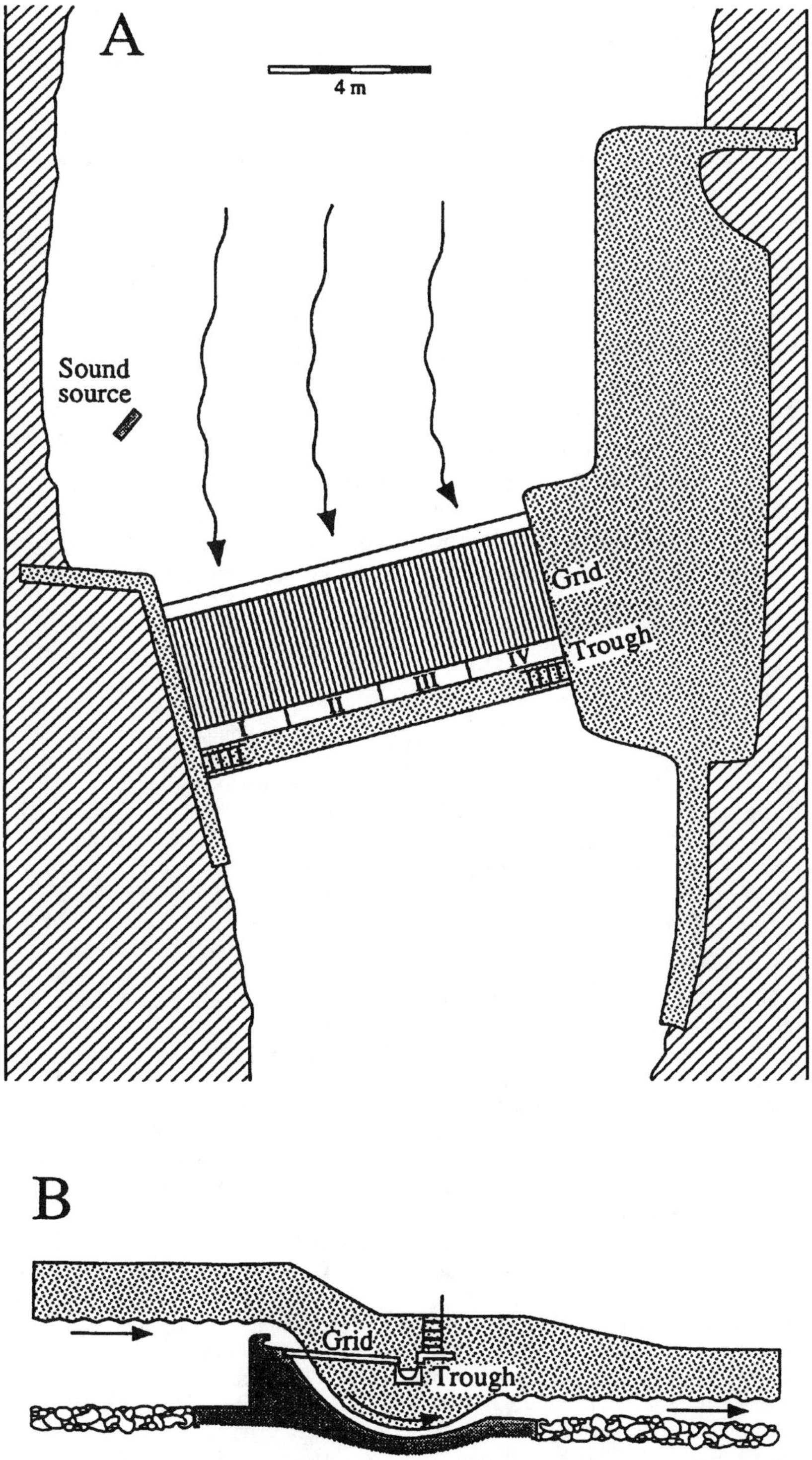

Figure 6. Sketch of the River Imsa at the site of the fish trap. (A) Horizontal view showing the position of the infrasound source relative to the trap. The collecting trough of the trap was separated in four equal sections (I-IV). (B) Longitudinal section of the trap, which catches all of the descending fish (from Sand et al. 2000).

12%, which is 43% of the corresponding proportion during the silent periods. The proportion of eels entering section IV increased to 26% when the infrasound source was switched on, which is 144% of the control value. The observed shift in the distribution of trapped eels is highly significant (Sand et al. 2000). This result is encouraging regarding the prospects of designing acoustic barriers for migrating silver eels.

Conclusion

Intense infrasound is an efficient deterrent for juvenile salmonids and silver eels, and has a great potential in acoustic fish barriers. We have constructed a submersible infrasound source designed to generate large near field particle acceleration. The effective range is about 3 m, making it possible to use several phase synchronized units to tailor the shape of the area from which the fish should be deflected. To be effective, an acoustic fish fence must be combined with an acceptable alternative route for the deflected fish.

References

Astrup, J., and B. Møhl. 1993. Detection of ultrasound by the cod *Gadus morhua*. Journal of Experimental Biology 182:71–80.

Carlson, T. J. 1994. Use of sound for fish protection at power production facilities: a historical perspective of the state of the art. U.S. Department of Energy, Project 92–071, Boneville Power Administration, Portland, Oregon.

Carlson, T. J., and A. N. Popper, editors. 1997. Using sound to modify fish behavior at power-production and water-control facilities: a workshop. U.S. Department of Energy, Project 92-071-01, Bonneville Power Administration, Portland, Oregon.

Chapman, C. J., and A. D. Hawkins. 1973. A field study of hearing in the cod, *Gadus morhau* L. Journal of Comparative Physiology 85:147–167.

de Vries, H. L. 1950. The mechanics of labyrinth otoliths. Acta Oto-Laryngology 38:262–273.

Dixon W. J. 1965. The up-and-down method for small samples. Journal of the American Statistical Association 60:967–978.

Dunning, D. J., Q. E. Ross, P. Geoghegan, J. Reichle, J. K. Menezes, and J. K. Watson. 1992. Alewives avoid high-frequency sound. North American Journal of Fisheries Management 12:407–416.

Fay, R. R. 1988. Hearing in vertebrates: a psychophysics databook. Hill-Fay Associates, Winnetka, Illinois.

Hadderingh, R. H., F. B. J. Koops, and J. W. van der Stoep. 1988. Research on fish protection at Dutch thermal and hydropower stations. Kema Scientific Technical Reports 6:57–68.

Harris, G. G. 1964. Considerations on the physics of sound production by fishes. Pages 233–247 *in* W. N. Tavolga, editor. Marine Bio-Acoustics. Pergamon, London.

Hawkins, A. D. 1981. The hearing abilities of fish. Pages 109–133 *in* W. N. Tavolga, A. N. Popper and R. R. Fay, editors. Hearing and Sound Communication in Fishes. Springer-Verlag, New York.

Hawkins, A. D., and A. D. F. Johnstone. 1978. The hearing of the Atlantic salmon, *Salmo salar*. Journal of Fish Biology 13:655–673.

Kalmijn, A. J. 1988. Hydrodynamic and acoustic field detection. Pages 83–130 *in* J. Atema, R. R. Fay, A. N. Popper, and W. N. Tavolga, editors. Sensory Biology of Aquatic Animals. Springer-Verlag, New York.

Kalmijn, A. J. 1989. Functional evolution of lateral line and inner ear sensory systems. Pages 187–215 *in* S. Coombs, P. Görner, and M. Münz, editors. The Mechanosensory Lateral Line—Neurobiology and Evolution. Springer-Verlag, Berlin.

Karlsen, H. E. 1992a. The inner ear is responsible for detection of infrasound in the perch (*Perca fluviatilis*). Journal of Experimental Biology 171:163–172.

Karlsen, H. E. 1992b. Infrasound sensitivity in the plaice (*Pleuronectes platessa*). Journal of Experimental Biology 171:173–187.

Knudsen, F. R., P. S. Enger, and O. Sand. 1992. Awareness reactions and avoidance responses to sound in juvenile Atlantic salmon, *Salmo salar* L. Journal of Fish Biology 40:523–534.

Knudsen, F. R., P. S. Enger, and O. Sand. 1994. Avoidance responses to low frequency sound in downstream migrating Atlantic salmon smolt, *Salmo salar* L. Journal of Fish Biology 45:227–233.

Knudsen, F. R., C. B. Schreck, S. M. Knapp, P. S. Enger, and O. Sand. 1997. Infrasound produces flight and avoidance responses in Pacific juvenile salmonids. Journal of Fish Biology 51:824–829.

Mann, D. A., Z. Lu, and A. N. Popper. 1997. A clupeid fish can detect ultrasound. Nature (London) 389:341.

Mueller, R. P., D. A. Neitzel, W. V. Mavros, and T. J. Carlson. 1998. Evaluation of low and high frequency sound for enhancing the capacity of fish screening facilities to protect outmigrating salmonids. Contract Number DE-A179-86BP62611, Bonneville Power Administration, Portland, Oregon.

Nestler, J. M., G. R. Ploskey, J. Pickens, J. K. Menezes, and C. Schilt. 1992. Responses of blueback herring to high-frequency sound and implications for reducing entrainment at hydropower dams. North American Journal of Fisheries Management 12:667–683.

OTA (Office of Technology Assessment). 1995. Fish passage technologies: protection at hydropower facilities. OTA-ENV-641. U.S. Government Printing Office, Washington, DC.

Popper, A. N., and T. J. Carlson. 1998. Application of sound and other stimuli to control fish behavior. Transactions of the American Fisheries Society 127:673–707.

Rogers, P. H., and M. Cox. 1988. Underwater sound as a biological stimulus. Pages 131–149 *in* J. Atema, R. R. Fay, A. N. Popper, and W. N. Tavolga, editors. Sensory biology of aquatic animals. Springer-Verlag, New York.

Ross, Q. E., D. J. Dunning, R. Thorne, J. K. Menezes, G. Tiller, and J. K. Watson. 1993. Response of alewives to high-frequency sound at a power plant on Lake Ontario. North American Journal of Fisheries Management 13:291–303.

Ross, Q. E., D. J. Dunning, J. K. Menezes, M. J. Kenna, and G. Tiller. 1996. Reducing impingement of alewives with high-frequency sound at a power plant intake on Lake Ontario. North American Journal of Fisheries Management 16:548–559.

Sand, O. 1981. The lateral-line and sound reception. Pages 459–480 *in* W. N. Tavolga, A. N. Popper, and R. R. Fay, editors. Hearing and Sound Communication in Fishes. Springer-Verlag, New York.

Sand, O., and A. D. Hawkins. 1973. Acoustic properties of the cod swim bladder. Journal of Experimental Biology 58:797–820.

Sand, O., and H. E. Karlsen. 1986. Detection of infrasound by the Atlantic cod. Journal of Experimental Biology 125:197–204.

Sand, O., and H. E. Karlsen. 2000. Detection of infrasound, and linear acceleration in fish. Philosophical Transactions of the Royal Society, series B 355:1295–1298.

Sand, O., P. S. Enger, H. E. Karlsen, F. R. Knudsen, and T. Kvernstuen. 2000. Avoidance responses to infrasound in downstream migrating European silver eels, *Anguilla anguilla*. Environmental Biology of Fishes 57:327–336.

Vøllestad, L. A., and B. Jonsson. 1988. A 13-year study of the population dynamics and growth of the European eel *Anguilla anguilla* in a Norwegian river: evidence for density-dependent mortality, and development of a model for predicting yield. Journal of Animal Ecology 57:983–997.

Vøllestad, L. A., B. Jonsson, N. A. Hvidsten, T. F. Næsje, Ø. Haraldstad, and J. Ruud-Hansen. 1986. Environmental factors regulating the seaward migration of European silver eels (*Anguilla anguilla*). Canadian Journal of Fisheries and Aquatic Sciences 43:1909–1916.

Wolf, P. 1951. A trap for the capture of fish and other organisms moving downstream. Transactions of the American Fisheries Society 80:41–45.